Natural History Collections in the Science of the 21st Century

SCIENCES

Biology, Field Director – Marie-Christine Maurel

Systematics, Phylogenomics and Taxonomy,
Subject Head – Philippe Grandcolas

Natural History Collections in the Science of the 21st Century

A Sustainable Resource for Open Science

Coordinated by
Roseli Pellens

iSTE

WILEY

First published 2021 in Great Britain and the United States by ISTE Ltd and John Wiley & Sons, Inc.

ISTE Ltd
27-37 St George's Road
London SW19 4EU
UK

www.iste.co.uk

John Wiley & Sons, Inc.
111 River Street
Hoboken, NJ 07030
USA

www.wiley.com

Library of Congress Control Number: 2021944047

British Library Cataloguing-in-Publication Data
A CIP record for this book is available from the British Library
ISBN 978-1-78945-049-1

ERC code:
SH5 Cultures and Cultural Production
SH5_7 Museums, exhibitions, conservation and restoration

MIX
Paper from responsible sources
FSC® C013604

Contents

Foreword

Bruno DAVID

Muséum national d'Histoire naturelle, Paris, France

The book you are holding in your hands offers an original and stimulating reflection on the role of natural history collections, their scientific interest, and their use in understanding the world we live in. On these themes, the *Muséum national d'Histoire naturelle* (French National Museum of Natural History), with its 68 million specimens and 2 million documents, archives and objects stored in its reserves and libraries, has had to make its voice heard through the contributions of many researchers from various academic backgrounds.

By shedding light on the different facets of the collections, from collection to conservation, Roseli Pellens, the coordinator of this ambitious editorial project, explores their close links with research, expertise and the dissemination of knowledge. These interfaces have been the *Muséum*'s *raison d'être* since its foundation and have contributed to its influence in France and abroad.

As Roseli Pellens rightly reminds us in Chapter 1, which is a general introduction, the collections do not speak for themselves; they are a medium that allows us to reconstitute the vast and complex puzzle of the history of life on Earth and open up broader avenues of reflection on the understanding of life from its most distant origins.

By underlining the essential link between collections and research, this book takes up a singular challenge: that of making the collections, witnesses of a long history, speak for themselves and of putting them at the center of the construction of a reliable and scientifically supported form of knowledge. But to fully understand the mysteries of this knowledge factory, a detour through the history of the *Muséum*'s collections is necessary.

When it was founded, the *Jardin du Roi* (the former name of the *Jardin des Plantes*) had an eponymous cabinet, built in 1633 on the site of its current central library. It housed the first curiosities that constituted the core of the *Muséum*'s collections: insects, plants, fossils, concretions and monster skeletons, as well as the beginnings of what was to become the first natural history library in France. This cabinet was the jewel of the institution, its standard, but also the foundation of its prestige. More than three centuries later, in 1996, Joseph Bonnier de la Mosson's famous cabinet of curiosities, purchased at auction by Buffon in 1774, was installed in the *Muséum*'s library and classified as a historical monument: a distant homage to the King's cabinet that had occupied the same premises. We can thus see that this first movement consisted of a series of accumulations of objects and natural history collections.

The priority was then to organize and give them a scientific meaning to build the story of life on Earth. The classification of the collections is indeed at the heart of the scientific activity of the *Muséum*: describing species and establishing their relationship and evolution are approaches that can be found in the classification of specimen reserves as well as in the library collections. But the scientific links between the specimens and the works do not necessarily go well with the physical or climatic precautions that their conservation requires. Where one would like to arrange by species, by theme, by author or collector, one is often led to other options by the constraints of space, fragility, hygrometry, etc. They can be arranged according to format, age or nature of the material. The spatial arrangement of collections has always reflected this tension between scientific logic and very practical considerations.

The *Muséum*'s primary mission can be summarized as that of a vast conservatoire of nature in all its diversity: meteorites, fossils, plants, animals, witnesses to the evolution of humankind and human cultures, etc. These collections have been – and still are – amassed by researchers. They have changed our understanding of the world by shedding light, for example, on the origin of the solar system (meteorites), the major crises of biodiversity (fossils) and the impact of anthropization over the last 10,000 years (archaeo-fauna, flora).

This book recalls the crucial role of collections in the production of knowledge and their indispensable window to research. They are a memory of the materiality of the world and serve to verify the accuracy of the names we give to living beings and things. They constitute an immense source of data, the archived specimens being bearers of triple information: that of their individual nature, accessible by increasingly powerful analytical techniques (imaging, bio-geochemistry, genomics, etc.), that of the history of science, which is interested in the motivations as well as the methods of their collection and archiving, and finally, that of the evolution of the state of the planet, because the collections are a privileged tool for documenting the evolutions of biodiversity, the environment and the climate. This reverse flow, which turns old collections into sources of data for future research, reinforces their heritage value and demonstrates that their continued enrichment fulfills a duty to archive.

Finally, and this is one of the strong points of this book, this collective reflection offers a fascinating tribute to the "people of the collections", these passionate men and women who bring the treasures of the *Muséum* to life on a daily basis and allow them to shine throughout the world by revealing their scientific interest. Behind each specimen and behind each article or book documenting the discovery of a taxon, there is an individual story, that of a person who made the decision to start a collection, who published the descriptions of objects, and who, sometimes, mixed their own name in with that of the object. Beyond this relationship established by the codes of taxonomic nomenclature, there is an intimacy between nature and humankind that animates the backstage and the professions of our museum.

The *Muséum* must be visionary in the enrichment and use of its collections, as well as in its public outreach activities. It must foster ethics of the planet oriented toward the preservation of its evolutionary potential and cultural diversity, while questioning the interactions between ecological and social vulnerability. It is this closeness and these strong links that are illustrated here.

August 2021

Acknowledgments

Roseli PELLENS
ISYEB – CNRS, Muséum national d'Histoire naturelle, Sorbonne Université, École Pratique des Hautes Études, Université des Antilles, Paris, France

A collective work cannot be achieved without a number of people embracing a common cause. I would therefore like to warmly thank each of the authors for the enthusiasm with which they have committed themselves to this project and for the rich ideas they have generously contributed. This project was a result of discussions during the module "Innovative uses of natural history collections in scientific research" that I organize for the École doctorale ED-227 "Sciences of nature and man: evolution and ecology" of the *Muséum national d'Histoire naturelle* (French National Museum of Natural History) and Sorbonne Université in Paris. These discussions, always very enriching, showed me how important it is to make visible the richness and diversity of innovative research based on collections. They also showed me the extent of our collective responsibility in protecting and enriching this heritage in order to allow future generations of researchers to make the best use of it in their quest for knowledge. I am therefore very grateful to all those who contributed directly or indirectly to these public discussions. I would like to thank Bruno David, president of the *Muséum national d'Histoire naturelle*, for agreeing to write the book's foreword and for providing historical insight into our heritage, Marie-Christine Maurel for supporting this project and for her always kind and benevolent advice, and Philippe Grandcolas for supporting this project and for his many suggestions during our rich discussions on this topic. Finally, I would like to thank Bruno Dastillung for his help with the layout and formatting corrections.

1

Natural History Collections: An Essential Resource for Science in the 21st Century

Roseli PELLENS

ISYEB – CNRS, Muséum national d'Histoire naturelle, Sorbonne Université, École Pratique des Hautes Études, Université des Antilles, Paris, France

As we find ourselves at the beginning of the 21st century, we are in the midst of a major environmental crisis (Hoekstra *et al.* 2012; Hoekstra and Wiedmann 2014; Steffen *et al.* 2015). Climate disruption, colossal loss of biological (Barnosky *et al.* 2011, 2012; Ceballos *et al.* 2015) and cultural diversity (Amano *et al.* 2014), and the emergence and spread of new diseases (Carlson 2020) are some of the most visible facets of this crisis. Because of its global scale, it is generating new momentum in international negotiations (e.g. Convention on Biological Diversity, United Nations Framework Convention on Climate Change), although it has not yet led to a significant shift toward a production and consumption model that can limit its effects.

This situation creates a significant demand for the scientific community to produce measurements and indicators to assess the causes, magnitude and the evolution trends of this crisis (IPCC 2018; IPBES 2019; Díaz *et al.* 2020; Rounsevell *et al.* 2020). It is in this context that bodies such as the IPCC (Intergovernmental Panel on Climate Change) and the IPBES (Intergovernmental Science-Policy Platform on Biodiversity and Ecosystem Services) at the

---

For a color version of all figures in this chapter, see www.iste.co.uk/pellens/natural.zip.

Natural History Collections in the Science of the 21st Century, coordinated by Roseli PELLENS. © ISTE Ltd 2021.

international level have been created. At the national level in France, numerous structures have also been created to analyze problems and implement appropriate solutions[1]. This context has led to the emergence of new scientific themes. For example, assessing the global state and trends of biological and cultural diversity and the environment requires a large amount of data and the organization of these data into temporal or geographical series. Natural history collections are an extraordinary resource in this context. In addition to their original function as repositories for systematics, they are now becoming a colossal source of information, contributing to the exploration of this problem. The economic benefits of their contribution, even if they are still little evaluated, could be of the order of several billion dollars (Suarez and Tsutsui 2004).

1.1. Collections in early 21st century science

Research on collections is gaining unprecedented scientific importance because of its potential to answer the major questions of our time. Collection specimens contain information that allows us to reconstruct the puzzle of the history of life, humankind and the universe, as well as provide clues for understanding environmental transformations. This broadens the scope of research on the collections. Collections remain indispensable for systematics. In turn, systematics contributes more than ever to the enrichment of collections because of the unprecedented development of identification methods and to phylogenetic and evolutionary inferences. The result is an increase in the number of identifications and the availability of new datasets that facilitate the study of living organisms as a whole. Systematics creates new forms of association between specimens and information that are increasingly numerous. Its role is essential and indispensable for any further use of the specimens.

Collections are increasingly used in two new areas: the research of new information on specimens and global analyses on large datasets of specimens. The principle of open science, on which systematics was built and which has always been the basis of collection data exchange, is thus shaping new forms of data provision. A wide range of tools has been developed to access information residing in collections. For example, new techniques are available to identify parasites and commensals (viruses, bacteria, protozoa, etc.) or to measure the chemical content

1 In France, law no. 2016-1087, known as "For the reconquest of biodiversity, nature and landscapes", as well as the creation of the *Office français pour la biodiversité* (French Office for Biodiversity) are direct consequences of the ratification of the Nagoya Protocol, the third main objective of the Convention on Biological Diversity, which refers to access to genetic resources and fair and equitable sharing of benefits arising from their use.

(carbon content, heavy metals, stable isotopes, etc.) of collection specimens. New computer approaches allow the availability of all collection data from single portals called "aggregators".

1.2. New explorations because of the magnitude and diversity of the collections' data

The magnitude of the taxonomic, spatial and temporal diversity of the data housed in the thousands of natural history collections is colossal. The collections host not only the name-bearing types that allow the use of scientific names but also, very often, numerous specimens presumed to belong to the same species. The richness of this content not only allows us to study questions of genetic diversity and species limits but also to understand global phenomena for which a large number of specimens are essential. The numerous specimens of the same species constitute genetic, spatial and temporal replicas that allow us to test many hypotheses.

The magnitude of the collections is also reflected in their gigantic spatial coverage. Most of the world's environments have been sampled and have corresponding specimens in the collections. For many localities, there is a wealth of data. Millions of specimens trace the history of thousands of institutional or individual missions that have allowed us to access the most diverse and challenging environments (Figure 1.1). They also tell the story of the explorations and many scientific questions asked during the last centuries. In fact, the collections allow us to have a global view of the distribution of life on the planet. No global expedition would be able to collect again these millions of pieces of information, especially since most of the places have been transformed and the natural environments of the past have been replaced by cities or cultivated areas.

Another important aspect of the collections is the fact that they contain specimens collected at various times in the past, even in very remote times. The most extraordinary examples mentioned in this volume are the remains of flower garlands found on the royal mummy of Ramses II (Chapter 12 by Quiles *et al.*), or the collection of mummies dated between the 2nd millennium BCE and the 19th century CE (Chapter 4 by Thomas *et al.*). In European institutions, it is not uncommon to have diverse collections on most of the biomes of the five continents dating from the 17th century. Some very old objects provide information on the way of life of the populations, their physical characteristics, their environment or their culture.

a)

b)

Figure 1.1. *Examples of environments that are difficult to sample. (a) The Canopy Raft in French Guiana, a device to collect and study tropical canopies (in the foreground is an entomologist's net to collect insects; inset, a general view of the device). (b) Mission to Spitsbergen, marine ecology and study of Carboniferous and Devonian rock outcrops. © R. Garrouste, ISYEB*

However, the main interest of large time series is to provide what is commonly called "zero points", that is, reference points on the state of populations or environments before certain human actions, for example, at the beginning of the industrial era. The distribution of organisms can thus be known before the destruction of natural habitats, and certain effects of climate change on biodiversity can be measured (see Chapter 15 by Robuchon *et al.* and Chapter 16 by Muller *et al.*). This knowledge is essential for refining proposals for the conservation of organisms, or even their reintroduction, as is the case with the germination of seeds of extinct plants from herbariums (see Chapter 10 by Muller *et al.*).

1.3. Research using and driving the constitution of natural history collections

Such a high number of positive results might lead one to imagine that the collection data speak for themselves. But nothing could be further from the truth. Extracting information is a scientific process that requires expertise, inventiveness, knowledge of tools and methods, as well as background knowledge of the issues being addressed. For this reason, very often, this research is inter- or trans-disciplinary and carried out in the framework of inter-institutional collaborations, where each party brings complementary expertise. These approaches benefit from international collaborative strategies, such as the creation of aggregators that bring together the databases and data used (e.g. GBIF), the creation of taxonomic repositories that make nomenclature and taxonomy available, the development of collaborative analysis systems (R Core Team) (see Chapter 18 by Monnet *et al.*), the creation of reference libraries for studies of ancient DNA (see Chapter 13 by Gagnevin *et al.*), and so on. The development or adaptation of methods, inherent to the scientific process, is part of the daily routine of this research.

The main objective of this book is to unveil this scientific universe based on natural history collections at the beginning of the 21st century. A single volume is not enough to deal with all of these scientific and societal issues, but it does allow us to present a representative sample. We will see that the collections constitute an important medium for knowledge. Their fundamental quality lies in the preservation and permanence of specimens that can be revisited in the light of new questions and new technologies. In this conception of science, the data obtained remain always available, structured and re-analyzable.

The chapters deal with various subjects and focus either on a scientific problem, a specimen, a group of organisms or a set of collections. What they all have in common is the recognition of the value of collections as revealed by science. On the one hand, the collections serve as a medium for the knowledge of life, humankind and the universe. On the other hand, the constitution of collections, their *raison d'être*, their management and their future are also objects of scientific and ethical questioning in perpetual renewal.

1.3.1. *Being able to return to the object: one of the major contributions of natural history collections*

In Chapter 2, Grandcolas summarizes and analyzes the concepts that led to the creation of collections and classification systems of life and minerals. In this conception, collections are representative of a science based on the unlimited possibility of returning to an object of study. His analysis highlights a paradox in the situation of collections in relation to current science. On the one hand, they have never been so necessary to shed light on scientific issues in an emergency. On the other hand, they risk being rapidly replaced by a science based on billions of data produced by methods that do not allow any return to the specimen, such as data from participatory sciences or metagenomics. The challenge is being able to produce knowledge without any traceability or possibility of revision or refutation, the opposite of what has been built over several centuries of natural history research. There is therefore an urgent need to find strategies for reconciling these different types of data, which some consider, respectively, as "quality data" versus "Big Data".

Being able to re-examine objects and associated materials is thus an inseparable aspect of the research on collections and is therefore part of most of the studies in this book (for a broader overview, see Chapter 15 by Robuchon *et al.*). Nevertheless, the following three chapters illustrate this element of the research on collections well. The rediscovery of the blue diamond, one of the two extraordinary gems of Louis XIV, represents an excellent example in this respect. Stolen in 1792, this diamond was considered lost until very recently. Detailed analysis of a wealth of documentation, the rediscovery of a clay cast and the ability to examine the modified diamond after the theft have confirmed earlier hypotheses about its whereabouts, and also allowed the reconstruction of a zircon copy including many of its unique characteristics (see Chapter 3 by Farges).

The second example is the study of the Chachapoya mummy (see Chapter 4 by Thomas *et al.*). This artifact, which is part of the museum and scenic history of the *Musée de l'Homme* and which inspired one of the most remarkable paintings in the history of art of the late 19th and early 20th centuries – Edvard Munch's *The Scream* – is now part of the history of modern science on Amerindian people. Recent studies and the use of new technologies have made it possible to examine the interior of the body and to understand how this people lived and the causes of their death.

The third example concerns new studies on human skulls, probably one of the most heavily examined artifacts in the history of science (see Chapter 5 by Friess and Galland). This study shows how new methods of 3D morphometry have provided new insights and contributed to a better understanding of important issues in the evolution of human diversity, including the relationship between skull shape and geographical distribution and linguistic diversity.

1.3.2. *Collections at the heart of highly innovative research thanks to new technologies*

Another aspect of innovative uses of collections is the importance of technologies to access information that could not be acquired before. Chapter 6 (by Invernón *et al.*) shows how a study of hyperaccumulators of metals and rare earths in herbarium plates was possible using X-ray fluorescence spectrometry. The results indicate how the herbarium can be used to identify not only useful species for the remediation of heavy metal contaminated sites but also to search for locations where the concentration of certain metals and rare earths may be significant. These fluorescence techniques as well as X-ray microtomography are also of great interest in the study of fossils, because they allow the visualization of structures invisible to the naked eye or through other traditional methods (see Chapter 7 by Charbonnier and Forel). Many diverse research are possible because of these new technologies (systematics, study of development or ancient paleogeography, etc.).

Another facet of the new questions and new technological possibilities is the creation of innovative collections. Collections that are initially intended to meet very specific demands open up a new universe of knowledge and experimentation. An excellent example is illustrated in Chapter 8 (by Duperron *et al.*) with the collection of live strains of cyanobacteria and microalgae, created to serve as a reference for ecotoxicological tests and environmental diagnostics. Duperron *et al.* show that this collection, in addition to its use in systematics, is of major interest in research facilitated by high-throughput "-omics" approaches. For example, it allows us to study the interactions between organisms, because the associated cyanosphere and phycosphere are preserved with the specimens. It also facilitates research into bioactive metabolites, mainly for pharmaceutical purposes, which are explored in an increasingly greater detail thanks to new technologies. Another example is presented in Chapter 9 (by Gerbault-Seureau and Dutrillaux) through the collections of cryopreserved cells and tissues of homeothermic vertebrates. This very rich, but still little exploited, collection represents an important source of cellular material, with its recent DNA and RNA of fibroblasts, available for research. Preserving tissues and cells of protected species is all the more interesting as these vulnerable species have small populations. Studies on living tissues of these species can be conducted repeatedly, after cell culture, without causing any harm to living individuals.

While cryopreserved cells and tissues do not enable the revival of homeothermic animals, seeds kept in natural history collections offer this possibility. Scientists have thus set themselves the goal of reviving extinct species whose seeds are preserved in the herbarium (see Chapter 10 by Muller *et al.*). This innovative approach obviously requires expertise in the reproductive physiology of the target species. This chapter therefore presents the first results of an ambitious project that is a precursor to a type of research that will probably become very widespread in the coming decades.

1.3.3. *A resource for global change research*

One of the major challenges in studying global change is establishing baselines from which we can see the magnitude and direction of ongoing changes. Global changes, and in particular climate change, are measured on time scales much longer than our contemporary observing systems that, for example, come from stations set up in the second half of the 20th century. Thus, studies on collections, whether paleontological, archaeological or current, seek to define and calibrate proxies to understand these changes.

Climate change research requires that these proxies be placed in the context of climate evolution over the last 20,000 to 100,000 years. In Chapter 11, Moreno and Bartolini demonstrate one such approach: the search for climate and salinity proxies in marine sediments (from a few tens of thousands of years ago to the last million years). This research is based on new collections of ocean cores, mainly from 1970 onwards. This chapter reveals the challenges inherent to the constitution and conservation, as well as to the analysis and understanding of data from such an extraordinary resource. In the same vein, Quiles *et al.* (Chapter 12) show how collections can provide access to atmospheric gas levels prior to a major landscape change. The aim is to calibrate the radiocarbon curve before the construction of the Aswan Dam in Egypt, which had a major impact on this region, and strongly affected the periodic flooding of the Nile. The analyses indicate that the herbarium plates of the 17th and 18th centuries allow for the detection of ^{14}C levels in the atmosphere of this period. This study thus contributes to the refinement of archaeological dating over a historical period of several millennia, a major challenge for the knowledge of the history of ancient Egypt. These two projects reveal how collections can be at the heart of highly transdisciplinary research.

The next two chapters discuss research on the dispersal of crop pathogens and invasive alien species, problems associated with the globalization of agriculture and the transportation of diverse agricultural products over very long distances. The study by Gagnevin *et al.* (Chapter 13) explores the use of herbaria to study *Xanthomonas citri* pv. *citri* which infects citrus and geminiviruses that attack manioc and sweet potato. These pathogens have a strong impact on agricultural production in the Indian Ocean and sub-Saharan Africa. Their results show how new molecular and genomic methods can be used to access information residing in herbaria, rich in species of agricultural interest. These analyses make it possible to trace the origin, the evolutionary history and the dispersal routes of diseases that can have a major impact on the regional economy. The role of collections in understanding biological invasions is illustrated by Villemant *et al.* (Chapter 14), concerning the Asian hornet *Vespa velutina*, the first exotic species of Vespidae acclimatized in Europe. Their study underlines the importance of gaining data from collections in order to discover the origin of invasive lineages, the structure of diversity in its area of origin, and thus to make more accurate predictions on potential areas of invasion.

As mentioned earlier, using collections to answer targeted questions requires methodological adjustments. Sometimes the data from the collections are not sufficient to answer a question. Nevertheless, taking them into account allows us to reframe the subject and adjust the protocol. This iterative aspect, so fundamental to scientific research, is underlined well by Robuchon *et al.*'s study (Chapter 15). The existing data do not allow us to reach clear conclusions. However, when supplemented by targeted collections of the same species at the same locations, they can represent a solid past reference point for research on community change and local extinctions.

Despite their number, the chapters presented in this book only offer a glimpse of the diversity of research on global changes for which collections are an indispensable source of data. The review in Chapter 16 (by Muller *et al.*) on herbaria confirms that there is already a surprising diversity of approaches, even for this type of collection alone. The contribution of collections to understanding the effects of global change is clearly still in its infancy.

1.3.4. *Designing the science of the future based on collections*

As all of these examples show, collections-based research is both a user of collections to answer specific questions and a driving force of new designs for collections. The chapters in this book highlight different know-how that also enrich collections.

In Chapter 17, Garrouste reflects on the work of a naturalist, with the association of 2D, 3D and even 4D (films, movements) digital images of the collected specimens. These images make it possible to easily preserve information on living specimens (precise colors, movements, behaviors, interactions). The association of this information to specimens of collections allows us to make a conceptual leap, so much so that they are essential to ecology, physics of movements, bio-inspiration and evolution studies. Despite these advantages, the author mentions two contexts where these uses are, respectively, exacerbated and neglected, and thus their scientific potential is reduced: the use of photos as types when describing new species and the limited association of images to identifications from participatory science projects.

Today, an important part of the research on databases from natural history collections is destined to validate the data or neutralize their biases with respect to the questions asked. In Chapter 18, Monnet *et al.* discuss the provision and use of data via data portals or aggregators, as part of a traceability system from the collection specimen to the last of a series of users. Indeed, over the last two decades, institutions have made enormous efforts to make their data available and the number of research studies that use them is increasing exponentially. Scientific journals also play a fundamental role in recommending or requiring the availability of the data used. Nevertheless, the whole practice needs to be reviewed with rigor, and data processing and use pipelines that are more in line with principles of validation and traceability need to be built. The monitoring of the evolution of knowledge on specimens, the association of the various information obtained for the same specimen (Gene Bank Accession Number, GBIF ID, etc.), as well as that of the recovery of resampled data by other research groups are still very limited. Considering the fact that database portals and aggregators are designed to last over time and cannot be reconstructed with each new study, it is clear that new procedures must be implemented in this regard.

Questioning the collections, their values, their needs for maintenance and continuity are part of the daily life of scientists in charge of collections. This link also leads to numerous questions not only about how to add information and ensure its traceability, as proposed in the studies in Chapters 17 (Garrouste) and 18 (Monnet *et al.*), but also about their *raison d'être* and their future. Bahuchet's study (Chapter 19) explores this issue through an analysis of biocultural collections. The basic question "do we need biocultural collections?" is posed with regard to the objects that are essential for testifying the relationship between people and nature. This question concerns all types of collections. It shows us that decisions concerning a collection cannot be taken without considering the scientific context that gave rise to their constitution.

The book concludes with a study of conservation, a research theme that is inseparable from the existence of collections. Rouchon (Chapter 20) details conservation practices and their changes over time, particularly in cases where the object to be conserved poses methodological and technical challenges. These examples reveal the magnitude of the conservation problem. How can we avoid alterations to the objects, whether they are of a physical nature, or the result of chemical reactions or biological activity? Through the question "why preserve?" the author underlines the different values that we give to collections or to certain objects, as well as the transformations that come as we gain more knowledge and understanding of the objects. Better conservation thus implies better knowledge of them.

As a conclusion, Chapter 21 questions what contributions collections will make to future science. In retrospect, it would have been difficult at the end of the 19th century to imagine much of the research currently being conducted on collections. The same is true for our ability to predict in detail the societal needs and scientific insights of the next century and so on. Nevertheless, two ideas can help us: on the one hand, if we consider the set of practices and principles that have allowed the conservation of the rich information residing in the collections; on the other hand, if we appreciate the evolution of systematics, the science that has served as a guideline for the establishment of a large number of the collections. These two elements already show us the future potential of research on collections.

1.4. References

Amano, T., Sandel, B., Eager, H., Bulteau, E., Svenning, J.-S., Dalsgaard, B., Rahbek, C., Davies, R.G., Sutherland, W.J. (2014). Global distribution and drivers of language extinction risk. *Proceedings of the Royal Society B: Biological Sciences*, 281(1793), 17–19.

Barnosky, A.D., Matzke, N., Tomiya, S., Wogan, G.O.U., Swartz, B., Quental, T.B., Marshall, C., McGuire, J.L., Lindsey, E.L., Maguire, K.C., Mersey, B., Ferrer, E.A. (2011). Has the Earth's sixth mass extinction already arrived? *Nature*, 471(7336), 51–57.

Barnosky, A.D., Hadly, E.A., Bascompte, J., Berlow, E.L., Brown, J.H., Fortelius, M., Getz, W.M., Harte, J., Hastings, A., Marquet, P.A., Martinez, N.D., Mooers, A., Roopnarine, P., Vermeij, G., Williams, J.W., Gillespie, R., Kitzes, J., Marshall, C., Matzke, N., Mindell, D.P., Revilla, E., Smith, A.B. (2012). Approaching a state shift in Earth's biosphere. *Nature*, 486 (7401), 52–58 [Online]. Available at: https://doi.org/10.1038/nature11018.

Carlson, C.J. (2020). From PREDICT to prevention, one pandemic later. *The Lancet* [Online]. Available at: https://doi.org/10.1016/S2666-5247(20)30002-1.

Ceballos, G., Ehrlich, P.R., Barnosky, A.D., García, A., Pringle, R.M., Palmer, T.M. (2015). Accelerated modern human-induced species losses: Entering the sixth mass extinction. *Science Advances*, 1:e1400253, 1–5.

Díaz, S., Zafra-Calvo, N., Purvis, A., Verburg, P.H., Obura, D., Leadley, P., Chaplin-Kramer, R., De Meester, L., Dulloo, E., Martín-López, B., Shaw, M.R., Visconti, P., Broadgate, W., Bruford, M.W., Burgess, N.D., Cavender-Bares, J., DeClerck, F., Fernández-Palacios, J.M., Garibaldi, L.A., Hill, S.L.L., Isbell, F., Khoury, C.K., Krug, C.B., Liu, J., Maron, M., McGowan, P.J.K., Pereira, H.M., Reyes-García, V., Rocha, J., Rondinini, C., Shannon, L., Shin, Y.J., Snelgrove, P.V.R., Spehn, E.M., Strassburg, B., Subramanian, S.M., Tewksbury, J.J., Watson, J.E.M., Zanne, A.E. (2020). Set ambitious goals for biodiversity and sustainability. *Science*, 370(6515), 411–413.

Gorenflo, L.J., Romaine, S., Mittermeier, R.A., Walker-Painemilla, K. (2012). Co-occurrence of linguistic and biological diversity in biodiversity hotspots and high biodiversity wilderness areas. *Proceedings of the National Academy of Sciences of the United States of America*, 109(21), 8032–8037.

Hoekstra, A.Y. and Wiedmann, T.O. (2014). Humanity's unsustainable environmental footprint. *Science*, 344, 1114–1117.

Hoekstra, A.Y., Mekonnen, M.M., Chapagain, A.K., Mathews, R.E., Richter, B.D. (2012). Global monthly water scarcity: Blue water footprints versus blue water availability. *PLoS One*, 7(2) e32688.

IPBES (2019). Summary for policymakers of the global assessment report on bio-diversity and ecosystem services. Report, Intergovernmental Science-Policy Platform on Biodiversity and Ecosystem Services, Bonn [Online]. Available at: https://doi.org/10.5281/zenodo.3553579.

IPCC (2018). Summary for policymakers. In *Global Warming of 1.5 °C*, Masson-Delmotte, V., Zhai, P., Pörtner, H.-O., Roberts, D., Skea, J., Shukla, P.R., Pirani, A., Moufouma-Okia, W., Péan, C., Pidcock, R. *et al.* (eds). World Meteorological Organization, Geneva.

Rounsevell, M.D.A., Harfoot, M., Harrison, P.A., Newbold, T., Gregory, R.D., Mace, G.M. (2020). A biodiversity target based on species extinctions. *Science*, 368(6496), 1193–1195 [Online]. Available at: https://doi.org/10.1126/science.aba6592.

Steffen, W., Richardson, K., Rockström, J., Cornell, S.E., Fetzer, I., Bennett, E.M., Biggs, R., Carpenter, S.R., De Vries, W., De Wit, C.A., Folke, C., Gerten, D., Heinke, J., Mace, G.M., Persson, L.M., Ramanathan, V., Reyers, B., Sörlin, S. (2015). Planetary boundaries: Guiding human development on a changing planet. *Science*, 347(6223), 1–10.

Suarez, A.V. and Tsutsui, N.D. (2004). The value of museum collections for research and society. *BioScience*, 54(1), 66–74 [Online]. Available at: https://doi.org/10.1641/0006-3568(2004)054[0066:TVOMCF]2.0.CO;2.

2

Natural History Collections: An Ancient Concept in a Present and Future Perspective

Philippe GRANDCOLAS
ISYEB – CNRS, Muséum national d'Histoire naturelle, Sorbonne Université, École Pratique des Hautes Études, Université des Antilles, Paris, France

Natural history collections must be considered from an innovative perspective of an open science, which adds to the traditional uses (taxonomies, identifications, etc.), with all the possible new uses linked to current problems (effects of global change, research of new natural resources and sustainable development, notably with bioinspiration, etc.). These new uses are possible because of the material status of collections that contain available specimens and their possible links with digital information. New practices that only generate virtual data (citizen science, metagenomics) that is analogous to collection data must be reconnected with specimens. The concern about biases in collection building, while understandable, does not represent a serious problem if one understands that any question requires *ad hoc* sampling and it is therefore important to "resample" collections to answer a question. Thus, it is therefore essential that the collections last and continue to increase in size, notably via the innumerable scientific studies that should make their data accessible as part of an open science policy.

Natural History Collections in the Science of the 21st Century,
coordinated by Roseli PELLENS.

2.1. Introduction

The role and importance of natural history collections are considered differently by different people. With their ancient origins, collections may seem obsolete to some, as if they were a testimony to another era when our natural history knowledge and tools were still insufficient and required intense descriptive activity and physical storage. And yet, in the context of global change, collections constitute a formidable archive of abused and endangered biodiversity. Indeed, society and very diverse parts of the scientific community are beginning to understand that collections not only have an undeniable utility, but even have a major role to play in today's and tomorrow's world (Suarez and Tsutsui 2004). At this critical juncture, these somewhat contradictory views of collections can blur their image and prevent us from assigning them an appropriate strategic trajectory.

It is therefore important to renew but also to clarify the way we consider collections. In such a context, we must not deny the reasons for their foundations and their current uses, which are extremely numerous (Funk 2004, 2018). We must therefore first try to understand their original *raison d'être*, their relationship with the classifications of living organisms and their evolution, with its constraints and assets. But it is also necessary to extricate ourselves from our habits and try to take a fresh look at them in order to integrate them in a world in environmental and technological revolution in the most profitable way. It is thus important to situate them in a general context where other types of collections – genetic, molecular, digital, etc. – or other modalities of biodiversity data acquisition – metagenomics, citizen sciences – are emerging.

With this approach, it is also important to be aware of the diversity of collections, encouraging interactions between biological and non-biological collections (mineralogy, human artifacts, etc.). This aspect is rarely mentioned and collections are almost always considered only in relation to biological specimens (Meineke *et al.* 2018). However, major concepts are common or analogous between all these collections, such as species or classification, representation of variability, availability, and so on. I will not systematically develop these analogies in the rest of my reflection, but I encourage the reader to take them into account in their approach.

2.2. A tribute to curiosity and coupling with classifications

For thousands of years, humans have shown a curiosity and knowledge of the natural environment, as evidenced by the parietal engravings or paintings of prehistoric populations or the vernacular classifications, known as "folk classifications" of the first people, often remarkably complete. This knowledge was based on daily practice, local but diversified use, and essentially oral tradition and transmission. Following sedentarization and agricultural and urban development, humankind's relationship with the world has changed, with both more distance and more focus on the main productions (a few cultivated or raised species, etc.). It was then necessary for them to keep a memory of the natural world from which they had distanced themself. The field of human intervention was then strongly extended with voyages of exploration with essentially commercial aims to distant countries. This memory then became more complicated to maintain, especially since many exotic living organisms were not visible *in natura* or *in vivo* but had to be evoked by specimens or specific writings. Collections thus assumed the role of an external memory for humanity, allowing a relationship with nature, in the same way as books, gardens or menageries do, the latter having the disadvantage of being materially more complicated to maintain.

It is thus possible to trace an evolution in the nature of collections from the collections of the schools of natural history, from that of Aristotle to those of the Renaissance, then to cabinets of curiosities, and finally those of the first great museums, places of European power in a world expanding toward new places of commercial activities (spices, silk, precious woods, coffee, etc.).

During this period, from Mediterranean Antiquity to the European Renaissance, collections and writings came into resonance. These collections had to be ordered according to a system or catalogued. Vernacular taxonomies were no longer sufficient to name organisms that were no longer encountered on a daily basis in the natural environment and whose numbers were beyond individual human memory. Natural histories such as Aristotle's testify to this process, which began a long time ago. However, it was not until the post-Renaissance period that the real beginning of a taxonomic literature based on binominal nomenclature (Linnaeus 1758) was noted. This literature took over from individual expertise and gradually gave shape to a collaborative system to which naturalists from different countries could contribute. This system still exists today because of natural history museums and the scientific literature in taxonomy.

The aim of natural history literature was to form a catalogue of Life, the mineral or the human. For practical reasons, this rapidly growing catalogue could not only be enumerative, it had to be classificatory. A classification allows us to find our way through an immense set of objects, with logical rules that propose a guiding structure in a too-long list by constituting groups of species that are subordinate to each other. The classifications have thus very quickly made it possible to navigate efficiently in a rapidly-growing sector of knowledge.

But what classification could have been constructed if not one that made sense, a meaning so obvious that everyone wished to use it for communication within society? This meaning, on which the first classifications were based, is the orderliness of Life that the great naturalists of the 18th century perceived and then interpreted as resulting from biological evolution (Perrier 1886; Le Guyader 2003).

Through this interaction between collections and classifications, the museums and their collections have thus helped the emergence of the idea of biological evolution that is central to biology today. Similar aspects have taken place for other types of collections, such as minerals or human artifacts. In the case of minerals, for example, classification is based on their chemical composition and thus developed with modern chemistry in the 19th century.

2.3. The structuring of our thoughts and actions by an ancient concept

The traditional vision of collections thus recapitulates these successive societal motivations at the historical level: the satisfaction of curiosity, the cataloguing of diversity with the formulation of the notion of species (essentialist, then biological for living organisms or chemical for minerals) and finally, recently, the organization of knowledge with the notion of evolution for living organisms. Curiosity for the natural world has, nevertheless, remained great, to the point of being the driving force behind the pedagogy offered by museums in their exhibitions.

The beginning of a more structured interest in biological diversity and the scientific refinement of the notion of species fueled very applied approaches, even if they proved to be extraordinarily powerful and innovative in many scientific respects. The naturalists of European museums actually supported the economic efforts of the great kingdoms with their ability to classify and to name important exotic resources. Subsequently, population biologists designed genetic tools for the improvement of domesticated, cultivated or bred species (Fischer 1930). The notion of species coupled with genetics thus became an important societal tool and all kinds of conservatoires and genetic data banks were created in that context. All this can also be said regarding the study of the diversity of mineral resources, which brought the Industrial Revolution by offering societies other materials and sources of energy than wood or earth, such as bronze, iron and their modern industrial companion, coal.

In the context of all these advances, the role of collections remained frozen at the stage of description of the diversity of Life, presumptively resulting from evolution. Therefore, collections are still a reference for the description and the identification of species. In this respect, 2.4 million species are known to science, while about

10 million probably exist in ecosystems today (estimates vary according to different authors but are rarely below this order of magnitude) May 2004; Costello *et al.* 2012)). Collections are growing steadily with taxonomic work (Tancoigne and Dubois 2013) and the continuation of field sampling, to the point that their overall size is difficult to estimate (Arino 2010).

However, for many decision-makers or non-natural history scientists, the reference role of collections is often perceived as limited, to the point that science policy articles regularly appear questioning the slow pace or even the quality of the taxonomic process of exploring living species and classifying them (Charles and Godfray 2002; Grandcolas 2017b, pp. 116–128). Many authors erroneously believed that living species are already well-known and we should just add coherence to the current system and rush to provide the necessary additions (May 2004; Padial *et al.* 2010; Costello *et al.* 2013). In reality, this view only takes into account the basic taxonomic description: we only know about 20% of living species, and still very imperfectly. It should be remembered that out of these 2.4 million described species, only a few tens of thousands are known for more than a taxonomic description and some elements of geographical distribution. In addition, many of them are already extinct or have become exceedingly rare in natural environments (Regnier *et al.* 2015). Moreover, even at the purely classificatory level, the taxonomic system is scientific and collaborative, and incessant corrections are made to the supposedly already acquired knowledge that is, in reality, very imperfect (Vaidya *et al.* 2018).

It is also important to understand that these living organisms that are still unknown to science do not only include rare "exotic" species or sisters of known species without much immediate societal interest: so-called exotic, unknown or unrecognized species remind us incessantly when we are confronted with a new "plague" or vector or pathogen for humans. For example, the recent discovery in France of exotic flatworms that have been introduced, are invasive and are predators of earthworms, leads to understanding of how much these organisms are totally misunderstood and scorned by everyone, including most scientists (Justine *et al.* 2018). Everyone then understands momentarily how valuable collections are and how they allow us to situate a species in relation to existing knowledge about living species in relation to taxon names.

We are therefore rather stuck in this traditional vision of collections as a matter of curiosity, repertoire and support for the study of evolution, in the narrow context of indispensable but very focused studies of taxonomy and evolution. We need to reintegrate natural history collections into more diverse scientific and societal issues by starting again from the fundamentals.

2.4. Collections: more than species catalogues

In terms of fundamentals, it is important to realize that collections are samples of living organisms that reflect not only specific richness but also genetic and phenotypic diversity among individuals. This goes against the traditional taxonomic conception that considers collections like Noah's ark, where the list of different species is proof of their importance. In fact, the sum of natural history collections in different institutions allows us to examine many individuals assigned to different species and not to study biological diversity with a simplistic view which would consider each species as invariant. Moreover, it must be understood that the notion of species is an operational facility to designate networks of relationships between individuals whose limits are somewhat imprecise. The species is not a natural kind or individual, and each identification and assignation to a species is an arbitrary scientific decision: a species cannot be enumerated on the basis of all its constituent individuals, but extensionally (Grandcolas 2017b, pp. 116–128). Not only is this enumeration not materially feasible and would require inordinate material resources, but it would also become impossible at the margin of each species in which individuals reproduce momentarily with those assigned to other species or are themselves of indefinite assignment.

When studying biodiversity, it is therefore important or even crucial that data can be linked to a specimen, especially in the case of poorly known "species", especially since the "species" taxon can be revised and its definition revised *a posteriori*, thus changing the meaning of the name and the assignment of individuals to a species group (Troudet *et al.* 2018).

This situation is very poorly understood by society as a whole and by a large part of the scientific community, which approaches the question of species with a very essentialist vision. In this vision, phenotypic discontinuities between species are wrongly taken as a confirmation of their reality, and attributed, without proof, to reduced gene flow. In this view, the species would evolve and be transgressed only by rare and unproductive hybridization events or by exceptional evolutionary changes. In reality, hybridization is often trivial and the species is a porous category at the genetic level; the discontinuities between species that we think we observe empirically are often a defect in the sampling of living organisms or the result of secondary extinctions, and not only the result of a splendid isolation coupled with a divergence phenomenon (Mallet 2008). Hence the importance of collections, which contribute to documenting a complex living world that is resistant to classification by multiplying the number of specimens stored, regardless of their momentary assignment to a particular taxon.

2.5. Big Data collections in space and time

Still in terms of fundamentals, specimens from collections have the great advantage of documenting a vast geographical coverage and an unparalleled temporal depth. As such, they can help answer all sorts of questions about changes in ecosystems in the context of the biodiversity crisis. Indeed, even if they were not acquired through a reasoned sampling program but erratically, as a result of various studies or opportunistic collections, the different specimens stored in museums ultimately document many different places and times and can often lend themselves to tests of scientific hypotheses formulated *a posteriori* (Dias Tarli *et al.* 2018). All global approaches with large numbers of individuals or species need this collection-resident information, though with the express condition that places and dates are accurately informed, which is not as obvious or as common as one might think. Toponyms can change or be inaccurate and even modern means of geolocation have their limits because we do not have a single exact reference ellipsoid for the whole planet.

Nevertheless, specimens in collections represent a biased sample of life on Earth, whether in terms of location, date, or taxonomic groups involved (Meyer *et al.* 2015; Amano *et al.* 2016; Troudet *et al.* 2017). Most of these biases are related to well-known "collector effects" (Pautasso and McKinney 2007), namely that biodiversity is best sampled in the most accessible or visited locations, and for the most societally and scientifically valuable taxa. These biases are often criticized and indeed it is often recommended that they are corrected by supplementing sampling with additional collection effort (Feeley and Silman 2011; Beck *et al.* 2012; Goodwin *et al.* 2015).

Of course, additional collection efforts are never negative, especially if they are concentrated on the most glaring deficiencies in terms of areas or taxa, since we only know 20% of all living organisms. Such efforts are often associated in people's minds with massive data acquisition operations (large expeditions, large monographs on fauna or flora). Nevertheless, it is illusory to hope to compensate for all the deficiencies by more massive sampling. Even if there are no longer obvious large "gaps" in terms of taxa or geographic areas, the available data will still be biased with respect to the multiplicity of questions addressed or accessible, and it is therefore at least of equivalent importance to design protocols for statistical correction of these biases (Dias Tarli *et al.* 2018).

Rather than relying solely on a few massive operations, it is also important to consider that all published scientific studies on biodiversity should make their specimens and observations available. The accumulation of all these specimens/ observations would potentially be colossal, precisely because of the immense number of published case studies of species on diverse evolutionary or ecological issues, and would dwarf the sum of the few large expeditions conducted each year. This is not at all the case at present, and there is a real culture change to be made in this area, in the very current context of an "open science" policy (Ayris *et al.* 2018). This is both a fundamental ethical issue for science and an extraordinary opportunity for the community to quickly increase the amount of indispensable data on biodiversity.

2.6. What future is there for the use of collections?

Collections have the great advantage of being material and having curated samples on which all sorts of retrospective and analytical approaches can be conducted (Rocha *et al.* 2014). In this way, one can go back to the source data and analyze them against completely different questions than those that governed their harvest in the past. The current trend, unfortunately, is to accentuate the acquisition of dissociated observations of specimens rather than continue to acquire many additional specimens (Troudet *et al.* 2018).

This trend is mainly related to the very strong development of initiatives of citizen science that generate millions of observations without specimens (Amano *et al.* 2016). Some scientists also argue ethical reasons for this lack of specimen collection (Minteer *et al.* 2014), but it should be noted that disciplines such as ecology, for example, have never really made available most of the data collected and used, regardless of the organisms involved (Schilthuizen *et al.* 2015; Mills *et al.* 2016). Finally, a few systematists oddly advocate taxonomic descriptions based on virtual data such as photography (Marshall and Evenhuis 2015). All of these trends are based on the same mode of thinking – derived from general biology – which implicitly considers our knowledge about biodiversity as already sufficient and organized enough to afford the creation of digital complementary data alone (Grandcolas 2017a).

The validity and the *a posteriori* exploration of these observations will be strongly limited in the great majority of cases, because it will be impossible to come back to the specimen, even in the form of a digital photograph or sound record associated with the observation. These observations, especially when they are linked to rare and poorly established names, which are sometimes doubtful or changing, will simply be null and void. Consider that, in metropolitan France alone, there are 40,000 species of insects; we cannot expect all of these taxa to be known or even stable in terms of their scientific significance, or to be the subject of unambiguous taxonomic assignments.

Concerning the specimens themselves, our analytical capabilities have increased considerably and allow for increasingly powerful and diverse studies through collections (Meineke *et al.* 2018). Access to and the study of specimens is facilitated by the digital data and images associated with them, especially through large digitization programs (Le Bras *et al.* 2017). Many future uses of collections will likely be stimulated in this way, by linking specimens to digital records or analytical results. It is therefore important that this linking is done correctly, which unfortunately is not always the case. For example, the millions of digital DNA or RNA sequences deposited in Genbank rarely include a link to the sequenced specimens which would be deposited in collections (Pleijel *et al.* 2008). Similarly, not all participatory science citizen programs, generating huge amounts of data, deposit their digital identification validation materials (photos, sounds, etc.) on portals like GBIF that are designed to link metadata (Troudet *et al.* 2018).

Collections are thus in a paradoxical situation. Their accessibility is greatly improved by modern techniques, but these same techniques can lead to collections becoming virtual in the minds of many (Grandcolas 2017a). Many biologists, and among them some taxonomists, consider that we have already mapped out the order of life and that we now just need to complete this picture with additions that are still numerous, but that will not upset the already established order. Therefore, the tree can hide the forest and digitization or participatory science can dissimulate the need that we have to keep material documents, the specimens, in still-increasing numbers.

Another future issue concerning collections is related to the development of metagenomics. Developed by microbiologists, this technique makes it possible to obtain global DNA samples from an environmental sample (soil, water, etc.) (Tringe and Rubin 2005). It does not distinguish between the individual organisms analyzed, but rather extracts the overall DNA of a given sample. It therefore does not allow us to return to individual specimens and to the material origin of the data obtained. When the sequenced taxa are unknown at the molecular level, they cannot be identified in a taxonomic sense and thus linked to the existing body of knowledge (Pellens *et al.* 2016, pp. 375–383). A huge study of plankton in the world's oceans, for example, reports over 100,000 operational taxonomic units (OTUs) detected by metagenomics, of which only a little over 11,000, at best, are described with a name in the scientific literature (Vargas *et al.* 2015). Initiatives with similar implications are developing with respect to fungi, for which molecular identification is preferred to the uneasy linkage to traditional taxonomy that is difficult to conduct on microscopic organisms with limited access to direct sampling (Nilsson *et al.* 2019).

Initially, all these developments occur without too many problems because they are based directly on existing knowledge and the taxon naming system. Nevertheless, it seems quite obvious that a second cycle of knowledge acquisition, on the occasion of future studies, with mixed reference to traditional taxa and molecular entities defined in the meantime, may lead to a great confusion, a lack of global coherence and a lack of capitalization in knowledge.

Unfortunately, there is no miracle solution to this duality between collections or traditional taxonomy and metagenomics, except by organizing crossover studies between these different fields of activity and thus reinforcing the links between species, taxa and molecular OTUs. Collections remain as indispensable as ever in this context and it is simply necessary to ensure that the links with molecular data, which are already very lax, are, at last, adequately rewoven.

2.7. Conclusion

Natural history collections are alive and well and guarantee us a rich future of diachronic or large-scale scientific studies. The usual criticisms of their deficiencies or biases (Beck *et al.* 2012) do not hold if we rightly consider that we need to resample collections for every new study we conduct instead of considering that they should be improved according to an ideal universal protocol.

Their future study can be very fruitful but their future enrichment can very quickly become a problem and slow down significantly if we do not modify our behavior by making our data properly available and if we do not create the necessary bridges between collections, taxonomy and new forms of sampling. These are open science objectives which must be rigorously implemented in relation to the FAIR data concept.

The permanence and coherence of knowledge on biodiversity come at this price. All the knowledge we have on biodiversity is scattered in a huge and diverse literature and only the taxonomic names and the specimens that are potentially associated with them allow for this knowledge to be united.

2.8. References

Amano, T., Lamming, J.D.L., Sutherland, W.J. (2016). Spatial gaps in global biodiversity information and the role of citizen science. *Bioscience*, 66, 393–400.

Arino, A.H. (2010). Approaches to estimating the universe of natural history collections data. *Biodiversity Informatics*, 12, 57–62.

Ayris, P., Lopez de San Roman, A., Maes, K., Labastida, I. (2018). Open science and its role in universities: A roadmap for cultural change. *LERU Advice Paper*, 24, 1–31.

Beck, J., Ballesteros-Mejia, L., Buchmann, C.M., Dengler, J., Fritz, S.A., Gruber, B., Hof, C., Jansen, F., Knapp, S., Kreft, H., Schneider, A.-K., Winter, M., Dormann, C.F. (2012). What's on the horizon for macroecology? *Ecography*, 35, 673–683.

Charles, H. and Godfray, J. (2002). Challenges for taxonomy. *Nature*, 417, 17–19.

Costello, M.J., Wilson, S., Houlding, B. (2012). Predicting total global species richness using rates of species description and estimates of taxonomic effort. *Systematic Biology*, 61, 871–883.

Costello, M.J., May, R.M., Stork, N.E. (2013). Can we name Earth's species before they go extinct? *Science*, 339(6118), 413–416.

Dias Tarli, V., Grandcolas, P., Pellens, R. (2018). The informative value of museum collections for ecology and conservation: A comparison with target sampling in the Brazilian Atlantic forest. *PLoS ONE*, 13, xe0205710.

Feeley, K.J. and Silman, M.R. (2011). Keep collecting: Accurate species distribution modelling requires more collections than previously thought. *Diversity and Distributions*, 17(6), 1132–1140.

Fischer, R.A. (1930). *The Genetical Theory of Natural Selection. A Complete Variorum Edition*. Oxford University Press, Oxford.

Funk, V.A. (2004). 100 uses for an herbarium. Document, Division of Botany, The Yale University Herbarium, CT.

Funk, V.A. (2018). Collections-based science in the 21st century. *Journal of Systematics and Evolution*, 56, 175–193.

Garnett, S.T. and Christidis, L. (2002). Taxonomy anarchy hampers conservation. *Nature*, 417, 17–19.

Goodwin, Z.A., Harris, D.J., Filer, D., Wood, J.R.I., Scotland, R.W. (2015). Widespread mistaken identity in tropical plant collections. *Current Biology*, 25(22), R1066–R1067.

Grandcolas, P. (2017a). Loosing the connection between the observation and the specimen: A by-product of the digital era or a trend inherited from general biology? *Bionomina*, 12(1), 57–62.

Grandcolas, P. (2017b). The living species is not a natural kind but an intellectual construction. In *Life Sciences, Information Sciences*, Gaudin, T., Lacroix, D., Maurel, M.C., Pomerol, J.C. (eds). ISTE Ltd, London, and John Wiley & Sons, New York.

Justine, J.-L., Winsor, L., Gey, D., Gros, P., Thévenot, J. (2018). Giant worms chez moi! Hammerhead flatworms (Platyhelminthes, Geoplanidae, *Bipalium* spp., *Diversibipalium* spp.) in metropolitan France and overseas French territories. *PeerJ*, 6, e4672.

Le Bras, G., Pignal, M., Jeanson, M.L., Muller, S., Aupic, C. (2017). The French Muséum national d'Histoire naturelle vascular plant herbarium collection dataset. *Sci. Data*, 4, 170016.

Le Guyader, H. (2003). *Classification and Evolution*. Le Pommier, Paris.

Linnaeus, C. (1758). *Systema Naturae per regna tria naturae, secundum classes, ordines, genera, species, cum characteribus, differentiis, synonymis, locis. Editio decima, reformata, Tomus I.* Holmiae, Laurentii Salvii.

Mallet, J. (2008). Hybridization, ecological races and the nature of species: Empirical evidence for the ease of speciation. *Philosophical Transactions of the Royal Society B: Biological Sciences*, 363(1506), 2971–2986.

Marshall, S.A. and Evenhuis, N.L. (2015). New species without dead bodies: A case for photo-based descriptions, illustrated by a striking new species of *Marleyimyia* Hesse (Diptera, Bombyliidae) from South Africa. *Zookeys*, 525, 117–127.

May, R.M. (1988). How many species are there on Earth? *Science*, 241 (4872), 1441–1449.

May, R.M. (2004). Tomorrow's taxonomy: Collecting new species in the field will remain the rate-limiting step. *Philosophical Transactions of the Royal Society of London B*, 359, 733–734.

Meineke, E.K., Davis, C.C., Davies, T.J. (2018). The unrealized potential of herbaria for global change biology. *Ecological Monographs*, 88, 505–525.

Meyer, C., Kreft, H., Guralnick, R., Jetz, W. (2015). Global priorities for an effective information basis of biodiversity distributions. *Nat. Commun.*, 6, 8221.

Mills, J.A., Teplitsky, C., Arroyo, B., Charmantier, A., Becker, P.H., Birkhead, T.R., Bize, P., Blumstein, D.T., Bonenfant, C., Boutin, S. *et al.* (2015). Archiving primary data: Solutions for long-term studies. *Trends in Ecology & Evolution*, 30, 581–589.

Minteer, B.A., Collins, J.P., Love, K.E., Puschendorf, R. (2014) Avoiding (re)extinction. *Science*, 344(6181), 260–261.

Nilsson, R.H., Larsson, K.-H., Taylor, A.F S., Bengtsson-Palme, J., Jeppesen, T.S., Schigel, D., Kennedy, P., Picard, K., Glöckner, F.O., Tedersoo, L., Saar, I., Kõljalg, U., Abarenkov, K. (2018). The UNITE database for molecular identification of fungi: Handling dark taxa and parallel taxonomic classifications. *Nucleic Acids Research*, gky1022-gky1022.

Padial, J.M., Miralles, A., De la Riva, I., Vences, M. (2010). The integrative future of taxonomy. *Frontiers in Zoology*, 7, 16.

Pautasso, M. and McKinney, M.L. (2007). The botanist effect revisited: Plant species richness, county area, and human population size in the United States. *Conservation Biology*, 21, 1333–1340.

Pellens, R., Faith, D.P., Grandcolas, P. (2016). The future of phylogenetic systematics in conservation biology: Linking biodiversity and society. In *Biodiversity Conservation and Phylogenetic Systematics: Preserving Our Evolutionary Heritage in an Extinction Crisis*, Pellens, R. and Grandcolas, P. (eds). Springer, Basingstoke.

Perrier, E. (1886). *La philosophie zoologique avant Darwin*. Félix Alcan, Paris.

Pleijel, F., Jondelius, U., Norlinder, E., Nygren, A., Oxelman, B., Schander, C., Sundberg, P., Thollesson, M. (2008). Phylogenies without roots? A plea for the use of vouchers in molecular phylogenetic studies. *Molecular Phylogenetics and Evolution*, 48, 369–371.

Regnier, C., Achaz, G., Lambert, A., Cowie, R.H., Bouchet, P., Fontaine, B. (2015). Mass extinction in poorly known taxa. *Proceedings of the National Academy of Sciences*, 112(25), 7761–7766.

Rocha, L.A., Aleixo, A., Allen, G., Almeda, F., Baldwin, C.C., Barclay, M.V.L., Bates, J.M., Bauer, A.M., Benzoni, F., Berns, C.M. *et al.* (2014). Specimen collection: An essential tool. *Science*, 344(6186), 814–815.

Schilthuizen, M., Vairappan, C.S., Slade, E.M., Mann, D.J., Miller, J.A. (2015). Specimens as primary data: Museums and "open science". *Trends in Ecology & Evolution*, 30(5), 237–238.

Suarez, A.V. and Tsutsui, N.D. (2004). The value of museum collections for research and society. *Bioscience*, 54(1), 66–74.

Tancoigne, E. and Dubois, A. (2013). Taxonomy: No decline, but inertia. *Cladistics*, 29(5), 567–570.

Tringe, S.G. and Rubin, E.M. (2005). Metagenomics: DNA sequencing of environmental samples. *Nature Reviews Genetics*, 6, 805–814.

Troudet, J., Grandcolas, P., Blin, A., Vignes-Lebbe, R., Legendre, F. (2017). Taxonomic bias in biodiversity data and societal preferences. *Scientific Reports*, 7, 9132.

Troudet, J., Vignes-Lebbe, R., Grandcolas, P., Legendre, F. (2018). The increasing disconnection of primary biodiversity data from specimens: How does it happen and how to handle it? *Systematic Biology*, 67, 1110–1119.

Vaidya, G., Lepage, D., Guralnick, R. (2018). The tempo and mode of the taxonomic correction process: How taxonomists have corrected and recorrected North American bird species over the last 127 years. *PLoS One*, 13(4), e0195736.

de Vargas, C., Audic, S., Henry, N., Decelle, J., Mahé, F., Logares, R., Lara, E., Berney, C., Le Bescot, N., Probert, I. *et al.* (2015). Eukaryotic plankton diversity in the sunlit ocean. *Science*, 348(6237), 1261605.

3

Louis XIV's Blue Gems: Exceptional Rediscoveries at the French National Museum of Natural History

François FARGES
IMPMC – CNRS, Muséum national d'Histoire naturelle, Institut universitaire de France, Paris, France

Rare minerals – and more particularly gems – were the preferred instruments of power for past rulers. In the 19th century, these political objects became scientific: gemology was born as a branch of mineralogy. Between thefts, sales, covetousness, fashions and other alienations, the majority of these masterpieces of lapidary art have not survived. The rare historical gems that remain today tell lost or fragmented stories: the history of art – French academic art history has long neglected the expertise of gems, relegating them to a craft without creative inspiration. In a worst-case scenario, its expensive trade would definitely disqualify it from aesthetic research, which is dedicated to a pure Bohemian nature and insensitive to markets. Entire sections of this history are, therefore, unknown and left to various "fashionable storytellers" and others who magnify myths through a literary inventiveness that distorts the work of current reasearchers. Among them is the diamond known in English as the "French Blue", a misleading name because the gem was never named as such in French but rather as the "*Grand Diamant Violet de Sa Majesté*" (Great Violet Diamond of his Majesty) or the "Great Blue Diamond of Louis XIV".

For a color version of all figures in this chapter, see www.iste.co.uk/pellens/natural.zip.

Natural History Collections in the Science of the 21st Century,
coordinated by Roseli PELLENS.

Science – be it the history of art or physical chemistry of geomaterials – helps us to restore these objects in a more rigorous way. Thus, surface morphometry and optical spectrometry combined with quantum mechanics has allowed the fabulous blue diamond of Louis XIV, which was stolen in 1792 without leaving any obvious trace, to be reconstituted exactly. Other methodologies of mineralogy allow historical but undocumented exchanges to be reconstituted. Nanotechnologies have even made it possible to recreate an exact replica of this lost diamond, which still remains an absolute natural rarity despite the opening of numerous and very productive mines since the 18th century (Brazil, Africa, Siberia, Australia, Canada, etc.). Unpublished archives also allow us to better understand the often tormented past of these gems and their presence in the French National Museum of Natural History (*Muséum national d'Histoire naturelle* (MNHN), also known just as the *Muséum*).

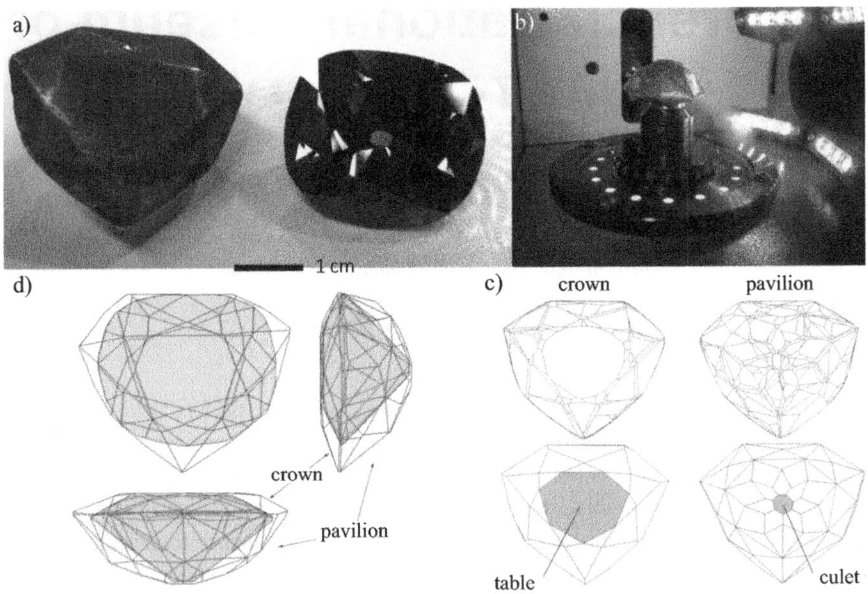

Figure 3.1. *From museum objects to digital models. © F. Farges/MNHN*

COMMENT ON FIGURE 3.1.– *(a) The blue diamond cast found at the MNHN in 2008 and the Hope diamond removed from its bezel. Note how the white paper stands out perfectly in the center of the Hope diamond, where it is thickest (12.9 mm), while in the periphery, the gem appears opaque while its thickness is reduced to 2 mm. The scan of the cast (b) allowed the arrangement of the polygons reflecting the rays of light to be located in terms of space. These data were then simplified on a computer*

(c) via decimation algorithms to reconstitute the effective facets of the gem, both for the top (crown) and the bottom (pavilion). This model was compared (d) to the one obtained on the Hope diamond in order to observe the correspondence between the two gems under three different angles.

3.1. Introduction

In the 17th century, Louis XIV took possession of two of the most beautiful blue gems known at the time: the *Grand Diamant Bleu* (Great Blue Diamond, also called the "French Blue") (69 carats, or almost 14 grams, Figure 3.1) and the *Grand Saphir* (Grand Sapphire) (135.7 carats, or more than 27 grams, Figure 3.2). They had the greatest honor as these two gems were listed at the top of the 1691 Crown Jewels inventory, items 2 and 3 respectively of the fabulous royal inventory listing hundreds of gems, amongst extraordinary gems and various jewels (Farges 2014) of which almost nothing remains today.

However, these two blue gems disappeared in September 1792, during the looting of the Hôtel du Garde-Meuble located on the present-day Place de la Concorde (Bapst 1889). Police investigations show that the jewel was probably taken by a certain Cadet Guillot, a young swindler who fled to London where he was eventually imprisoned (Bapst 1889). The diamond disappeared forever. As for the *Grand Saphir*, it must also have disappeared into the hands of thieves because the gem was not found amongst the few remaining gems. On the other hand, the sapphire was found shortly afterward, according to an inventory from December 1792 (Farges *et al.* 2015). It was deposited at the Hôtel de la Monnaie where it was recovered in 1796 by Daubenton (1716–1799), then a professor at the *Muséum*. Indeed, the 1789 inventory of the Crown Jewels described the item as a "sapphire piece" because the mineralogist Romé de l'Isle (1736–1790) published – in his *Essai de Cristallographie* (1772) – that the gem was, according to his studies, a sapphire crystal (an erroneous conclusion as the object was definitively faceted by humans, i.e. a gem).

Our research in the French National Archives (*Archives nationales*, AN, Minutier MC/ET/XXXIX/619) led us to the will of this little-known crystallographer. This document informed us of the existence of crystal models developed by this pioneer of the science of crystals, which he indicated as constituting the most precious legacy of his collection. Forgotten up until now, this legacy was found, identified and recollected (MIN000 5020) within the mineralogy collections of the *Muséum*: these hundreds of clay casts of unknown origin and age correspond in fact to the oldest models (in compacted clay of the finest texture) of Romé de l'Isle's crystals known to date and published in the opus of 1772 (Farges *et al.* 2015). Among them, No. 71 (Figure 3.2) corresponds, according to Romé, to the crystallographic model of the *Grand Saphir*, which he considered natural. As for the original, declared "uncut" in 1789 (AN, O[1] 3362) and therefore not very valuable, it found its "rough" acolytes at the *Muséum* under the tutelage of Daubenton in 1796, who understood well that he was making a bargain on behalf of his museum by acquiring such an exceptional gem. Even though Romé questioned his conclusion in 1783 in the second edition of his *Essai de Cristallographie*, he continued to suspect the *Grand Saphir* of being natural in a 1787 letter to Delamétherie (1743–1817), another French mineralogist of that period who sponsored Romé. Indeed, it was in fact a gem, probably polished by a Mughal hand in 17th-century India. Its most original faceting had fooled the experts of the Enlightenment, but not those of the Grand Siècle[1], a hundred years earlier... .

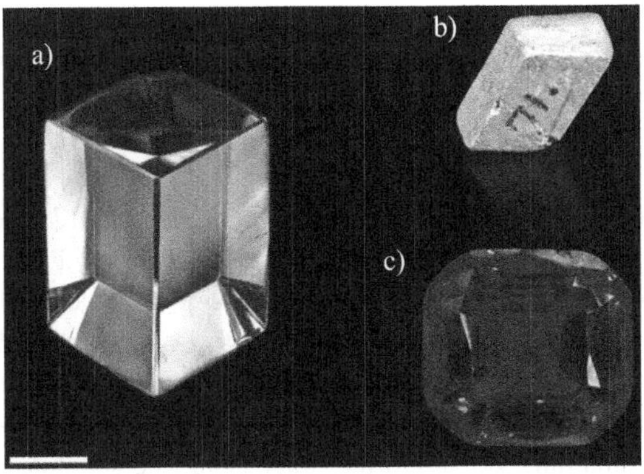

Figure 3.2. *Gems and models. © F. Farges/MNHN*

COMMENT ON FIGURE 3.2.– *(a) Louis XIV's Grand Saphir (MNHN inv. A.67); (b) ceramic model no. 71 by Romé de l'Isle (circa 1772) representing the Grand Saphir as a natural crystal of "adamantine spar" (corundum); (c) rhinestone model of the Ruspoli sapphire also found in the MNHN collections (inv. 50.167) and given by the same Parisian gem cutter as the model of the blue diamond (see Figure 3.1(a)): a certain M. Achard. The scale is 1 cm.*

More recently, it was said that this sapphire should be named the *Ruspoli*, after the eponymous Roman prince who is said to have owned it before Louis XIV (see references in Farges *et al.* 2015). The mechanics of storytelling are so effective that this sapphire is actually better known by this nickname than by its original name *Grand Saphir*. In fact, a misreading of a paragraph by Mr Barbot (see Farges *et al.*

1 A period in French history under the rule of Louis XIII and Louis XIV. The period was known for the development of art and literature.

2015) was at the origin of the confusion between the *Grand Saphir* and the *Ruspoli*, which are actually two different gems. This other sapphire made the headlines – from both a society and legal point of view – in 1813 for obscure reasons, due to the mischievous baiting that jewelers liked to throw back and forth. In fact, the *Ruspoli* was rediscovered at the *Muséum* (inv. 50.167), at least its historical replica in cobalt blue rhinestones, dating back to the 1820s (Figure 3.2). Its faceting is largely dissimilar to that of the *Grand Saphir*. The *Ruspoli* was still traceable until 1952 when it was likely purchased by Cartier in New York (Farges *et al.* 2015). We do not know what has happened to this gem since then.

For its part, Louis XIV's *Grand Saphir* remains strangely absent from the MNHN's brand image, even though it is certainly the most precious piece in its collections: for example, consider the popularity of the *Hope* diamond, the "muse" of the Smithsonian Institution in Washington and, more generally, of the Americans who recognize in this blue diamond no less than their own *Mona Lisa*. Indeed, some semi-confidential gossip in the jewelry world has long suspected the *Hope* diamond of being Louis XIV's blue diamond, cut by thieves or their collectors. Without proof, the story will remain as such. France quickly forgot its masterpiece since no painting of this French jewel is known. Until 2007 that is, when we found – in the collections of the *Muséum national d'Histoire naturelle* – the only cast of this gem stolen in 1792 (Figure 3.1(a)).

3.2. A scientific investigation of color

The first investigation consisted of deciding between the two opponents claiming to be the representation of the stolen French Blue diamond. On the one hand, we had the Russian clan which, through the *Terestchenko* diamond of 42.92 carats, claimed prestigious royal paternity by its ovoid shape. On the other hand, the American clan had the *Hope* diamond of 45.52 carats. A laser scan of the lead model was performed to obtain a 3D mesh of thousands of polygons of the lead cast of the French Blue diamond to find the truth of these two hypotheses. Numerical methods such as edge collapse decimation enabled the restoration of the original faceting of the gem. This result, when compared to the two competing representations, was clear: the *Terestchenko* was too long and not wide enough. The American won hands down (Figure 3.1(d)). The thieves of 1792 thus savagely recut the three main corners of the French triangular diamond to give it the present ovoid shape of the *Hope* diamond.

Then arose the question of the color of this mythical gem. Confusion reigned: the diamond was described as "violet" in the royal inventories of 1691, as "sapphire blue" in 1787 and then as "light blue" in 1813 (Farges *et al.* 2012). The study began with the *Hope* diamond, which was removed from its Cartier necklace for the occasion.

An optic oddity – never noted or interpreted before – was revealed as I placed the mythical diamond on a light background: in the center of the gem, where it is thickest, the object became very light in color, which contrasted with the rest of the volume, which was of a navy blue almost black. It became obvious that the gem behaved like a lightly colored parallel prism (Figure 3.1(c)), which no one had noticed. On the other hand, at the periphery, the diamond appeared dark even though it is thinner. To understand this curious anomaly, an optical absorption study was performed on the gem (Figure 3.3).

3.3. The digital decoding of the creative genius of the royal gem cutter

Considerable *ab initio* calculations of the color of the *Hope* diamond were undertaken to elucidate the anomaly (Farges *et al.* 2012). For this purpose, the theoretical dielectric function (ψ) of a diamond was calculated by considering a doping of its cubic atomic structure with boron atoms, present in trace amounts in the carbon atomic structure of this mineral and supposedly the source of its blue color (Fritsch 1998). We then solved the Bethe–Salpeter equation (Farges *et al.* 2012) that quantified the interaction between light and this mineral to determine its absorption spectrum between ultraviolet and near-infrared (400–900 nm). In summary, these complex calculations allowed us to reproduce and therefore understand the color anomaly of this diamond. In fact, the *Hope* diamond was not dark blue as it appeared at first sight, but pale blue! The calculation was able to show that its faceting amplified the pale color of the mineral constituting the *Hope* diamond to the point where the latter appeared blue (almost black) to us (Figure 3.3).

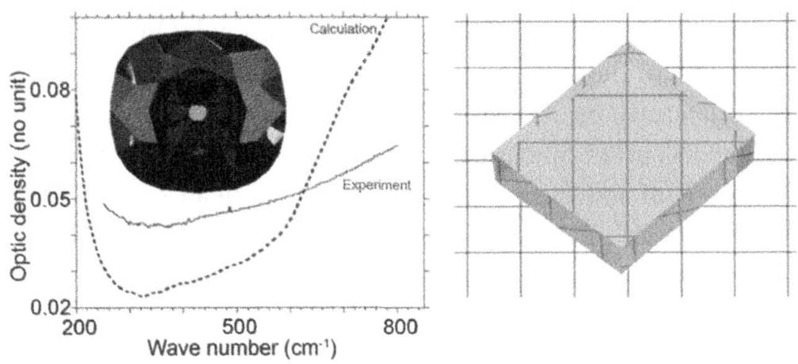

a) b)

Figure 3.3. *(a) Results of* ab initio *calculations on the color of the Hope diamond (dashed line) compared to the experiment (solid line) and photo-realistic simulation of this diamond reproducing the color contrast at the center of the gem (compare with Figure 3.1(a)); (b) photo-realistic simulation of this diamond without its cushion-faceting. Images © F. Farges/MNHN*

As the faceting of the *Hope* diamond differed drastically from that of its ancestor, we then had to recalculate the effective color of the French Blue diamond based on the scanned and decimated cast. The application of this concept to the Louis XIV blue diamond was based on the concept of critical angle (θ_2), which is related to the angle of total reflection itself as a function of the diamond's refractive index, worth 2.42 on average (i.e. $\theta_2 \sim 26°$, Figure 3.4(a)). The light path (ray 1 in Figure 3.4(b)) passed through the diamond given its angle of incidence (θ_1) zero with respect to the perpendicular angle of the diamond table. It passed through the diamond/gold diopter and returned the image of the underlying gold setting. The path (radius 2 in Figure 3.4(b)) corresponds to the incidence of light on the slightly off-center diamond (θ_1). Again, the angle of incidence was less than the critical angle and the light was still reflected by the gold. As for the third path (ray 3 in Figure 3.4(b)), it corresponded to the case where the light arrived on the facets of the pavilion with an angle greater than the critical angle: the light was then reflected in the diamond which increased its absorption and, consequently, the color perceived by the observer in return. Clearly, Louis XIV's gem cutter had mastered, knowingly or not, the founding concepts of linear optics published by Descartes (1596–1650) in *La Dioptrique* (Descartes 1637). The computer simulation of the Louis XIV blue diamond gave a diamond with blue hues centered on a sapphire blue. In the 17th century, the term "violet" was used to designate both dark blue and the current violet. As far as the Louis XIV diamond is concerned, it is the "sapphire blue variant" of "violet" that is the correct color for this diamond.

The 1691 Crown Diamond Inventory (AN, O^1 3360) tells us that the diamond was mounted on a "*baton d'or émaillé par derrière*" (gold stick enameled from behind). A second computer simulation (Figure 3.4(c)) of this diamond set in this manner was calculated following these archival indications. The result showed a central seven-pointed gold figure set in a deep blue background and studded with white spot chips. Clearly, this star was the emblem of the Sun King, Louis XIV, and, more generally, of the monarchical France of the time, whose symbolic colors were celestial Marian blue and solar gold (Farges 2014). The seven golden facets located around the central facet (culet) symbolize the seven planets then considered (Mercury, Venus, Earth, Mars, Jupiter and Saturn, plus the Moon). Depending on the inclination of the diamond, these facets lit up or went out while the culet remained "always lit with gold". We postulate that this corolla of seven facets represented an allegory of the graces and disgraces of the king toward the heavens, which were subject to his good will and in precession around the solar star. This "royal sun" shone in a blue background studded with stars representing the universe. These parables can be found in the allegories contemporary to the making of this decidedly extraordinary diamond, such as those of the Ballet de Flore (1669) where the king danced before the Court in the presence of the four white horses of Apollo's quadriga: the libretto by Isaac de Benserade (1612–1691) multiplied the allusions to the benefits of the sun on the Universe while the king was returning victorious from his first European conflict, the War of Devolution (1667–1668).

This heliocentric vision contrasted with the Platonic dogmas then recognized by the Church in which the Sun was supposed to orbit the Earth. It was not until 1755 that the Vatican admitted this "modernist" version of the mechanics of the solar system. Clearly, Louis XIV did not wait for the papacy: Descartes had already been published in his post-humous opus *The World* (originally dated 1633; Descartes 1664) where he took up the cause of Galileo and his assertions on heliocentrism. Through his "modernist" vision, Louis XIV asserted his Gallicanism above all against the ultramontanes (and other "Vatican-ultras" in France like the *Compagnie du Saint-Sacrement* (Company of the Blessed Sacrament) who were more subservient to the Vatican than to the power of the young sovereign, which the latter wanted to be absolute.

As for the *Grand Saphir* (Figure 3.2), it was also set in gold as the royal inventories indicate. We have been able to undertake a gemological study since the original gem has been preserved today: it constitutes, along with the *Hortensia* diamond of the Louvre Museum, the only two remains of the fabulous gem acquisitions by Louis XIV that have been kept intact to date. UV-vis-NIR spectroscopic analysis (Farges *et al.* 2015) of this gem shows that the sapphire originated in Sri Lanka. We can even see growth zones and inclusions that allow us to conclude that this sapphire has not been heated to accentuate its blue color. The zonations show that the original crystal was much larger, but these extra volumes must have been less transparent and were removed by recutting.

In the end, it is astonishing that the two blue gems of Louis XIV ended up in the *Muséum national d'Histoire naturelle*: one intact and the other in the form of a lead cast, the original of which has definitely disappeared. To a certain extent, this long-ignored lead cast has become the "original" and its precise study shows that its creator – still unknown at this stage – wanted to keep the trace of exceptional expertise by making this historical replica. My very recent observation (end of 2018) of the surface of this lead cast would seem to indicate that the date "1806" is inscribed on it. This type of lead casting is usually done prior to a sale or re-cut so that the gem cutter can keep track of his excellence for future customers. If this hypothesis is true, this information would indicate the date of the recutting of the Louis XIV blue diamond, which corresponds exactly to the time window between the theft of the French diamond in Paris and the first appearance in London (1792–1812) of what would later be named the *Hope* diamond, now kept at the Smithsonian.

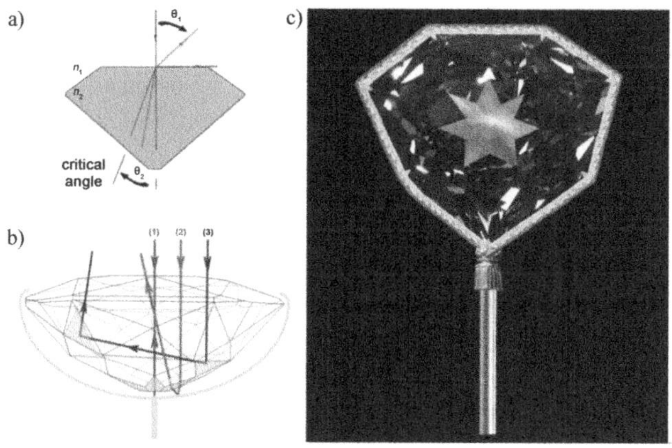

Figure 3.4. *Optical principles of a generic faceted diamond explaining definition and measurement. © F. Farges/MNHN*

COMMENT ON FIGURE 3.4.– *(a) Measurement of the critical angle, θ2, for an air (1)/gem (2) interface (also called a diopter). For a light incidence (θ1) smaller than this angle, the diamond will behave like a transparent diopter. Otherwise, the facets of the gem will behave like mirrors: the light will be more absorbed and the color will be perceived as more intense (darker); (b) optical principles for three typical light paths; (c) computer simulation of the Louis XIV blue diamond on the basis of the MNHN inv. 50.165 cast and the color of the* Hope *diamond and then placed in a model of a setting according to the 1691 inventory.*

3.4. Epilogue: toward a renaissance...

To free ourselves from the virtual, we have reconstructed the "almost real" diamond. The challenge is potentially expensive: if this mineral had actually existed, the price would have been stratospheric. However, another path was considered: cubic zirconia, ZrO_2, is a crystallized zirconium oxide that has optical properties (refractive index and optical dispersion) close enough to those of a diamond without costing a fortune. But no cubic zirconia has the "cold blue" color so typical of blue diamonds. On the contrary, they show "warm" blues matted with purplish red, which gives us the wrong idea of Louis XIV's blue diamond. We then opted for a "physical" solution. Indeed, it is possible to give the illusion of a color by sputtering, that is, depositing metallic particles "nano-pulverized" at high temperature and under ultra-high vacuum on a substrate of colorless zirconia. The color – obtained by light scattering by this thin layer of nanoparticles – is adjustable according to the operating conditions (ultra-high vacuum, temperature, nature of the metal to be deposited, etc.). After 2 years of testing, a colorless zirconia cut with the exact shape and faceting of Louis XIV's blue diamond was "nano-coated" on its surface: it almost perfectly reproduces the theoretical simulation of the original blue diamond (Escobar 2017). Moreover, this replica will soon be set in gold so that we can see this magnificent diamond again as it once appeared in the eyes of the Sun King: a political instrument in the colors of a brilliant monarch. As amazing as it stands, no gem cutter – even one with modern, digitally enhanced skills – would have dared to imagine such a level of sophistication.

3.5. References

Bapst, G. (1889). *Histoire des Joyaux de la Couronne de France*. Hachette, Paris.

Descartes, R. (1633). *Traité du monde et de la lumière*. Bobin et Le Gras, Paris.

Descartes, R. (1637). *La Dioptrique*. Maire, Leyden.

Escobar, N. (2017). Now there are near-perfect copies of the hope diamond. *Smithsonian Magazine* [Online]. Available at: https://www.smithsonianmag.com/smithsonian-institution/now-there-are-near-perfect-copies-hope-diamond-180967372/.

Farges, F. (2014). Les grands diamants de la Couronne de François Ier à Louis XVI. *Versalia*, 17, 55–78.

Farges, F., Sucher, S., Horovitz, H., Fourcault, J.-M. (2008). Deux découvertes majeures autour du "diamant bleu de la Couronne". *Revue de Gemmologie*, 165, 17–24.

Farges, F., Vinson, J., Rehr, J.R., Post, J.E. (2012). The rediscovery of the "French Blue" diamond. *Europhysics News*, 43, 22–25.

Farges, F., Panczer, G., Benbalagh, N., Riondet, G. (2015). The Grand Sapphire of Louis XIV and the Ruspoli Sapphire. *Gems & Gemology*, 51, 392–409.

Fritsch, E. (1998). The nature of color in diamonds. In *The Nature of Diamonds*, Harlow, G. (ed.). Cambridge University Press, Cambridge.

Romé de l'Isle, J.M L. (1772). *Essai de cristallographie*. Didot jeune, Paris.

4

Rediscovering Human Mummies: Unpublished data on the Chachapoya Mummy Exhibited at the *Musée de l'Homme*

Aline THOMAS[1], Arnaud ANSART[2], Christophe BOU[3], Jean-Bernard HUCHET[3,4], Véronique LABORDE[5], Samuel MERIGEAUD[6] and Éloïse QUETEL[7]

[1] *EA – CNRS, Muséum national d'Histoire naturelle, Université Paris Diderot, France*

[2] *Consultant*

[3] *PACEA, UFR Sciences Odontologiques, CNRS – Université de Bordeaux, Pessac, France*

[4] *AASPE, ISYEB – CNRS, Muséum national d'Histoire naturelle, Sorbonne Université, École Pratique des Hautes Études, Université des Antilles, Paris, France*

[5] *DGDcoll, Muséum national d'Histoire naturelle, Paris, France*

[6] *Clinique du Parc, Castelnau-le-Lez, France*

[7] *Pôle Patrimoine, Sorbonne Université, Paris, France*

As a biological, archaeological and ethnographic vestige, the mummy is at the intersection of multidisciplinary issues and requires the use of adapted tools and techniques. By presenting the work carried out on the Chachapoya mummy, an emblematic figure of the *Musée de l'Homme*, this chapter presents the means used to

---

For a color version of all figures in this chapter, see www.iste.co.uk/pellens/natural.zip.

Natural History Collections in the Science of the 21st Century,
coordinated by Roseli PELLENS. © ISTE Ltd 2021.

study the mummified human remains of the *Muséum's* anthropological collections. After a brief historical review (chrono–cultural origin of the mummy, context of its discovery, its entry into collections, up to its current exhibition), this chapter puts forward for the first time an illustrated scientific assessment of the mummy, obtained by means of various analyses: bio-medical exploration (virtual autopsy by scanner imaging), comparative experimentation (cranial trepanation) and archeo-entomological study.

4.1. Introduction

The scientific appeal of the *Musée de l'Homme*'s Chachapoya mummy on the cultural, funereal, paleoanthropological and even medical levels is closely linked to museum and conservation issues. Exhibited at the Palais de Chaillot for more than a century, its long museographic history mirrors the scientific and scenographic evolution of the *Muséum*. It claimed its place in the "history of art" as the inspiration for Edward Munch's *The Scream* and several other works by Paul Gauguin. It fascinates people far beyond the realm of natural sciences, and continues to be a key figure in the *Muséum's* anthropological collections. In spite of its fame, the mummy MNHN-HA-30187 (formerly 1879.1.22) had only been the subject of occasional, scattered expert appraisals, which have remained unpublished to date. This chapter aims to bring together these different analyses and to present their main results. Medical imaging, which gives non-invasive access to the internal anatomy of the body – organic and skeletal remains – allows us to determine the biological and health identity of the deceased. Combined with experimental research aimed at reproducing the cranial trepanation seen on the individual, as well as a specific analysis of insect remains discovered on the subject's skin, this data accurately documents the mortuary practices applied to the body of the deceased prior to mummification.

4.1.1. *The* Muséum's *collection of human mummies*

Because of the quantity and chrono/geographical diversity of its specimens, the *Muséum's* anthropology collection is one of the four largest of its kind in the world. It consists of approximately 200 complete or partial bodies from the five continents, dated between the 2nd millennium BCE and the 19th century CE. More broadly, the category of mummies extends to all human remains that include soft tissue: eviscerated and bandaged bodies, autopsy and skeletonized mummies, shrunken heads ("tsantzas"), anatomical preparations, dried skin, and even scalps. The term "mummy" does not presume the preservation method of the body, which may be the result of intentional preparation or of a favorable natural environment (cold and anaerobic, for example). It was from the 19th century onwards that the current collection of anthropological pieces of the *Muséum*, mummified or not, was essentially created (Froment and Mennecier 2014)[1]. However, some pieces were collected well before this date. The oldest complete mummy preserved in the collections, for example, was given to the *Cabinet du Roi* in 1756[2]. The anthropology collection, which has grown through expeditions, donations, acquisitions and deposits, has been housed at the *Musée de l'Homme* since its creation in 1937, after having been merged with, on this same site, the collections of the former *Musée d'ethnographie du Trocadéro*.

In the context of the renovation of the *Musée de l'Homme* (carried out from 2009 to 2015), a major inventory work, coupled with a phase of protection and restoration of the mummified remains, was undertaken in 2006 by A. Raggi[3]. This work made it possible to update a detailed inventory of the collection to produce an illustrated condition report and to ensure an optimal conservation framework for each specimen through an adapted individual reconditioning. In 2018, the mummy collection, temporarily stored on the *Jardin des Plantes* site, was returned to the *Musée de l'Homme*, along with the rest of the *Muséum's* anthropology collections. It now occupies a dedicated room, with controlled temperature and humidity, and can now be consulted again.

The laborious work of "technically" repairing the mummies was the *sine qua non* of any research work pertaining to the collection. The CNRS research group (GDR 3446) "*Centre d'investigation et de recherche sur les momies*" (CIRM), led by A. Froment[4] from 2011 to 2015, laid the groundwork for a multidisciplinary scientific experiment, in collaboration with various research institutes, museums and health professionals. The definitive return of the collection to the *Musée de l'Homme* and its "reopening" to researchers was for us, former participants of the GDR and current *Muséum* staff attached to the mummy collection, an opportunity to initiate a new phase of collective work aiming, first of all, to synthesize the analyses carried out on the Chachapoya mummy MNHN-HA-30187.

---

1 The first inventory register of anthropology collections was created in 1857 (Mennecier 2006).

2 Mummy described in: Buffon, Georges-Louis Leclerc, *Histoire naturelle, générale et particulière avec la description du Cabinet du Roy*, volume XV.

3 Restorer, in charge of the conservation and restoration of the mummies of the Museum until 2013. A. Raggi was assisted in this work by C. Lair (MNHN) and various other people, both volunteers and on a contractual basis. This work was continued between 2013 and 2017 by E. Quétel, curator–restorer of heritage, specialising in human remains and organic materials.

4 Scientific manager of the Museum's anthropobiology collections until 2016, succeeded by S. Bahuchet.

4.1.2. *Origin, discovery, donation and exhibition: a brief history of the mummy*

The Incas named Chachapoyas, or "warriors of the clouds", large populations with white skin who lived in the Inca province. Organized into autonomous regions between the 9th and 15th centuries CE, they were colonized by the soldiers of the Inca emperor Tupac Yupanqui between the end of the 15th and the beginning of the 16th century (de Cieza de León 1984). Their territory extended approximately 65,000 km^2, in the north of the Andes in modern-day Peru, between 2,000 and 3,200 m of altitude (Cornejo García 2002). For pre-Hispanic Andean societies, the dead participated in creation myths and in defining local families or *ayllu* (Doyle 1998), over whom they could exercise power, beneficial or otherwise (Mantha 2009; Lau 2013). These influential ancestors thus required care and attention from the living (de Arriaga 1968). The Chachapoya dead were found in a variety of contexts and funereal structures: in individual anthropomorphic or *purunmachus* sarcophagi, inside collective structures with the appearance of limestone houses, *chullpas*, and sometimes in caves (Kauffmann *et al.* 1989; Nystrom 2003; Schjellerup 2005; Knutson 2006; Fabre *et al.* 2008). Corpses could be buried until the flesh decomposed, then exhumed so that their bones could be gathered and wrapped in cloth (Nystrom 2003; Friedrich *et al.* 2010). Other bodies were, on the contrary, voluntarily preserved through embalming. Evisceration was performed through an anatomical orifice of the corpse (vagina, anus), which was then plugged with a rolled up cloth pad. In order to preserve a certain volume of the body, cotton balls were inserted between the teeth, cheeks, and nose (Guillén 1998; Wild *et al.* 2007). The body was finally placed in a folded position, knees brought to the face, fingers bound and tied to the head, and then wrapped with several layers of cloth made up in *fardo*.

Figure 4.1. *Current state of the discovery site of the Chachapoya mummy MNHN-HA-30187, photo © S. Ziemendorff*

It was in 1877 that the explorer P. Vidal-Senèze discovered the Chachapoya mummy, in the heart of the Utcubamba valley of the Piedra Grande mountain. According to the descriptions given by P. Vidal-Senèze, this one was laid to rest, with others, in a tomb in the shape of a "spherical cap, or of a beehive of bees" (Vidal-Senèze 1877), made of stone and clay, overhung by a sculpted top and nestled in the side of cliff (Figure 4.1). The mummies were wrapped in cloth and rope at the time of their discovery and accompanied by objects, bags and pottery. Four of them were sent to France (Vidal-Seneze 1877; Vidal-Senèze and Noctzli 1877).

The mummy MNHN-HA-30187 entered the collections of the *Musée d'ethnographie du Trocadéro*'s (MET) Department of America in 1879, two years after its discovery; it was registered in the inventories under the number 1879.1.22 ("Vidal-Senèze collection"). Less than 20 years later, E.-T. Hamy, director of the MET, gave a description of it in a book dedicated to the American gallery of the *Muséum* (Hamy 1897). We learn here that the mummy was presented to the public, stripped of its cloth envelope. It was cordoned off and accompanied by a tapestry bag with geometric designs. Next to it was another mummy from the Vidal-Senèze collection (no. 1879.1.21) swaddled in its original wrapping (Figure 4.2(a)). The mummy remained on public display after the creation of the *Musée de l'Homme*, in the section dedicated to "New America" (Figures 4.2(b) and (c)). It became part of the *Muséum*'s anthropology collection in 1981.

The closing of the *Musée de l'Homme* for its refurbishment, which lasted between 2009 and 2015, was an opportunity for a detailed assessment of the mummy's state of conservation. It was found that it had not undergone any substantial degradation since the description made by E.-T. Hamy in 1897. However, since that date, the bag that had accompanied it has been removed and kept separately. In the absence of any alteration other than soiling due to the years spent in the exhibition spaces, the mummy was subjected to a superficial dusting and cleaning[5]. A pedestal system was created in order to allow good maintenance without compromising its initial position. Since the reopening of the *Musée de l'Homme* in 2015, the mummy has been displayed in the showcases of the permanent exhibition space dedicated to funereal treatment, "Living, Dying and Being Aware" (Figure 4.2(d)).

5 Dusting with a soft brush and a HEPA filter vacuum cleaner equipped with a micro-suction kit; cleaning with a latex resin and RubGum®, resin by E. Quétel.

Figure 4.2. *Museum history of the Chachapoya mummy MNHN-HA-30187 at the Palais de Chaillot*

COMMENT ON FIGURE 4.2.– *(a) 1895: Exhibited at the* Musée d'ethnographie du Trocadéro, *photo* © Musée du quai Branly – Jacques Chirac, *Dist. RMN-Grand Palais/ image* Musée du quai Branly – Jacques Chirac. *(b) 1939: Inauguration of a new section of the* Musée de l'Homme, *dedicated to "New America" by Jean Zay, Minister of National Education, in the presence of Paul Rivet, a French physician and ethnologist who initiated the creation of the* Musée de l'Homme *in 1937, photo:*

© *Excelsior – L'Équipe/Roger-Viollet. (c) Same showcase, 1939, photo* © Musée du quai Branly – Jacques Chirac, *Dist. RMN-Grand Palais/image* Musée du quai Branly – Jacques Chirac. *(d) 2015: Exhibited in the permanent exhibition of the new* Musée de l'Homme, *showcase devoted to funereal treatment, photo © MNHN – JC Domenech.*

4.2. Materials and methods

4.2.1. The MNHN-HA-30187 mummy: position of the body, measurements and external appearance

The mummy is sitting on its buttocks, in the "fetal position" (Figure 4.3(a)). The lower limbs are bent, the knees brought up toward the chin, the legs bent over the thighs and the thighs up to the chest. The feet are positioned on top of each other, the left over the right, brought back against the pubis. The upper limbs are folded, the arms along the body and the forearms against the chest. The closed hands come in contact with the cheeks, with the palms against the mandible. The head is slightly tilted to the right, slumped into the shoulders, and the mouth is open.

A tie made of four strands, wound in pairs, maintains the folded position of the limbs by surrounding the mummy (Figure 4.3(c)). It passes to the front below the knees, opposite the middle of the forearms, then backwards at the level of the scapula and the upper thoracic spine. In this "hyperflexed" position, the mummy occupies a minimum space of 73 centimeters in height, 36 centimeters in bilateral width, and 38 centimeters in anteroposterior width.

The skin is gray-beige in color. It appears distended, especially on the face, where it covers part of the upper right orbit, or the occlusal surface of the lower teeth. The skin is eroded in places, and part of the skeleton is sometimes visible, especially on the extremities of the limbs: ankles/feet and hands (Figure 4.3(d)). On most of the body, the skin shows multiple traces of activity of necrophagous insects.

On the back of the skull, the skin is cut away to reveal a cranial cavity, which retains the trace of a large perforation of anthropic origin (Figure 4.3(b)). This craniectomy, commonly called cranial trepanation, is located at the level of the left lambdoid suture, straddling the left parietal bone and the occipital bone. It is subcircular in shape and has a "serrated" appearance at its edges, which suggests a multi-drilling technique for the extraction of a bone flap.

Figure 4.3. *Current state of the mummy MNHN-HA-30187. (a) General photographs of the mummy. (b) Detail of the trepanation at the back of the skull. (c) Detail of the rope that binds the mummy. (d) Detail of the erosion of the skin observed on the feet, photos © MNHN – JC Domenech*

4.2.2. *Medical imaging protocol and technique*

To transport the mummy to the hospital center for a CT scan[6], and in anticipation of the examination itself, the mummy was packaged using radio-transparent material for two main reasons. The first was of a conservationist nature in order to limit direct contact between the mummy and physical variations in the atmosphere (temperature, hygrometry), potential sources of degradation. The second was sanitary, with respect to the hospital space and the scanner room. It was indeed imperative to avoid any risk of transmission of nosocomial infection via equipment used for patients who are sometimes immunocompromised. The use of such a container made it possible to eliminate the risks of contaminating the atmosphere with possible germs originating from the mummy, spores in particular. A protocol for cleaning and disinfecting the equipment was also deployed at the end of the procedure, following the recommendations of the *Comité de lutte contre les infections nosocomiales* (CLIN). The examination consisted of a multidetector scan[7] of the whole body in standard resolution[8], then a targeted scan of the teeth in high resolution[9].

4.2.3. *Protocol for experimental reproduction of trepanation*

In order to study the operating mode of the perforation carried out at the back of the mummy's skull, to estimate the drilling technique, the tools used and the duration of the operation, a protocol for experimental reproduction on an actual skull was developed (Bordeaux Forensic Odontology Laboratory). This protocol was based on the theory that an opening of the cranial cavity created by removing a bone-skin flap through successive and confluent micro-drilling was carried out (see method no. 3 of Lisowski (1967)). The use of flint points – created for this experiment – was the preferred option here with respect to the archaeological context of the Chachapoyas (Toyne 2015). Following CT imaging, segmentation and 3D surface reconstruction of the trepanation (Figure 4.4(a)), a resin duplicate was created by proto-typing (b) for

6 The scans were acquired on February 19, 2011 at La Pitié Salpêtrière (Department of Pr. Grenier). The authors thank Dr. Isabelle Huynh-Charlier for her technical assistance during these scans.

7 MDCT Philips Brilliance 64-channel, The Netherlands.

8 "Standard" acquisition parameters. Collimation: 64 × 0.625; 120 kV/50 mAs. Slice thickness: 1.5 mm; increment: 0.75 mm; filter B; matrix: 512. Cutting thickness: 0.67 mm; increment: 0.33 mm; YB filter; matrix: 768.

9 "Ultrahigh resolution" acquisition parameters. Collimation: 20 × 0.625; 140 kV/100 mAs. Slice thickness: 0.625 mm; increment: 0.5 mm; D filter; matrix: 512.

the creation of a matrix (c), which was then positioned on an anatomical model to replicate the perforation (d).

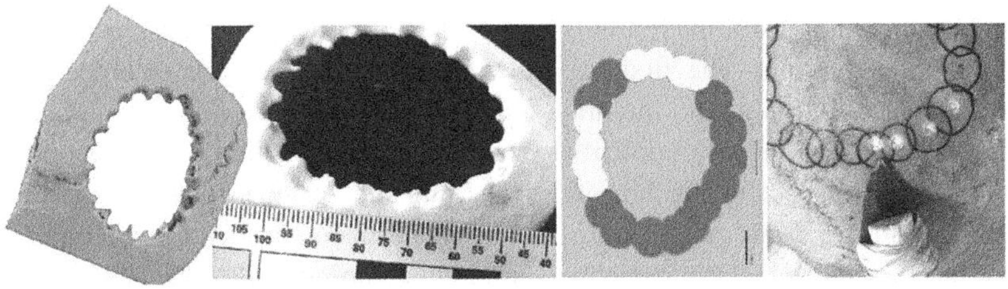

Figure 4.4. *Experimental reproduction protocol of the cranial trepanation of the mummy MNHN-HA-30187*

COMMENT ON FIGURE 4.4.– *(a) 3D surface reconstruction of the trepanation. (b) Resin duplicate. (c) Matrix for the reproduction of the peripheral perforations, 10 mm diameter (in red) or 70 mm diameter (in yellow). (d) Tracing of the matrix on the actual skull and production of flint points. Images and photos: © Dr. C. Bou*

4.3. Results

The CT imaging of the mummy enable us to see all of the individual, generally and in detail, from all angles (Figure 4.5). Depending on the image processing, the skeleton or the soft tissue can be examined. These images complement surface observations and allow for an initial biological assessment.

4.3.1. *Basic biological identity*

Despite the small volume occupied by the mummy due to its constricted position, its size is comparable to that of an adult. This is confirmed by the scan data, which allows a detailed visualization of the bone and dental state of the subject (Figure 4.6). The individual shows complete maturation of the long bones (fused conjugation cartilages), closure of the spheno-occipital synostosis and complete calcification of the third molars, indicating an age of over 20 years. The symphyseal pubic articular surfaces are relatively difficult to study in 3D because of the presence of fibrous cartilage. However, a not completely fused nucleus on cross-sectional images (Figure 4.6(c)) suggests an age of less than 40 years (Suchey and Brooks 1986). This age category is consistent with low dental attrition, seen even on the first molar (Figure 4.6(e)).

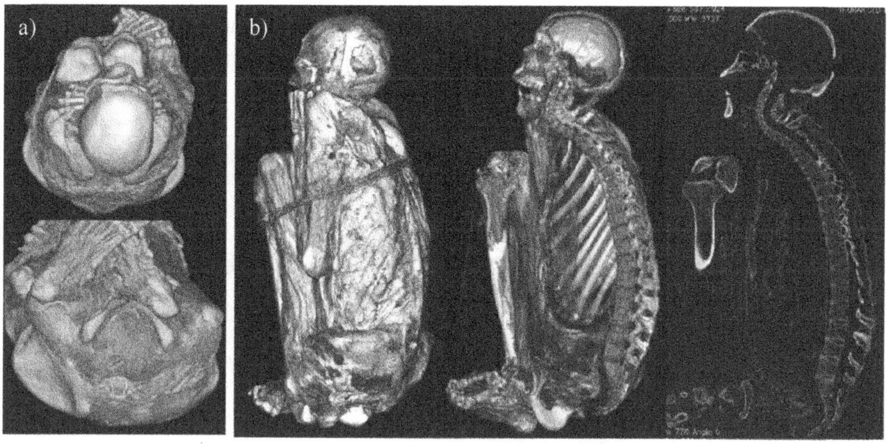

Figure 4.5. *Examples of images of the mummy obtained from CT imaging. (a) Views from above and below. (b) Left profile views, and sagittal sections progressively shifted to the right profile, according to the application of various color filters. Images: (a) © Dr. S. Mérigeaud MD/Tridilogy; (b) © Dr. C. Bou*

CT images are consistent with the presence of a penis, but structures at this level are difficult to recognize. Coxal bone measurements allow for a "probabilistic sexual diagnosis" based on a known global sex reference corpus (Murail *et al.* 2005). Of the 10 coxal bone measurements proposed by the method, six[10] could be taken from the scanned images, classifying this specimen as "male" with a probability of 0.99. The maximum length of the humerus (317 mm), tibia (371 mm) and femur (463 mm) were measured. Based on femur length, which is best correlated with overall stature, their height is between 1.71 and 1.75 m, according to regressions based on current populations (Trotter and Gleser 1958; Cleuvenot and Houët 1993).

4.3.2. *Osteo-dental status*

There are no acquired skeletal abnormalities. The bone is not altered and there is no sign of arthrosis. However, there is a congenital transitional anomaly of the lumbosacral spinal hinge, commonly seen in modern populations: a lumbarization of the first sacral vertebra (S1) giving the impression that there are six lumbar vertebrae instead of five (Figure 4.7). The osteodental pathologies detected through

10 *Pum* (pubic length) = 73.0 mm; *spu* (pubo-acetabular width) = 28.0 mm; *Dcox* (maximum height of coxal bone) = 209.8 mm; *Iimt* (height of greater ischial incisure) = 32.2 mm; *Scox* (ilium width) = 146.7 mm; *sa* (spino-auricular length) = 72.6 mm.

imaging are the three carious lesions present on the contact faces between the upper left first molar (distal face) and second molar (mesial face) (Figure 4.6(c)) and on the lower left third molar.

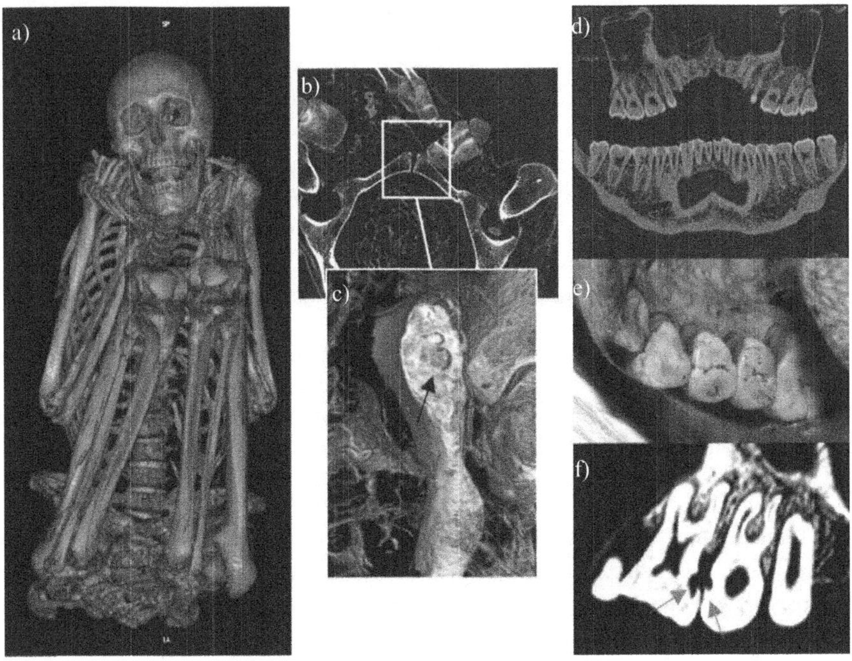

Figure 4.6. *Osteo-dental condition of mummy MNHN-HA-30187*

COMMENT ON FIGURE 4.6.– *(a) General view of the skeleton in anterior view after CT image processing. (b) Axial section and (c) 3D virtual reconstruction of the articular surface of the symphysis pubis. (d) General view of the dental arches. (e) Photograph of the maxillary teeth. (f) Detailed CT section of the caries present on the upper left molars 1 and 2. Images a, f and e: © Dr. C. Bou; images b, c and d: © Dr. S. Mérigeaud MD/Tridilogy*

4.3.3. *Internal organs*

The soft tissues are well preserved externally and internally, except for the brain which is not visible. Only the dura mater is present inside the skull and the spinal canal (Figure 4.8(a)). The soft parts of the face and neck are also present (soft palate, tongue, larynx) (Figure 4.8(b)). At the thoracic level, the trachea, lungs, heart and diaphragm are clearly visible (Figures 4.9(a) and (c)). The presence of mediastinal and pulmonary calcifications (Figure 4.9(d)) is suggestive of tuberculosis. It could also be granulomatosis. The abdomen and pelvis show visceral structures that are difficult to recognize (possible mesentery), apart from the liver and kidneys (Figures 4.9(e) and (f)), which are atrophic due to dehydration.

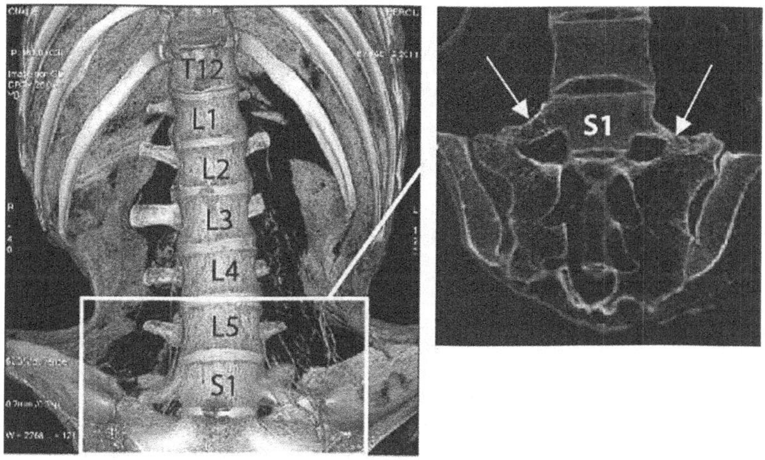

Figure 4.7. *Coronal section and 3D reconstruction of the lumbosacral junction, showing the transitional anomaly. Images © Dr. S. Mérigeaud MD/Tridilogy*

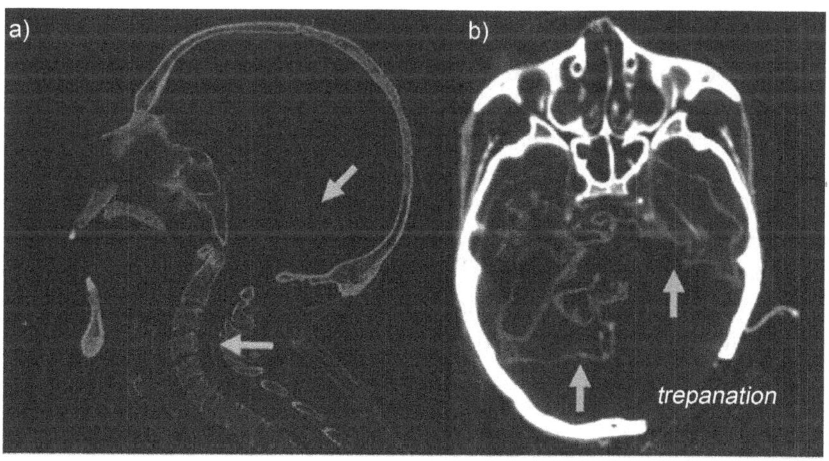

Figure 4.8. *(a) Sagittal and (b) axial sections showing intracranial and intraspinal dural remains and preservation of oropharyngeal-laryngeal soft tissue. Images © Dr. S. Mérigeaud MD/Tridilogy*

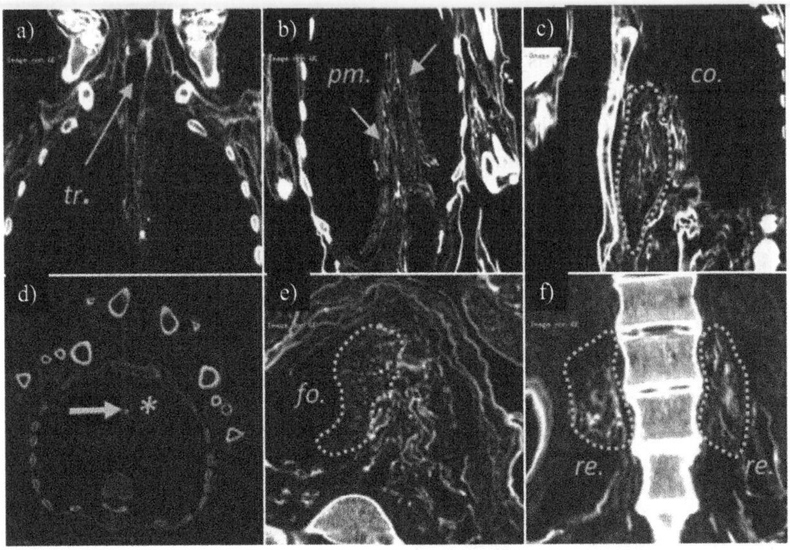

Figure 4.9. *Chest CT images. (a) Coronal section of the trachea [tr.]. (b) Coronal section of the lungs [pm.]. (c) Sagittal section of the heart [co.]. (d) Axial section of the thorax: mediastinal and pulmonary calcifications [*]. (e) Axial section of the liver [fo.]. (f) Coronal section of the two kidneys [re.]. Images: © Dr. S. Mérigeaud MD/Tridilogy*

4.3.4. *Archeoentomology*

External examination of the mummy revealed multiple traces of activity of necrophagous insects. Although visible on a large part of the body, these stigmata are mainly located on the lower limbs and the perianal region. However, to a lesser degree, similar marks can be seen on the collar region and the face. These circular perforations result from the action of the digestive juices of dipteran larvae (flies), which have consumed part of the muscles, viscera and skin. Close examination of various parts of the body, especially the thighs and the perianal region, reveals the presence of a large number of extremely well preserved *puparia*[11], attributable to a single species of Calliphoridae fly (Figure 4.10(a)).

---

11 During pupation (passage from larva to adult), the external envelope of the "maggot" hardens and constitutes a small rigid barrel (*puparium/plur. puparia*) inside which the insect enters into metamorphosis.

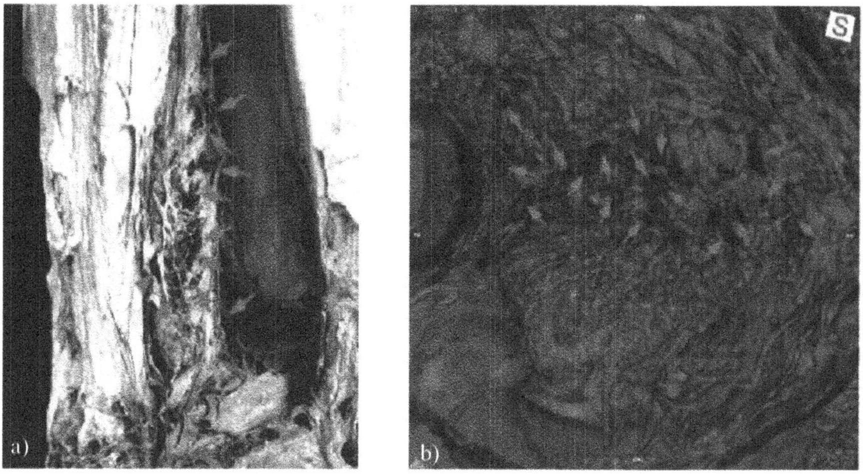

Figure 4.10. *Location of Calliphoridae diptera* puparia *(red arrows) (a) on the left thigh of the mummy (photo: © J.B. Huchet), and (b) in the abdominal region, internal view. Image: © Dr. C. Bou*

In parallel, CT examination of the abdomen revealed the presence of innumerable thin-walled oval structures with aerial contents of identical size in the abdominal cavity (Figure 4.10(b)). These structures are unmistakably related to *puparia* of necrophagous Diptera and indicate an internal colonization of larvae that pupated *in situ*.

From a taxonomic point of view, the conformation of the *puparia* (notably the arrangement of the taste buds and the shape of the respiratory stigmata) indicates that they correspond to immature stages of Calliphoridae diptera (Chrysomyinae), a "pioneer" family of flies of forensic interest, which lay eggs on corpses in the moments following death. Morphologically, these *puparia* are similar to those of the genus *Compsomyiops* Townsend, 1918 (Figure 4.11). This taxon, mostly neotropical, has six species, three of which are found in Peru (Gonzalez-Mora *et al.* 1998). Unlike most of the synanthropic Calliphoridae flies used in forensic science for estimating the *postmortem* interval (PMI), species of the genus *Compsomyiops* show a strong preference for environments little or not influenced by human activities (i.e. asynanthropy) (Figueroa-Roa and Linhares 2002; Mariluis *et al.* 2008).

Greenberg and Szyska (1984), who studied the biology and ethology of two species found in Peru (*Compsomyiops verena* (Walker) and *C. arequipensis*

(Mello)[12]), attest that the lifecycle of *C. verena* (egg to adult) varies between 15.2 and 19.7 days. At significantly higher temperatures, however, Dale and Prudot (1986) report a duration of 12.5 to 15 days for *C. arequipensis*. Based on observations by Greenberg and Szyska (1984), supplemented by Baumgartner and Greenberg (1985), *C. arequipensis* preferentially occupies high-altitude environments and *C. verena* low valley environments. Data from Mariluis and Schnack (1996) on the biology of representatives of the genus *Compsomyiops* in Argentina reveal that these species are particularly fond of forest and rural environments and are only active during the warmest months of the year. Finally, the experiments conducted by Figueroa-Roa and Linhares (2002) underline a very marked heliophilia[13] in the species *Compsomyiops fulvicrura* (Robineau-Desvoidy) in Chile.

Although, as Greenberg and Szyska (1984) point out, the similarities between the immature stages of these two species cannot be denied, data from their biology and from the environmental characteristics of the site where the mummy was found (Reichlen and Reichlen 1950) allow us to hypothesize that the species could correspond to *C. arequipensis* (Mello) (Figure 4.11).

Figure 4.11. *Remarkably preserved subfossil puparium of a Calliphoridae fly (*Compsomyops *cf.* arequipensis *(Mello)) from the mummy's left thigh. Photo © J.B. Huchet*

4.3.5. *Cranial trepanation: location, size and mode of operation*

The cranial trepanation present in the mummy, subcircular in shape and with a diameter of approximately 5 cm, extends over an area of more than 20 cm^2 (Figures 4.12(a) and (b)). Endocranial 3D reconstructions show that this trepanation avoids

12 Species cited as *C. boliviana* (Mello).

13 Who seeks the light of the sun.

the main pericerebral venous (superior longitudinal sinus, torcular, left lateral sinus) and arterial vascular structures (the imprints of the main branches of the meningeal arteries, visible in the endocranial view, are remote) (Figure 4.12(c)).

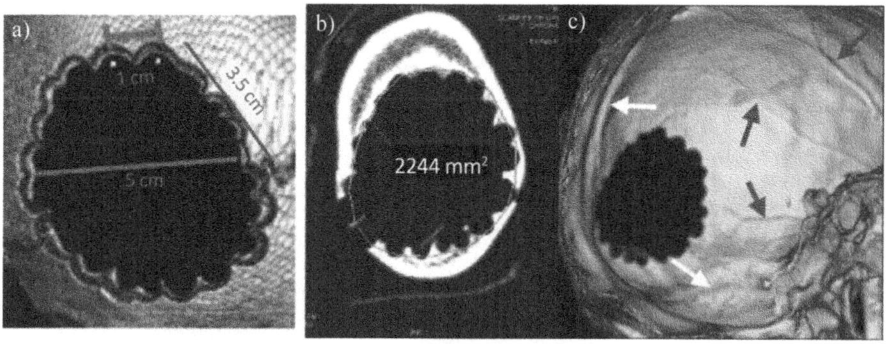

Figure 4.12. *(a and b) Trepanation measurements. (c) Endocranial view of the trepanation site away from the main venous (white arrows) and arterial (red arrows) vascular structures. Image: (a) © Dr. C. Bou; images: (b and c) © Dr. S. Mérigeaud MD/Tridilogy*

The edges of the trepanation preserve the trace of 20 microperforations, with a diameter of 0.7 and 1 cm, arranged in a circle, in a regular and confluent manner. A distance of approximately 1 cm between the center of each microperforation is maintained along most of the perimeter of the trepanation, with the exception of an area limited to 3.5 cm where the microperforations are positioned in a more irregular and constricted manner (Figure 4.12(a)).

In the experimental trepanation, the 20 microperforations were reproduced by rotating a flint tip along an axis vertical to the plane of the cranial vault. The rotations were first performed manually to initiate the perforation, and then with a bow until all perforations were joined (Figure 4.13(a)). The duration of the experiment, to obtain complete trepanation, was over four hours. At the end of the flap removal, the "jagged" morphology of the edges of the experimental trepanation was consistent with that of the mummy. More specifically, the similarity of the marks left by the rotation of the flint at the level of the external and internal tables and the diploe of the cranial vault with those observed on the archaeological case (Figures 4.13(b) and (c)) seems to confirm the technique used.

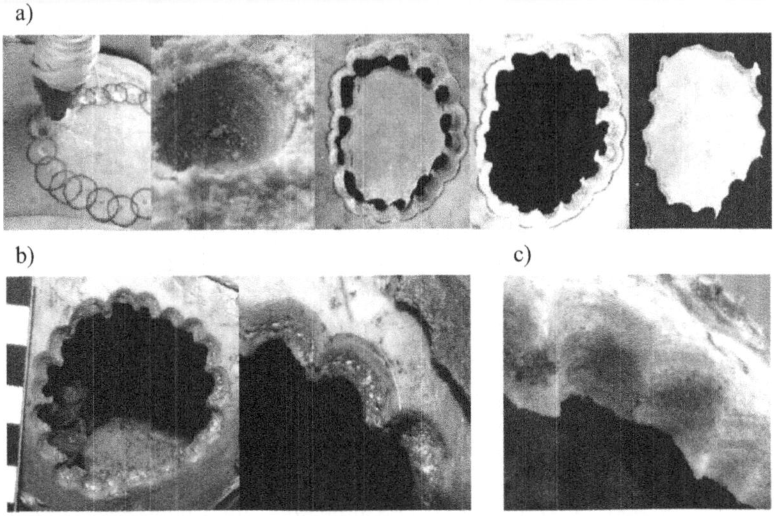

Figure 4.13. *Experimental extraction of the trepanation flap. (a) Successive steps of the protocol until the bone flap was obtained. (b) Photo of the edges of the trepanation of MNHN-HA-30187. (c) Comparison of the edge obtained experimentally on dry skull. Photos: © Dr. C. Bou*

4.4. Discussion

4.4.1. *Identity of the deceased and health status*

The Chachapoya mummy MNHN-HA-30187 is that of a male who died between 20 and 30 years of age, approximately 1.73 m in size (±2 cm). Consistent with an early age of death, the individual shows an overall "good" skeletal condition: no pathology, degenerative or traumatic, is noted, and despite the presence of three small carious inclusions, the oral condition is satisfactory, with moderate attrition and no *ante mortem* tooth loss. However, soft tissue imaging suggests that the male had an infectious lung disease. The presence of mediastinal calcifications suggests tuberculosis or granulomatosis. Tuberculosis is a disease known to have affected expanding Andean populations (Guillén 2012). Several cases were documented at the Laguna de los Condórès Chachapoya site, where 25% of 188 mummies studied were diagnosed with tuberculosis (Friedrich *et al.* 2010). At the scale of the Piedra Grande region, where mummy MNHN-HA-30187 originated, no epidemiological study has been conducted to date. If the diagnosis were confirmed, we do not know at this stage if the health status of the deceased is correlated with a specific burial treatment. In any case, if the young male was a carrier of this disease at the time of his death, there is no evidence that it was the cause of death. The evidence found on the corpse, which bears witness to the *perimortem* treatment to which it was subjected, nevertheless raises questions about the link between health status and burial practices.

4.4.2. *Treatment of the corpse and embalming*

The experimental approach which tested the trepanation method practiced on the mummy's skull supports the hypothesis of a craniectomy by successive circular micro-drilling, obtained by the rotation of a tool applied perpendicularly to the bone surface. The flint tool, in the form of a driven point, used manually or with a bow, is compatible with the wear observed macroscopically on the edges of the trepanation. However, at the end of this experiment, an important point must be made. The flint tool wears out rapidly and must be replaced regularly (eight were used in this experiment). The need for more and more tools, whose morphology evolves over the course of the experiment, does not fit with the "standardization" effect implied by the regularity of the diameters of each drill phase carried out on the mummy. No doubt new tests, using different materials – always in keeping with what the cultural framework allows us to envisage – should be attempted in the future. On dry bone, obtaining a bone flap with a diameter of 20 cm^2, made up of 20 confluent holes, required four hours of work. It is certainly probable that in the hands of experienced operators, the opening of the cranial cavity of the Chachapoya male would have required less time, but the methodical operation to which it testifies could not have been carried out quickly.

On the mummy, the edges of the trepanation are sharp and show no signs of healing. The constituent structures of the cranial vault, the external table, the diploe and the internal table are very clearly legible along the entire contour of the opening, and show no evidence of reworking. The edges are sharp and draw the contour of each drilling with precision. The trepanation of the mummy is thus comparable in every way to that carried out on dry bone, and thus implies a *post-mortem* operation. However, the absence of healing does not mean we can fully exclude the hypothesis of an *in vivo* trepanation, which would have been followed rapidly by the subject's death. The location of the trepanation, which spares the main pericerebral vascular structures, suggests a strategic choice aimed at limiting the risk of bleeding, in contradiction with a *post-mortem* procedure. Nevertheless, it should be emphasized that the opening is also located astride the lambdoid suture, which suggests a high risk of complication. In addition, the skull's skin, which is absent at the trepanation site, has the same serrated cut on some of its edges as the edges of the trepanation itself (Figure 4.3(b)). This indicates that the drilling was carried out directly from the outside of the scalp, which seems unlikely on a living subject.

The hypothesis of a *post-mortem* craniectomy remains the most parsimonious. The sophisticated arrangement of this large opening in the cranial cavity of the corpse of the young Chachapoya male can be related to the absence of the brain. The organ would have been removed at an early stage of the funereal process, at the time of the body's burial. This hypothesis is consistent with the very first observations made by P. Broca for the mummies of Piedra Grande of Utcabamba, several of which bear the trace of unhealed cranial openings. He concluded, before of his colleagues of the *Société d'anthropologie* (French Society of Anthropology), that they had been subjected to a process of mummification whose objective was "[to] introduce aromatic substances into the skull to prevent putrefaction" (Broca quoted by Hamy (1897, p. 66)). With regard to the case of MNHN-HA-30187, this hypothesis is supported by P. Vidal-Senèze who affirmed that he had extracted "a sponge that took the place of the encephalon and must have been soaked with an antiseptic liquid" from the mummy's skull (Vidal-Senèse in Hamy (1897, p. 66)). If the trepanation of this Chachapoya mummy was a stage of embalming, this must be distinguished from other cases of cranial perforations known in this same chrono-cultural context, whose traces of healing indicate an *ante-mortem* surgical gesture, and where the hypothesis of a cranioplasty practiced after trauma has been advanced (Kurin 2013).

Although the brain was removed from the corpse of the young Chachapoya male, the body was not completely eviscerated. At the abdominal level, the supra-diaphragmatic anatomical structures (lung and heart) are clearly visible. Under the diaphragm, the kidneys and liver are also present. On the other hand, the intestines and the stomach are not clearly identifiable. Have these tissues, which promote the decomposition processes, been extracted? Were the methods of evisceration of the mummy carried out in a "partial" manner, was this the norm or an alternative applied to particular deceased people? The existence of other non-eviscerated Chachapoya mummies is attested, notably at Laguna Huayabamba (Muscutt 2003; Nystrom 2003). The possibly infectious nature of the corpse of MNHN-HA-30187 could explain a specific evisceration procedure that would have avoided contact with certain organs. An epidemiological study conducted on a larger scale, in conjunction with the meticulous restitution of mortuary gestures, would make it possible to test various embalming practices according to the health and social status of the deceased.

4.4.3. *Chronology of mortuary gestures*

Biological data on the insect species associated with the mummy allow us to suggest different hypotheses on the *post-mortem* "history" of this individual and to infer new data on Chachapoya funereal gestures. *Calliphoridae diptera* belong to the first waves of colonizers and lay their eggs on corpses within minutes of death (Huchet 2014; Huchet and Greenberg 2010). Oviposition occurs most often in natural openings (eyes, nostrils, mouth, anus, etc.) but is also possible in open wounds. In this case, the very high concentration of *puparia* in the lower part of the body strongly suggests that the infestation started in this area of the body. The feet, thighs and perianal area show severe damage and degradation of soft tissue attributable to the action of the larvae. The larvae then colonized the interior of the abdomen as clearly indicated by medical imaging (Figure 4.10). It cannot be ruled out that a possible *post-mortem* treatment of rectal evisceration such as that described on the Chachapoya mummies of La Laguna de los Cóndores (Guillén 1998; Wild *et al.* 2007) could have favored the infestation by Calliphoridae flies. Another hypothesis is that the individual was colonized by these Diptera during his lifetime and died of myiasis[14]. Although the main cases of human myiasis in the Neotropics are due to various species of Calliphoridae (in particular the native species *Cochliomyia hominivorax* (Coquerel)), Dear (1985) indicates that certain species of the genus *Compsomyiops* can also cause myiasis in infected wounds. This assertion was recently confirmed by Guimarães and Papavero (1999) who cite a case of myiasis due to *Compsomyiops fulvicrura* in Argentina.

If, however, colonization is considered more likely to have occurred *post-mortem*, the fact that only one species of Diptera is present within the mummy indicates that the envelopment phase in the layers of tissue constituting the funereal bundle or *fardo* was carried out rapidly, probably within a few days of death. As Calliphoridae larvae leave the body to pupate at a distance, the presence of *puparia* inside the mummy indicates that this migration, inherent to the biology of these Diptera, was annihilated by the presence of the numerous layers of tissue. Depending on the species and environmental conditions, the length of the larval cycle in the feeding phase (i.e. before migration) of species of the genus *Compsomyiops* is between 4 and 10 days (Greenberg and Szyska 1984; Dale and Prudot 1986). The biological data reported by Dale and Prudot (1986) for *C. arequipensis* being between 4 and 5.9 days means that we can estimate that the body was packed in a time interval of less than 6 days.

Data from the study of insect remains present on the mummy indicates a relatively rapid funereal treatment and packaging of the body in *fardo* within less than 6 days. This conjecture is corroborated by the fact that the binding of the mummy in the fetal position was probably carried out only after complete disappearance of *rigor mortis,* that is, between 2 and 3 days after death. The hypothesis of myiasis having been the

14 Parasitic disease due to the infestation of tissues by insect larvae (various species of Diptera in particular).

cause of death by toxaemia or *sepsis* is not to be totally ruled out. Finally, the phenology (periods of activity) of the species of the genus *Compsomyiops* would suggest that the death occurred during the hottest months of the year (i.e. between mid-August and early October for the Amazonas region).

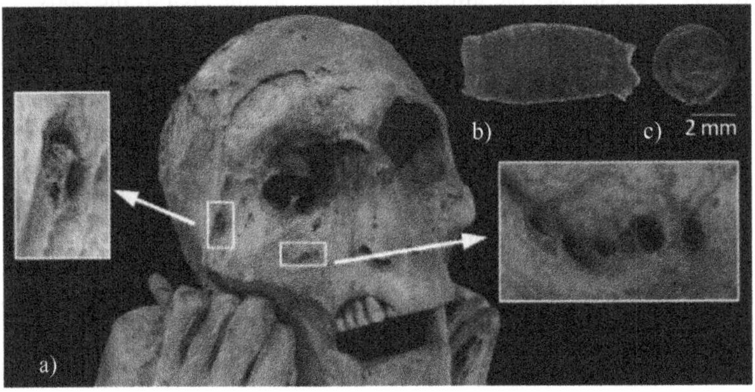

Figure 4.14. *(a) Location of Calliphoridae dipteran* puparia *on the mummy's face (white arrows). (b) Hatched* puparium *of* Compsomyiops *cf.* arequipensis, *lateral view. (c) Idem, caudal view, posterior respiratory stigmata (center) circled by tubercles (integumentary sensory papillae). Photos © J.B. Huchet*

4.5. Conclusion

Among archeological human remains, mummies constitute an exceptional legacy, which allow us to examine the corpses of disappeared populations. Beyond the emotion that this proximity can generate in the observer, mummies represent true archives in which a great quantity of information, relative to the way of life, to the cultural and funereal practices of ancient populations, is registered. Rare and fragile, they require the development of the least invasive procedures possible. In this context, medical imagery proves to be a great asset, as it gives access to the internal anatomy of the body, organs and skeleton without affecting its integrity. This data, which substantially builds upon surface observations, constitutes, moreover, a real iconographic and conservative stake. Finally, it opens up the prospect of additional reasoned analyses, backed up by solidly supported hypotheses. For mummies, as for all thc anthropological hcritage collections of the *Muséum*, particular attention is paid to research aiming at taking a fraction of organic remains (skin, bone, tooth, hair, insects, parasites), however small. Exploratory analyses such as those carried out on the Chachapoya mummy MNHN-HA-30187 make it possible to envisage an optimal scientific framework for future paleo-genetic and isotopic analyses by ensuring that relevant questions are asked. It will be necessary in the long run that the results be put into perspective by means of meta-analysis, so that the studies do not become anecdotal and deliver a coherent synthesis of the *Muséum's* collection of mummies. It is important to encourage, in the continuation of the GDR, a national network of specialists in this field, and that this network can acquire international visibility.

4.6. References

de Arriaga, P.J. (1968). *Extirpacion de la idolatria del Piru.* Atlas, Madrid.

Baumgartner, D.L. and Greenberg, B. (1985). Distribution and medical ecology of the blowflies (Diptera: Calliphoridae) of Peru. *Annals of the Entomological Society of America*, 78(5), 565–587.

de Cieza de León, P. (1984). *Primera Parte de la Crónica del Perú.* Université pontificale catholique du Pérou/Académie nationale d'histoire, Lima.

Cleuvenot, E. and Houët, F. (1993). Proposition de nouvelles équations d'estimation de la stature applicables pour un sexe indéterminé et basées sur les échantillons de Trotter et Gleser. *Bulletins et Mémoires de la Société d'Anthropologie de Paris*, 5(1–2), 245–255.

Cornejo García, M. (2002). Prospección y excavación arqueológica en la cuenca de los ríos Huayabamba y Huayabacu. Informe final de investigación. Report, INC-Trujillo, National University of Trujillo, Trujillo.

Dale, W.E. and Prudot, E.A. (1986). Apuntes sobre la biología de las moscas Calliphoridae en la costa central peruana. *Revista Peruana de Entomología*, 29, 105–111.

Dear, J.P. (1985). A revision of the new world Chrysomyini (Diptera: Calliphoridae). *Revista Brasileira de Zoologia*, 13, 109–169.

Doyle, M.E. (1998). Ancestor cult and burial ritual in seventeenth and eighteenth century central Peru. PhD Thesis, University of California, Los Angeles, CA.

Fabre, O., Guyot, J.-L., Salas-Gismondi, R., Malaver Pizarro, M., Maniero, E. (2008). Los chachapoya de la región de Soloco: Chaquil, del sitio de hábitat a la cueva funeraria. *Bulletin de l'Institut français d'études andines*, 37(2), 271–292.

Figueroa-Roa, L. and Linhares, A.X. (2002). Systematics, morphology and physiology. Sinantropia de los Calliphoridae (Diptera) de Valdívia, Chile. *Neotropical Entomology*, 31(2), 233–239.

Friedrich, K., Nemec, S., Czerny, C., Fischer, H., Plischke, S., Gahleitner, A., Bence Viola, T., Imhof, H., Seidler, H., Guillén, S. (2010). The story of 12 Chachapoyan mummies through multidetector computed tomography. *European Journal of Radiology*, 76, 143–150.

Froment, A. and Mennecier, P. (2014). Les collections de restes humains archives de l'Humanité. In *Penser, classer, administrer: pour une histoire croisée des collections scientifiques*, Daugeron, B. and Le Goff, A. (eds). Muséum national d'Histoire naturelle, Paris.

Gonzalez-Mora, D., Peris, S.V., Mariluis, J.C. (1998). Notas sobre la taxonomia y distribucion del género *Compsomyiops* Townsend, 1918 (Diptera, Calliphoridae). *Boletín de la Real Sociedad Española de Historia Natural*, 94(1–2), 15–21.

Greenberg, B. and Szyska, M.L. (1984). Immature stages and biology of fifteen species of Peruvian Calliphoridae (Diptera). *Annals of the Entomological Society of America*, 77, 488–517.

Guillén, S.E. (1998). Laguna de los Cóndores: Donde viven los muertos. *Bien Venida*, 6, 43–48.

Guillén, S.E. (2012). Momies chachapoyas du Pérou ancien. In *La préhistoire des autres*, Schlanger, N. and Taylor, A.-C. (eds). La Découverte, Paris.

Guimarães, J.H. and Papavero, N. (1999). *Myiasis in Man and Animals in the Neotropical Region: Bibliographic Database*. Editora Plêiade/FAPESP, São Paulo.

Hamy, E.-T. (1897). *Galerie américaine du Musée d'ethnographie du Trocadéro, choix de pièces archéologique et ethnographiques*. Ernest Leroux, Paris.

Huchet, J.-B. (2014). L'archéoentomologie funéraire. In *Insectes, cadavres et scènes de crime: principes et applications de l'entomologie médico-légale*, Charabidze, D. and Gosselin, M. (eds). De Boeck, Louvain-la-Neuve, Belgium.

Huchet, J.-B. and Greenberg, B. (2010). Flies, Mochicas and burial practices: A case study from Huaca de la Luna, Peru. *Journal of Archaeological Science*, 37(11), 2846–2856.

Kauffmann, D.F., Salazar, M., Morales, D., Mackay, I., Sacay, O. (1989). Andes amazónicos: Sitios intervenidos por la Expedición Antisuyo/86. *Arqueológicas*, 20, 6–57.

Knutson, S. (2006). A cave exploring culture from antiquity. *NSS News*, 64(2), 4–10.

Kurin, D.S. (2013). Trepanation in South-Central Peru during the early late intermediate period (ca. AD 1000–1250). *American Journal of Physical Anthropology*, 152, 484–494.

Lau, G.F. (2013). *Ancient Alterity in the Andes*. Routledge Press, New York.

Lisowski, F.P. (1967). Prehistoric and early historic trepanation. In *Diseases in Antiquity: A Survey of the Diseases, Injuries and Surgery of Early Populations*, Brothwell, D. and Sandison, A.T. (eds). C.C. Thomas Publisher, Springfield, IL.

Mantha, A. (2009). Territoriality, social boundaries and ancestor veneration in the central Andes of Peru. *Journal of Anthropological Archaeology*, 28(2), 158–176.

Mariluis, J.C. and Schnack, J.A. (1996). Elenco específico y aspectos ecológicos de Calliphoridae (Insecta, Diptera) de San Carlos de Bariloche, Argentina. *Boletín de la Real Sociedad Española de Historia Natural*, 92(1–4), 203–213.

Mariluis, J.C., Schnack, J.A., Mulieri, P.P., Patitucci, L.D. (2008). Calliphoridae (Diptera) from wild, suburban, and urban sites at three Southeast Patagonian localities. *Revista de la Sociedad Entomológica Argentina*, 67(1–2), 107–114.

Mennecier, P. (2006). Les collections d'anthropologie du musée de l'Homme (Muséum national d'Histoire naturelle, Paris). In *Les collections ostéologiques humaines : gestion, valorisation et perspectives*, Ardagna, Y., Bizot, B., Boëtsch, G., Delestre, X. (eds). Bulletins archéologiques de Provence, France.

Murail, P., Bruzek, J., Houët, F., Cunha, E. (2005). DSP: A probabilistic sex diagnosis tool using worldwide variation of pelvic measurements. *Bulletins et Mémoires de la Société d'Anthropologie de Paris*, 17, 167–176.

Muscutt, K. (2003). History channel site report on Vira Vira, Perú. Report, Department of Anthropology, The University of New Mexico, Albuquerque, NM.

Nystrom, K.C. (2003). Report on the human remains recovered from the Laguna Huayabamba, Chachapoyas, Perú. Report, Department of Anthropology, The University of New Mexico, Albuquerque, NM.

Nystrom, K.C., Goff, A., Goff, L. (2005). Mortuary behaviour reconstruction through palaeoentomology: A case study from Chachapoya, Peru. *International Journal of Osteoarchaeology*, 15, 1–11.

Nystrom, K.C., Buikstra, J.E., Muscutt, K. (2010). Chachapoya mortuary behavior: A consideration of method and meaning. *Chungara Revista de Antropología Chilena*, 42, 477–495.

Reichlen, P. and Reichlen, H. (1950). Recherches archéologiques dans les Andes du Haut Utcubamba. *Journal de la Société des Américanistes*, 39, 219–246.

Schjellerup, I. (2005). *Incas y españoles en la conquista de los chachapoya*. Institut français d'études andines/Pontifical Catholic University of Peru, Lima.

Suchey, J.M. and Brooks, S.T. (1986). *Instruction for Use of Suchey-Brooks System for Age Determination of the Male Os Pubic. Instructional Materials Accompanying Male Pubic Symphyseal Models of the Suchey-Brooks System.* Diane France, Fort Collins, CO.

Toyne, J.M. (2015). Tibial surgery in ancient Peru. *International Journal of Paleopathology*, 8, 29–35.

Trotter, M. and Gleser, G. (1958). A re-evaluation of estimation of stature based on measurements of stature taken during life and of long bones after death. *American Journal of Physical Anthropology*, 16(1), 79–124.

Vidal-Senèze, P. (1877). Perforations crâniennes sur d'anciens crânes du Haut-Pérou. *Bulletins de la Société d'anthropologie de Paris*, II(12), 561–563.

Vidal-Senèze, P. and Noctzli, J. (1877). Sur des momies découvertes dans le Haut Pérou. *Bulletins de la Société d'anthropologie de Paris*, II(12), 640–641.

Wild, E.M., Guillén, S., Kutschera, W., Seidler, H., Steier, P. (2007). Radiocarbon dating of the Peruvian Chachapoya/Inca site at the Laguna de los Cóndores. *Nuclear Instruments and Methods in Physics Research*, 259, 378–383.

5

Reconstructing the History of Human Populations: A Challenge for Biological Anthropology

Martin FRIESS and Manon GALLAND
UMR 7206 "Éco-Anthropologie", Muséum national d'Histoire naturelle, Université Paris Diderot, France

The thousands of human remains that make up the MNHN's anthropological collection, accumulated between the 17th and early 20th centuries, still hold many secrets. As true archives of human biocultural history, they offer us a look at human diversity and the evolutionary mechanisms that produced it. Through two examples, this chapter illustrates what the study of skulls can tell us about our recent past.

5.1. Introduction

5.1.1. *How human remains have also become scientific objects*

Biological anthropology is the science of the physical variation of humans in time and space. As such, it seeks to reconstruct our evolutionary history over the last 6 million years, and to understand the biological variability of modern-day humans across the globe. While the beginnings of this discipline go back to the time of the

For a color version of all figures in this chapter, see www.iste.co.uk/pellens/natural.zip.

great explorations and the discovery of the Other, that is to say, indigenous populations with their traditions and customs, but also their physical appearance, often regarded with disdain by Euro-centric thought, it has relatively quickly included the humans of the past in its considerations, in other words, skeletons.

One of the first natural history explorers to build up a true anthropological bone collection was Johann Friedrich Blumenbach (1752–1840), often considered to be the father of biological anthropology. Originally a physician, he became interested in human variability with an anatomist's eye and built up a collection of several hundred skulls from around the world (Schultz and Reich 2013). Others followed, such as Samuel Morton (1799–1851) in Philadelphia and Paul Broca (1824–1880) in Paris, whose original collection is now integrated into that of the MNHN.

5.1.2. *The MNHN biological anthropology collection*

With nearly 30,000 entries, the anthropological collection housed at the *Musée de l'Homme* is one of the largest in Europe, and it continues to be solicited by numerous researchers from all over the world. This is mainly due to its diversity, both chronologically and geographically. Chronologically, the oldest specimens date from the last ice age, about 38,000 years before our era, and correspond to the Neanderthal remains of La Ferrassie (Heim 1978). The most recent specimens are from the 19th century, when collecting human remains reached its peak in Europe. Because of this chronological diversity, researchers who come to study Neanderthal remains are able to collect comparative data from modern-day humans during their visit, an elementary step in any paleoanthropological research. Geographically, the modern human skeletal remains come from over 150 countries, from the United States to China, and from South Africa to Greenland. However, the constitution of this collection was not carried out according to a strict and rigorous scientific reasoning as would probably be done today. Its composition rather reflects the vagaries of the history of science, with a vast majority of specimens of French origin, mainly from archeological excavations and donations. Others were obtained through the great expeditions, such as that of Créqui-Montfort and Sénéchal de la Grange in South America (Chervin 2007), and still others were the object of exchanges between museums or between researchers. Another bias, due to the history of the discipline, is that of the 30,000 entries, more than 18,000 consist of skulls, rather than complete skeletons, a phenomenon easily attributable to the cephalic obsession of many researchers (Sibeud 2012). Because of these limitations, the composition of the collection probably does not fully satisfy the criterion of randomness, in the sense of strict scientific sampling.

Nevertheless, today the anthropological community recognizes the inestimable scientific value of this collection, because each specimen constitutes a true archive of biocultural history, usable by a multitude of analytical methods. Research in biological anthropology is, nowadays, extremely diversified and explains aspects such as functional anatomy, growth, sexual dimorphism, the history of populations, their diet or the history of certain diseases that leave traces on bones.

5.1.3. *Cranial morphology as an indication of biocultural processes*

While the first biological anthropologists of the 18th and 19th centuries were mainly concerned with documenting and measuring morphological diversity, often focusing on the shape of the skull (Blanckaert 1987), the advent of genetics during the second half of the 20th century has strongly shifted the main point of the discipline from the simple description of variability to the analysis of the mechanisms involved in this variability (Washburn 1951). On a biological level, these mechanisms are, above all, the evolutionary forces, namely mutation, selection, gene flow and genetic drift. The last two are often associated with the migratory movements of populations in the past.

Because of advances in ancient DNA[1] analysis, it is now possible to identify traces of migrations between different regions of the world and to quantify the genetic proximity of past populations. However, these analyses are invasive, remain relatively expensive, especially on a large scale, and are not always successful, especially in cases of poor preservation of biomolecules. Morphology, on the other hand, can testify to the proximity of populations, even in cases where the bone no longer contains sufficient collagen to extract DNA. Thus, morphological resemblance is used as an indication for genetic relationships, although morphology also depends on other factors besides genetics (Mielke *et al.* 2011). It is in this sense that anthropologists seek to reconstruct the history of human populations from prehistory to the present day across the globe by inferring phylogenetic relationships from craniometric data. Since the early days of biological anthropology, much effort has been made to improve morphometric methods. Calipers and other mechanical measuring instruments have long been replaced by digital technologies based on 3D imaging (Figure 5.1), and linear dimensions (lengths, widths, angles, etc.) by statistical analysis of Cartesian coordinates (Slice 2005) (see note "Craniometrics in the 21st century"). The purpose of these methods is to facilitate the geometric analysis of the object of study in order to be able to compare specimens objectively and to test hypotheses regarding the history of their migrations and their ancestors. A first example illustrates this type of research by looking at the settlement of the Americas. In a second step, we will look at the climatic impact on cranial diversity on a global scale.

1 See Chapter 13 of this book for further examples and details of the various uses of ancient DNA.

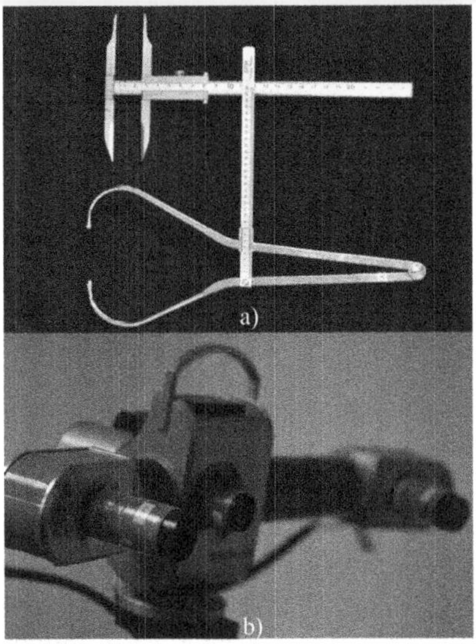

Figure 5.1. *"Evolutions" in biological anthropology: as the 21st century has gone on, the caliper (a) has largely given way to the the 3D scanner (b). Source: Friess, MNHN*

5.2. Cranial morphology and settlement history

The first settlement of the Americas has always attracted the attention of many researchers from various disciplines, such as prehistory, linguistics, genetics and biological anthropology. While a consensus seems to exist suggesting that the first modern humans arrived, through the Bering Strait, on the American continent at least 15,000 years ago, different scenarios on the settlement pattern continue to be debated. According to a long-standing anthropological view, the morphological difference between the oldest American fossils and present-day Native American populations suggests two successive waves, of which only the most recent was successful, and from which present-day Native Americans are descended (Neves *et al.* 1991). Genetic data, on the other hand, have led some researchers to suggest that present-day and fossil Native Americans are descended from a single migratory wave (Fagundes *et al.* 2008). In addition to this basic dichotomy, a number of studies have made the interpretation of this settlement process more complex. For example, studies focusing on cranial morphology have proposed that the morphological difference between Paleoamericans and present-day Native Americans would be consistent with a single wave, followed by processes of drift and local adaptation (Azevedo *et al.* 2011). In contrast, new genetic evidence from present-day Native Americans is consistent with the existence of at least two waves (Malaspinas *et al.* 2014).

NOTE.– *Craniometrics in the 21st century.*

While some morphological studies continue to be performed with calipers and other instruments invented mostly during the 19th century, today's biomedical research benefits most often from progress in two areas. On the one hand, 3D imaging (X-ray tomography, surface scanners) allows researchers to examine any specimen virtually, making it permanently accessible and allowing examinations that are impossible on the physical specimen. On the other hand, the quantitative analysis of shape (in the generic sense of morphometrics or in the specific sense of craniometrics) has undergone what was dubbed in the 1990s a revolution: the development of geometric morphometrics coupled with imaging allows the comparison of anatomical landmarks, or even entire surfaces, rather than linear dimensions. On the algebraic level, it proceeds by a decomposition of the form in terms of size and conformation, which allows for better understanding of the possible differences between these two aspects. It does this by normalizing the coordinates. To make them statistically comparable, it is necessary to eliminate all differences related to the size, position and orientation of the specimens at the time of recording. This is achieved by the so-called Procrustes superimposition, which corresponds to a common scaling, a superimposition of the specimens in their center of gravity, then their rotation in order to minimize the distance between corresponding points. The adjusted coordinates then allow any statistical analysis of size, shape and their covariation in the same space in which the coordinates were recorded. This guarantees not only more consistent analyses, but also a visualization of the results closer to the object of study and easier to interpret.

Morphometric methods help us to better understand the diversity of Native American populations, and, to put it in a broader context in order to better identify the factors contributing to morphological diversity.

For this purpose, the MNHN's anthropological collections, supplemented by those of other museums around the world, serve as an analytical corpus. Figure 5.2 summarizes the geographical composition of the sample of modern skulls used in this study. In total, we included 889 modern skulls, representing 30 populations from the Americas and Australasia. In addition, 89 fossil skulls from 25 archeological sites were included. These fossils often represent isolated cases, rather than populations, with the exception of three groups: Tam Hang, Laos, a presumed Terminal Pleistocene site (seven individuals; 15,700 BP; Fromaget 1940); Lagoa Santa, Brazil, an Early Holocene site (nine individuals; 8,200–8,500 BP; Neves *et al.* 2007); Indian Knoll, Kentucky, a Middle Holocene site (29 individuals, 4,530–5,390 BP; Morey *et al.* 2007).

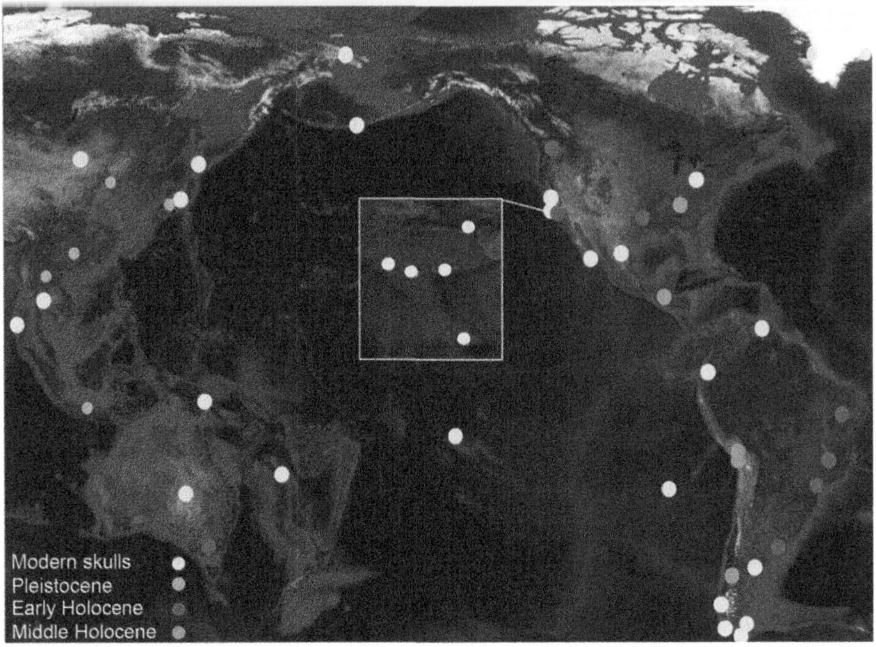

Figure 5.2. *Map showing the provenance of the Native American and Australasian cranial sample used to reconstruct the settlement history of the Americas*

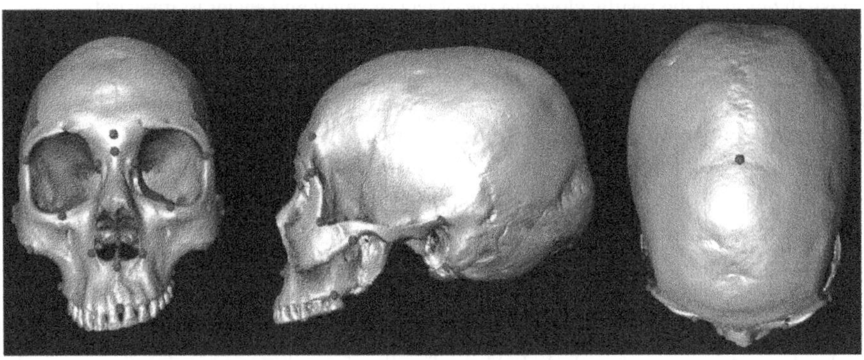

Figure 5.3. *Diagram, in frontal, lateral and superior views, representing the craniometric landmarks used*

The skulls were digitized using a surface scanner to produce a virtual 3D model. Once the virtual rendering was obtained, the Cartesian coordinates of 39 anatomical markers were recorded using Landmark editor software (Wiley 2005) (Figure 5.3).

These landmarks represent the general form of each skull and, after specific processing (see note "Craniometrics in the 21st century"), can be used for statistical analyses to compare populations in terms of their cranial variation.

5.2.1. *A new look at the diversity of Native Americans*

At the global scale (Figure 5.4), we observe, on the one hand, that all circum-Arctic populations are grouped together, and, on the other hand, the Australasian populations form a group. These two groups are relatively distant from the Native American populations. As for the latter, they are largely separated into continental populations (Mexico, Ecuador, etc.) and geographically isolated populations (Tierra del Fuego, Channel Islands of California, etc.).

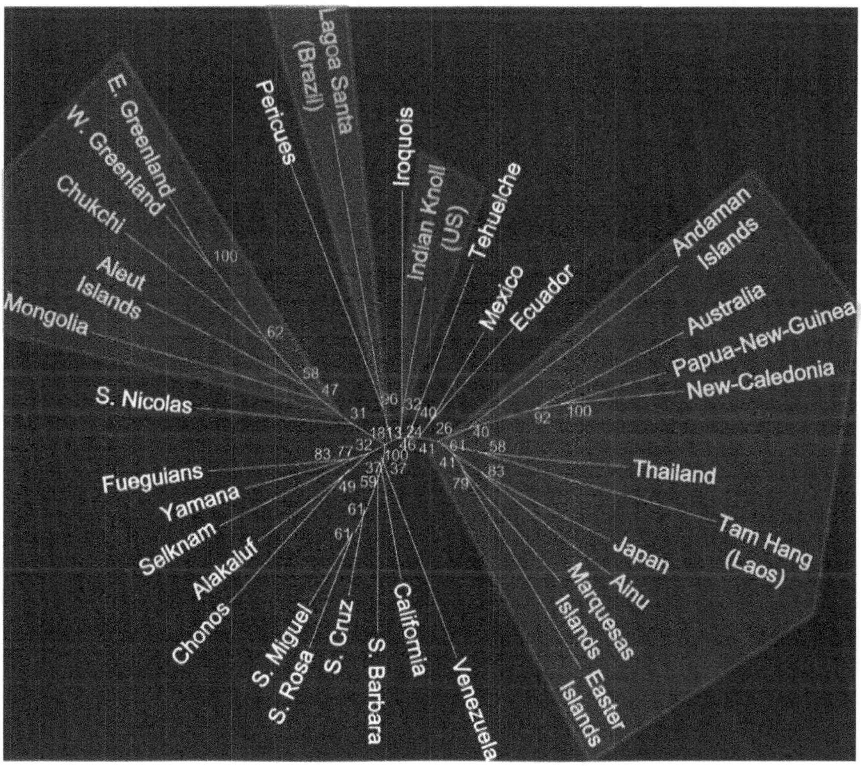

Figure 5.4. *Phenetic tree using the neighbor-joining method, computed from Procrustes distances, full sample, and showing the phenetic proximity of the populations*

Within Native Americans, a principal component analysis (Figure 5.5), allowing for the inclusion of fossil individuals, confirms that they are scattered among other populations of indigenous peoples of the Americas, but do not form a coherent whole. The notion of a simple dichotomy between the oldest Native Americans, on the one hand, and the current populations, on the other hand, is therefore not supported.

In order to better identify what influences cranial variability, the sample is grouped according to extrinsic factors: geographical region, chronology, climatic zone, diet and linguistic families. The importance of these factors for cranial diversity is then assessed by a MANOVA (multivariate analysis of variance). This analysis assigns a weight to each factor according to the portion of variance it can explain. The analysis reveals that membership in a particular language family explains by far the largest portion of cranial diversity, while chronology explains the smallest.

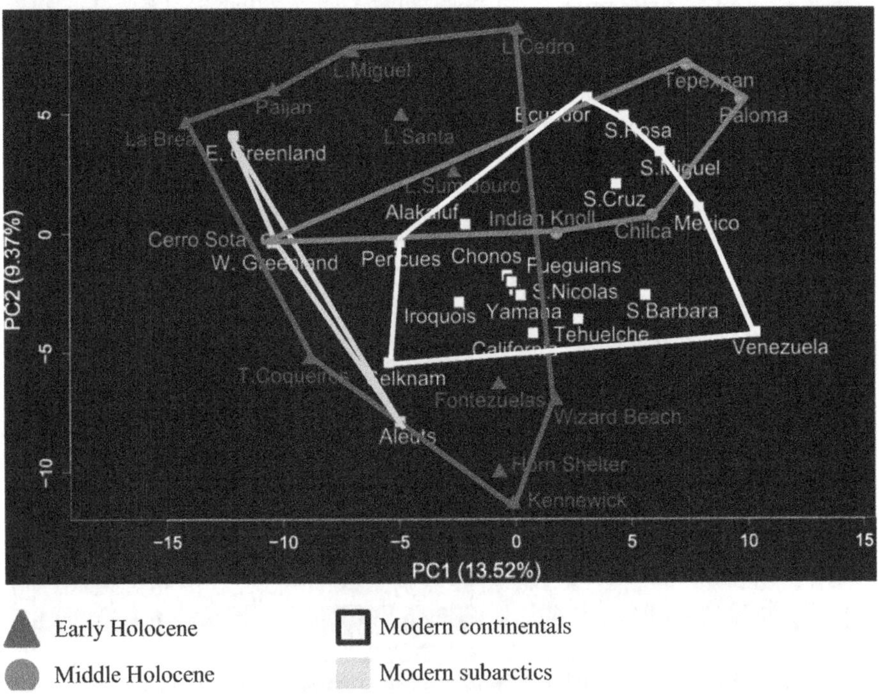

Figure 5.5. *Principal component analysis calculated on Procrustes coordinates of the present-day and fossil Native American samples, showing the major axes of variation*

In other words, being separated by thousands of years of history generates less morphological divergence than being separated by a language barrier. Other important factors contribute to present-day and fossil cranial diversity, such as climate and diet, and thus may obscure the evolutionary history of populations. It is therefore important for any study of population history to quantify the effect of these factors and to account for them in the interpretation, for example, by removing such effects from the phylogenetic signal (Galland and Friess 2016). However, this requires large samples of diverse populations with respect to these factors. For example, samples covering all climatic zones, representing populations with various diets, languages, etc., would be required. The greatest scientific value of an anthropological collection, such as that of the MNHN, lies in its diversity, because it is precisely this diversity that allows us to better examine the sources of morphological variation and the dispersion mechanisms of populations.

5.3. Cranial morphology and adaptation to the environment

NOTE.– *The neutral theory of evolution.*

The synthetic theory of evolution combines the Darwinian notion of natural selection with the knowledge of population genetics. It results in four main forces being responsible for evolution: mutation, selection, gene flow and drift. While mutation is the mechanism that produces new alleles, selection, gene flow and drift act by altering the frequencies of existing alleles. The effect of a mutation on the phenotype depends on natural selection, and this mutation can be beneficial or deleterious in terms of reproductive fitness, which influences its frequency in the following generations. On the other hand, gene flow (migration) and genetic drift (random loss of part of the alleles) are stochastic events, which do not necessarily have an effect on reproductive fitness. Their effect on allele frequencies is not directional, like that of selection, but random or neutral. A controversial question is which of the two main mechanisms, selection or randomness, plays a more important role in the evolution of a gene pool. The dominant theory in favor of selection is now replaced by the so-called neutral theory (Kimura 1968), which states that at the molecular level, population diversity is largely the cumulative result of random events, that is, drift. Many molecular studies, including human studies, tend to confirm this theory, and it has been suggested that the neutral theory also applies to cranial diversity within our species.

Numerous studies have suggested the existence, in modern humans, of morphological characteristics that are linked to climatic adaptation. Thus, the distribution of body size, trunk/extremity proportions or skin color varies clinally along the latitudes. The question of adaptive signs in the skull has also been studied, but the answers remain ambiguous. According to some authors (e.g. Betti *et al.* 2009), cranial variability within *Homo sapiens* is largely the result of drift, that is, randomness. This would be consistent with the neutral theory of evolution (see note "The neutral theory of evolution"), which states that the genetic differences accumulated between present-day populations are mainly the result of drift.

Indeed, if we extend the neutral theory to the phenotype, the cranial diversity of human populations is supposed to depend largely on the geographical distance that separates them from each other as well as from their common geographical origin, namely eastern sub-Saharan Africa. The extraordinary richness, in terms of geographical diversity, of the anthropological collections of the MNHN allows us to provide some answers to this question.

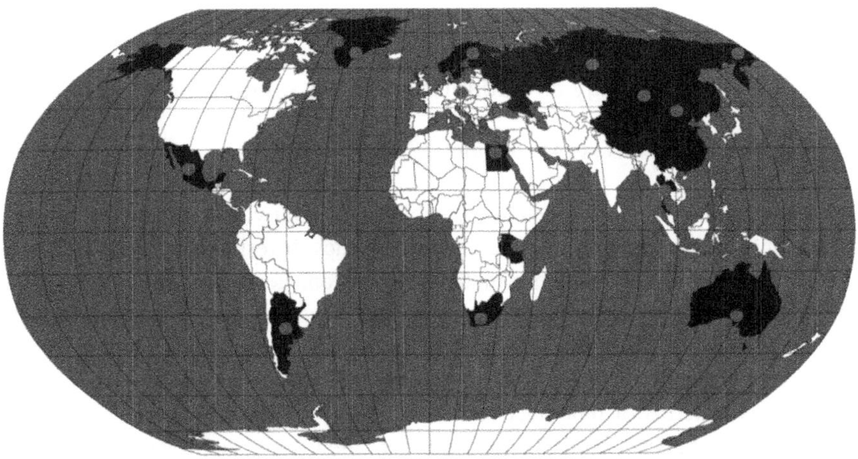

Figure 5.6. *Map showing the provenance of the sample of 525 skulls used to test the neutral theory of evolution*

To this end, we assembled a large sample, with 525 geographically representative human skulls (Figure 5.6). On this sample, 25 Cartesian coordinates were recorded on each specimen to capture the general shape of the skulls. This sample included 15 populations from all continents (except Antarctica), including the most extreme latitudes inhabited before modern times. It also represents the major eco-geographic zones occupied by humanity, namely polar, continental, temperate, tropical and arid zones (Kottek *et al.* 2006). For each individual, mean geodetic coordinates were determined, as well as the distances between them and from the hypothetical common origin in East Africa. The calculation of these distances was done by imputing land crossing points, according to the presumed migration routes from Africa. The registration of the landmarks was done using a *Microscribe*[2]. The Cartesian coordinates were subjected to Procrustes superimposition, and the Procrustes distances between populations were calculated.

5.3.1. *Cranial diversity beyond randomness*

Under the assumption of neutral evolution, the Procrustes distances should closely reflect the geographic distances, which is not quite the case (Figure 5.7). Indeed, there is no statistical correlation between these two distance matrices (Mantel test, $p = 0.3$ after 10,000 permutations), suggesting that factors other than drift explain a non-negligible part of the cranial diversity in *H. sapiens*, or contribute enough to disrupt the signal expected under the neutral theory. To remove any variation due to human migration history, Procrustes coordinates were regressed on migration distance from East Africa, and the residuals of this regression, which were by definition independent of distance, were analyzed. A discriminant analysis (CVA) was calculated using the five eco-geographic zones as a grouping variable. The purpose of this analysis was to test and visualize the major shape differences between populations inhabiting these zones. As shown in Figure 5.8, the first axis, accounting for 44% of variance between groups, tends to separate the polar and continental zones from the tropical and arid zones. The difference in shape associated with this axis corresponds mainly to two aspects of cranial morphology: sphericity of the neurocranium, as well as orthognathism, both of which are more accentuated in the polar and continental zones. The sphericity of the neurocranium translates into a reduced length and height relative to the width, while orthognathism goes hand in hand with a more projected position of the zygomatic bone, which contributes to the flattened aspect of the face, notably in Inuit populations (see Steegmann 1970).

The increased sphericity in extreme latitude populations is consistent with observations that brain size is relatively larger in extreme latitudes than in equatorial latitudes (Beals *et al.* 1984; Ash and Gallup 2007; Bailey and Geary 2009). Given that a sphere has a larger volume than an ellipsoid for the same surface area, this increased sphericity can be interpreted as a reflection of the increased endocranial volume at extreme latitudes, which correspond to colder eco-geographic zones. As for a more orthognathic and flattened face, the interpretation remains obscure. The suggestion that it would lead to less exposure to cold has been rejected in physiological experiments (Steegmann 1970). In the absence of evidence, this may be a random result, due to drift rather than selection. Alternatively, diet appears to be a possible source of variation of the face, specifically the masticatory apparatus, which it may influence (Sardi *et al.* 2006; Katz *et al.* 2017).

2 Available at: http://3d-microscribe.com.

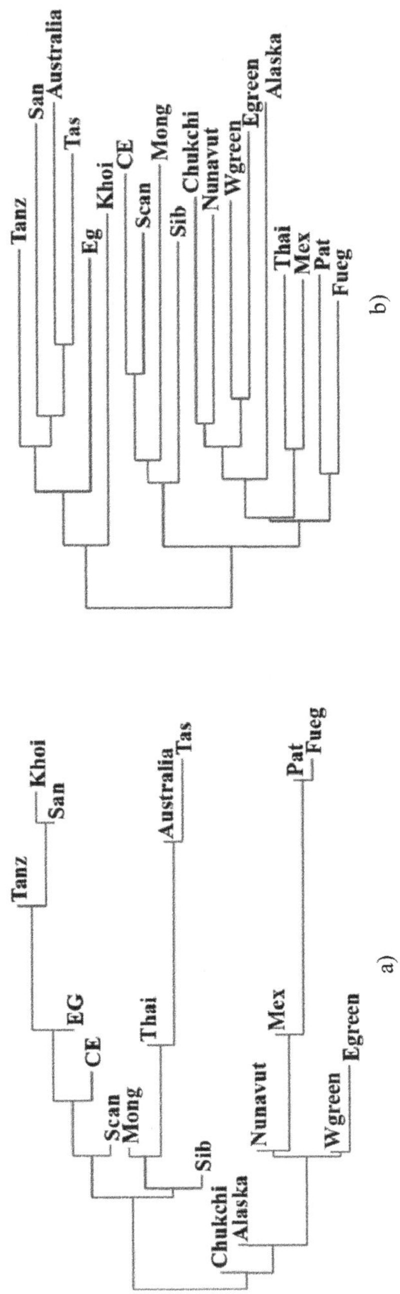

Figure 5.7. *Comparison between geographical distances (a) and shape distances, estimated with the Procrustes method (b), in the form of a phenetic tree by neighbor-joining. Each terminal point corresponds to one of the sampled populations (see Figure 5.5)*

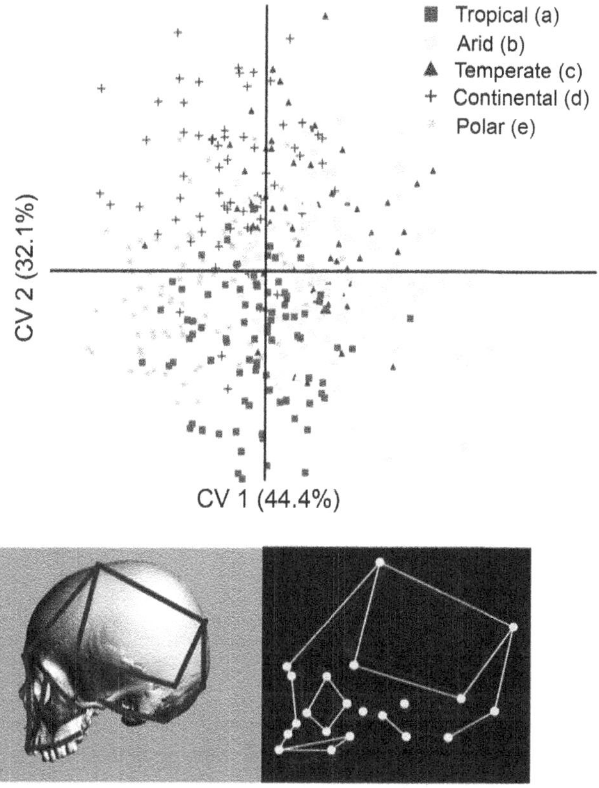

Figure 5.8. *Discriminant analysis (CVA) performed on the Procrustes coordinates, after removing the effect of geographic distance. The groups correspond to the five eco-geographic zones of the Köppen–Geiger classification*

5.4. The importance of cranial collection for the advancement of research in biological anthropology

Whatever the cause, this type of analysis is a relevant way to decipher the extent to which evolutionary history has shaped the phenetic diversity of our species. However, not all morphological diversity is the result of evolutionary forces. Some environmental factors can alter our appearance without affecting the genes, and thus our descendants. When we are exposed to intense sunlight, our skin temporarily increases melanin production to protect us from the harmful effects of UV rays. When we reach a high altitude, our body increases the production of red blood cells to compensate for the lack of oxygen. However, soon after we return to our usual environment, these effects cease. Other environmental factors can have a permanent effect on the humans, as long as they alter their growth. One of the particularities of our species, *H. sapiens*, is that these environmental factors can be cultural. This is the case, for example, for body mutilations, which exist under very divergent forms and in varied geochronological contexts (Flower 1981). Teeth filed or encrusted with precious stones, labrets, elongated necks of "giraffe women" or cranial deformations constitute sources of non-hereditary phenotypic variation. Some of these biocultural phenomena inform us about the plasticity of human growth while others can serve as social markers, reflecting membership of a certain group, which gives us information about the social structures of past populations.

No matter how much progress is made in the anthropological sciences, the majority of the research will always depend on the immense amount of biocultural information contained in skeletal remains. Recent advances in invasive analyses, such as DNA and environmental and dietary stable isotopes, have thus far only scratched the surface of this information.

5.5. References

Ash, J. and Gallup Jr., G. (2007). Paleoclimatic variation and brain expansion during human evolution. *Human Nature*, 18, 109–124.

de Azevedo, S., Nocera, A., Paschetta, C., Castillo, L., Gonzalez, M., Gonzalez-Jose, R. (2011). Evaluating microevolutionary models for the early settlement of the New World: The importance of recurrent gene flow with Asia. *Am. J. Phys. Anthropol.*, 146, 539–552.

Bailey, D. and Geary, D. (2009). Hominid brain evolution. *Human Nature*, 20, 67–79.

Beals, K., Smith, C.L., Dodd, S.M. (1984). Brain size, cranial morphology, and time machines. *Curr. Anthropol.*, 25(3), 301–330.

Betti, L., Balloux, F., Aros, W., Hanihara, T., Manica, A. (2009). Distance from Africa, not climate, explains within-population phenotypic diversity in humans. *Proc. R. Soc. B*, 276, 809–814.

Blanckaert, C. (1987). Les vicissitudes de l'angle facial et les débuts de la craniométrie (1765–1875). *Revue de synthèse*, 108(3–4), 417–453.

Chervin, A. (1907). *Anthropologie bolivienne. Mission scientifique de G. de Créqui Montfort et E. Sénéchal de la Grange*. Imprimerie nationale, Paris.

Fagundes, N.J.R., Kanitz, R., Bonatto, S.L. (2008). A reevaluation of the Native American mtDNA genome diversity and its bearing on the models of early colonization of Beringia. *PLoS ONE*, 3, e3157.

Flower, W.H. (1881). *Fashion in Deformity*. MacMillan, London.

Fromaget, J. (1940). La stratigraphie des dépôts préhistoriques de *Tam-Hang* (chaîne Annamitique septentrionale) et ses difficultés. In *Proceedings of the Third Congress of Prehistorians of the Far East, Singapore, 24th January–30th January, 1938*, Nutter Chasen, F. and Tweedle, M.W.F (eds). Government Printing Office, Singapore.

Galland, M. and Friess, M. (2016). A three-dimensional geometric morphometrics view of the cranial shape variation and population history in the New World. *Am. J. Hum. Biol.*, 28(5), 641–646.

Heim, J.L. (1978). *Les hommes fossiles de La Ferrassie*. Masson, Paris.

Katz, D.C., Grote, M.N., Weaver, T.D. (2017). Changes in human skull morphology across the agricultural transition are consistent with softer diets in pre-industrial farming groups. *Proceedings of the National Academy of Sciences USA*, 114, 9050–9055.

Kimura, M. (1968). Evolutionary rate at the molecular level. *Nature*, 217, 624–626.

Kottek, M., Grieser, J., Beck, C., Rudolf, B., Rubel, F. (2006). World map of the Köppen-Geiger climate classification updated. *Meteorol. Z.*, 15, 259–263.

Malaspinas, A-S., Lao, O., Schroeder, H., Rasmussen, M., Raghavan, M., Moltke, I., Campos, P.F., Sagredo, F.S., Rasmussen, S., Gonçalves, V.F. *et al.* (2014). Two ancient human genomes reveal Polynesian ancestry among the indigenous Botocudos of Brazil. *Current Biology*, 24(21), R1035–R1037.

Mielke, J.H., Konigsberg, L.W., Relethford, J. (2011). *Human Biological Variation*. Oxford University Press, Oxford.

Morey, D.F., Crothers, G.M., Stein, J.K., Fenton, J.P., Hermann, N.P. (2002). The fluvial and geomorphic context of Indian Knoll, an Archaic shell midden in West-Central Kentucky. *Geoarchaeology*, 17(6), 521–553.

Neves, W.A. and Pucciarelli, H.M. (1991). Morphological affinities of the first Americans: An exploratory analysis based on early South American remains. *J. Hum. Evol.*, 21, 261–173.

Neves, W.A., Hubbe, M., Pilo, L.B. (2007). Early Holocene human skeletal remains from Sumidouro Cave, Lagoa Santa, Brazil: History of discoveries, geological and chronological context, and comparative cranial morphology. *J. Hum. Evol.*, 52, 16–30.

Sardi, M.L., Novellino, P.S., Pucciarelli, H.M. (2006). Craniofacial morphology in the Argentine Center-West: Consequences of the transition to food production. *Am. J. Phys. Anthropol.*, 130, 333–343.

Schultz, M. and Reich, M. (2013). Blumenbachsche Schädelsammlung. In *Die Sammlungen, Museen und Gärten der Universität Göttingen*, Beisiegel, U., Reich, M., Pietzner, K. (eds). Universitätsverlag Göttingen, Göttingen.

Sibeud, E. (2012). A useless colonial science? Practicing anthropology in the French Colonial Empire, circa 1880–1960. *Curr. Anthopol.*, 53, S83–S94.

Slice, D. (ed.) (2005). *Modern Morphometrics in Physical Anthropology*. Springer, New York.

Steegmann Jr., A.T. (1970). Cold adaptation and the human face. *Am. J. Phys. Anthropol.*, 32(2), 243–250.

Washburn, S.L. (1951). The new physical anthropology. *Transactions of the New York Academy of Science*, 13, 298–304.

Wiley, D.F. (2005). Landmark v 3.0. document. Institute for Data Analysis and Visualization, University of California, CA.

6

The Discovery of New Metal-Hyperaccumulating Plant Species in Herbaria

Vanessa R. INVERNÓN[1], Romane TISSERAND[2], Pierre JOUANNAIS[3], Dulce M. NAVARRETE GUTIÉRREZ[2], Serge MULLER[1], Yohan PILLON[4], Guillaume ECHEVARRIA[2] and Sylvain MERLOT[3]

[1] *ISYEB – CNRS, Muséum national d'Histoire naturelle, Sorbonne Université, École Pratique des Hautes Études, Université des Antilles, Paris, France*

[2] *LSE, INRAE, Université de Lorraine, Nancy, France*

[3] *I2BC – CEA – CNRS, AgroParisTech, Université Paris-Saclay, Gif-sur-Yvette, France*

[4] *LSTM, IRD – INRAE – CIRAD, Institut Agro, Université Montpellier, France*

As the sixth extinction has begun, describing the diversity of plant species is a major challenge for the scientific community. A small group of species, called hyperaccumulators, have the unique ability to accumulate quantities of metals, such as nickel, manganese or zinc, 10–100 times higher than the toxic level for most plants. These species are currently receiving particular attention because they can be used for restoring mining or polluted sites, and their biomass can be used to recycle the accumulated metals. This chapter presents some results of recent research projects that are being developed in several herbaria around the world to identify new hyperaccumulating species, through elemental analysis of specimens

For a color version of all figures in this chapter, see www.iste.co.uk/pellens/natural.zip.

by X-ray spectrometry. The first results generated, and in particular those obtained at the MNHN herbarium, demonstrate that herbaria represent a precious source and are also an extremely efficient tool to identify new metal hyperaccumulators species in regions insufficiently studied until now, such as the Americas or Madagascar. This work also highlights the great importance of collecting geographic and edaphic data associated with herbarium specimens, which are essential to effectively conduct this type of research.

6.1. Metal-hyperaccumulating plants

In order to grow, plants need various mineral elements such as nitrogen, potassium and phosphorus, as well as many metallic elements such as iron, zinc, manganese and copper, which are involved in the functioning of enzymes that play essential roles in photosynthesis and respiration. These elements, necessary in small amounts, can be toxic when present in too large quantities in the soil. This is, for example, the case of substrates naturally rich in metals, such as serpentine (a rock rich in nickel), mining sites or polluted soils (by industries, batteries, etc.). The majority of plants growing in these environments limit the entry of metals in their roots or their transfer to the aerial parts. However, a more limited number of species will absorb and store large quantities of these metals in their leaves. When these concentrations in the leaves exceed a defined threshold, these plants are called metal hyperaccumulators. The threshold at which a plant is considered as hyperaccumulator varies according to the element considered. Some authors have suggested that a plant is a hyperaccumulator of an element if the measured levels are 100–1,000 times higher than in a "typical" plant. For example, a plant is considered to be hyperaccumulating nickel if its leaf concentration exceed 1,000 $mg \cdot kg^{-1}$ (or 1,000 parts per million [ppm], or 0.1% of dry mass) under natural conditions, 3,000 ppm for zinc and 10,000 ppm for manganese.

Abnormally high concentrations of nickel were first reported in *Alyssum bertolonii*, a herbaceous plant of the Brassicaceae family native to Italy (Minguzzi and Vergnano 1948). Then, in the 1970s, new examples of nickel hyperaccumulating species were discovered in Australia (Severne and Brooks 1972) and New Caledonia (Jaffré and Schmid 1974; Brooks *et al.* 1977). In New Caledonia, the "blue sap" tree, *Pycnandra acuminata* (Figure 6.1(a)), still holds the record for accumulation; its latex can contain up to 25% nickel (in dry mass), which gives it a characteristic blue color (Jaffré *et al.* 1976). Currently, more than 700 hyperaccumulating species of metals are known (Reeves *et al.* 2018), including more than 500 for nickel, a few dozen for copper, cobalt, manganese, zinc, and less than a dozen for lead, cadmium, or rare earth elements. While more than 50 botanical families contains hyperaccumulating species, one lineage of flowering plants in particular, the "COM clade" (orders Celastrales, Oxalidales and Malpighiales), concentrates more than 40% of the known hyperaccumulating species of nickel alone, including *Geissois lanceolata* (Figure 6.1(b)) (Pillon *et al.* 2010).

Figure 6.1. *(a) Blue latex of* Pycnandra acuminata *(image by S. Merlot). (b)* Geissois lanceolata, *a nickel hyperaccumulating species, endemic to New Caledonia (image by Y. Pillon). (c)* Phyllanthus rufuschaneyi *under cultivation, producing more than 250 kg of nickel per hectare per year (Sabah, Malaysia) (image by G. Echevarria)*

There are several hypotheses that could explain this characteristic (Macnair 2003). Accumulation could protect plants from herbivores (insects, vertebrates) and pathogens (fungi, bacteria). Other hypotheses, such as drought resistance or competition with other plants (allelopathy), have also been put forward.

Human needs for metals are increasingly important (Pitron 2018), and the metal cycle has accelerated considerably during the Anthropocene, from their extraction to their more or less controlled release into nature. Hyperaccumulating plants are interesting because they are able to develop in metal-rich environments, but especially because of their ability to extract these metals. They can thus be used for ecological restoration and phytoremediation of mining or metal contaminated sites. The metal-rich biomass can then be used to produce, with a reduced ecological footprint, pure metal salts with high added value (Figure 6.1(c)) (van der Ent *et al.* 2015), or bio-sourced catalysts used for the synthesis of flavors or drugs in the chemical industry (Grison *et al.* 2015).

Today, our knowledge of metal hyperaccumulating plants is still very patchy. Due to a growing interest in these species, it has become essential to complete the inventory of them, in order to identify the best candidates for applications in phytoextraction, for example, but also to study the complex physiological mechanisms involved and their place in the plant kingdom.

Tropical regions, with their extraordinary biodiversity, should harbor a large number of new hyperaccumulating species. New Caledonia and Cuba are two regions whose serpentine flora have been studied in detail for several decades (Jaffré 1980; Borhidi 1996). As metal hyperaccumulation is most frequently associated with serpentines and other ultramafic rocks, it is not surprising that these two islands concentrate more than half of the known hyperaccumulating species. Other regions of the world have not received as much attention and hyperaccumulation may also occur on other types of substrates (Figure 6.2).

Tropical America, Africa and the Indochinese peninsula are areas where hyperaccumulating plants have been little researched, despite the significant biodiversity of these regions. Herbaria are an ideal place to look for new candidates because they allow the study of a large number of species at low cost, including extinct species, without having to travel to areas that are sometimes difficult to access (extreme climate, conflicts, etc.). Today, new techniques for non-destructive analyses allow us to use these collections accumulated during the past centuries to address very modern problems in a different way.

Figure 6.2. *Distribution of nickel hyperaccumulating species worldwide, from the Global Hyperaccumulator Database (retrieved January 02, 2019)*

6.2. The screening of herbarium collections: from atomic absorption to X-ray fluorescence

The use of herbaria (see Chapter 16 by Muller *et al.*) to discover metal hyperaccumulating plants dates back to the 1970s, when Brooks *et al.* studied the herbarium specimens of tropical plants to identify nickel hyperaccumulators (Brooks *et al.* 1977; Jaffré *et al.* 1979; Macnair 2003). Until the 1980s, atomic absorption spectrometry (AAS) was used to quantify metals in plants. However, this technique was not very sensitive at the time and could not easily measure multiple elements in parallel, which limited the discovery of plants accumulating metals other than nickel. The study of nickel-hyperaccumulating plants was also aided by a simple colorimetric test that could be performed in the field by botanists, using the ability of dimethylglyoxime (DMG) to form a red complex in the presence of nickel (Figure 6.3(a)). Currently, inductively coupled plasma spectrometry (ICP-AES or ICP-MS) is predominantly used for quantitative elemental analysis of plant samples. ICP techniques allow several elements to be measured simultaneously, and with a very high sensitivity. However, they require a time-consuming, relatively expensive and destructive preparation of samples, which is not well adapted to the large-scale analysis of herbarium specimens. Conversely, X-ray fluorescence spectrometry (XRF) is a non-destructive method that allows the simultaneous measurement of the concentration of several elements in a plant matrix, without prior preparation of the sample. Its sensitivity is lower than ICP techniques (a few tens of ppm) but largely sufficient for the analysis of metal hyperaccumulating species (Jaffré *et al.* 2013). In addition, the absence of water in the herbarium specimens improves the sensitivity of XRF measurements. The principle of XRF relies on the use of an X-ray source to excite the electrons of the elements, inducing the emission of fluorescence radiation whose energy is characteristic of the electronic orbitals of each element. This fluorescence radiation is analyzed by an X-ray detector, allowing the identification of the different elements present in the sample. Today, the miniaturization of X-ray sources and detectors allows us to build handheld XRF spectrometers, which are very efficient tools for the analysis of herbarium specimens without damaging them (Figure 6.3(b)). Indeed, the analysis of a herbarium specimen by XRF is carried out very rapidly (less than 1 minute), which makes it possible to analyze more than 300 specimens per day (i.e. the equivalent of 6,000 per month).

a) b)

Figure 6.3. *(a) Nickel transport (turquoise green) in conducting tissues of* Blepharidium guatemalense *(Chiapas, Mexico) – DMG test (image by G. Echevarria). (b) Screening of the MNHN herbarium by portable XRF (image by S. Merlot)*

The screening of herbariums with portable XRF machines to identify new metal hyperaccumulating species has some limitations, however. This technology is not yet able to accurately measure the lightest elements, such as aluminum, or the heaviest elements, such as uranium. The presence of cadmium is also masked in plant matrices by the radiation of potassium present in large quantities. The quantification of elements obtained with XRF machines must also be taken with caution because it uses mathematical models that have not been developed for plant material. A calibration of the portable XRF machines with plant samples of known composition is therefore essential. Finally, herbarium specimens can be contaminated by soil-borne dust, by elements such as mercury and arsenic that were used in the past as insecticides and fungicides in herbaria, or by zinc present in some glues used in the assembly of the sheets, as observed at the Sandakan herbarium in Malaysia (van der Ent *et al.* 2019b).

6.3. The discovery of new metal-hyperaccumulating plants at the MNHN herbarium

Several herbarium screens using portable XRF equipment have recently been conducted, including in Noumea (New Caledonia), in the Malaysian state of Sabah (Borneo) and in New Guinea (Do *et al.* 2020). In Noumea, this screening identified approximately 90 new hyperaccumulating taxa of manganese, nickel, cobalt or zinc (Gei *et al.* 2018). The Sabah region of Malaysia is known to be a high biodiversity area hosting many endemic nickel hyperaccumulating plants. The screening of more than 6,000 specimens from the Sandakan Forest Research Centre herbarium resulted in the discovery of new species that hyperaccumulate nickel, manganese and cobalt simultaneously or not (van der Ent *et al.* 2019b). In total, more than 100 new cobalt, manganese, nickel and zinc hyperaccumulating species have been discovered in the last four years (Reeves *et al.* 2018; van der Ent *et al.* 2019b). XRF herbarium analysis is currently underway around the world to identify new metal hyperaccumulating plants. In this chapter, we will present the work we are currently conducting at the MNHN herbarium, which is mainly aimed at discovering new metal-hyperaccumulating species in the Americas and Madagascar.

6.3.1. *The interest of the MNHN herbarium for the research of metal-hyperaccumulating plants*

The MNHN's herbarium in Paris contains a collection of about eight million specimens. This natural history collection gathers specimens from all over the world, collected for nearly four centuries, and representing almost all botanical families of seed plants (angiosperms and gymnosperms), ferns and other spore plants, bryophytes, algae, as well as fungi and lichens. All of this makes the herbarium of Paris the largest herbarium in the world, but also one of the oldest, even if it continues to be enriched with several thousand new specimens each year.

To facilitate scientific research on the MNHN herbarium, all of the vascular plant specimens, as well as some of the algae and bryophytes, were digitized during its complete renovation between 2008 and 2013. Thus, nearly six million images of specimens are now available in the SONNERAT database[1], which is unparalleled for

1 Available at: https://science.mnhn.fr/institution/mnhn/collection/p/item/search.

such a natural history collection (Le Bras *et al.* 2017). Each image has associated metadata, such as its scientific names (family, genus, species), as well as the large geographical area. All other data (locality and date of collection, name of the collector, etc.) has only been computerized for a small fraction of the collections and remains to be transcribed for more than 80% of the specimens. A citizen science program, Les Herbonautes[2], was launched in 2012 by the *Muséum* to accelerate the transcription and recording of all herbarium label data from images of the corresponding specimens (Rouhan *et al.* 2014).

Because of its richness, the herbarium of Paris is a very interesting collection to screen a maximum of plant species with XRF technology and discover new species that are hyperaccumulators of metals. The choice of the herbarium of Paris is all the more relevant as it is one of the richest in the world for plants from New Caledonia (200,000 specimens), Madagascar (more than 400,000 specimens), Brazil (more than 300,000 specimens) and Mexico (more than 135,000 specimens). These territories, which are areas of significant diversity and possess important surfaces covered by metalliferous soils, have been little studied, until now, in the search for metalhyperaccumulating plants.

The collection of vascular plants at the herbarium is organized according to phylogenetic classification (Angiosperm Phylogeny Group 2009) up to the taxonomic rank of the families, then the genera are ordered according to a specific classification based on the evolutionary relationships between these taxa (when they are known). Each genus is then divided into a few large geographical areas, and the species are then classified in alphabetical order. Faced with this complexity for access to specific specimens, and the number of specimens available for the American and Madagascar sectors, we adopted two methodological approaches to sampling for analysis: (1) a systematic approach by screening species (five specimens per species) of plant families known to contain hyperaccumulating species, (2) a geographic/geological approach, by screening specimens of species collected in areas with metalliferous soils. The first results of our research show that the systematic approach is faster and at least as effective as the geographical approach because of the taxonomic classification of species in the herbaria. However, the geographical/geological approach had the advantage of being able to identify hyperaccumulating species in families where this character was still unknown. The geographical approach could even be improved by acquiring the metadata associated with the herbarium specimens (precise geographical location and/or type of soil),

2 Available at: http://lesherbonautes.mnhn.fr.

which would allow a better targeting of the specimens, by privileging those collected on metalliferous soils. However, this relevant information is often absent (not provided) for old specimens. For the acquisition of data present on the labels of screened specimens, a specific mission, *Fan de metal* (literally Metal Fan), was launched in 2019 on the citizen science program Les Herbonautes[3].

6.3.2. *From the herbarium to the field: new nickel hyperaccumulators in the genus* Orthion

The study of serpentine flora in different regions of the world, such as Cuba and New Caledonia, has identified more than 500 nickel-hyperaccumulating species, representing the largest contingent of metalhyperaccumulating plants. However, there are many other nickel-rich outcrops around the world that have not yet been studied, suggesting that new nickel hyperaccumulating species may be identified.

The presence of nickel hyperaccumulating specimens belonging to the genus *Psychotria* (Rubiaceae), collected at Los Tuxtlas Tropical Biological Station in the state of Veracruz, Mexico, was first reported to us by Joseph Pollard, professor at Furman University in the United States (McCartha *et al.* 2019). These findings suggested that the soil in this region is rich enough in nickel to observe the expression of the hyperaccumulation trait in some species. A scientific mission in this region confirmed the hyperaccumulation of nickel by the local *Psychotria* species, and further identified *Orthion veracruzense*, of the Violaceae family, as a new species hyperaccumulating this metal. This discovery was followed by a systematic XRF analysis of all specimens belonging to the genus *Orthion* in the MNHN herbarium in Paris and in the MEXU herbarium in Mexico. These analyses indicated that all species of the genus *Orthion* are capable of hyperaccumulating nickel and that one of them, *Orthion subsessile*, can reach concentrations up to 47,200 ppm. This discovery is remarkable since it is the first example of nickel hyperaccumulation in the Violaceae family in the Neotropics. In addition, some hyperaccumulating specimens of *Orthion subsessile* were collected in the Chiapas region of southeastern Mexico. Since then, their presence has been verified in the field through new collections, suggesting that this region contains soils rich enough in nickel to allow the expression of this character. This result illustrates the interest of searching for metal hyperaccumulating plants in herbarium collections to identify new geographical areas with metal-rich soils.

3 Available at: http://lesherbonautes.mnhn.fr.

a) b)

Figure 6.4. *(a) Specimen of* Orthion subsessile, *collected in the state of Chiapas by E. Martinez in 2004 (MEXU herbarium 1151696, Departamento de Botánica, IBUNAM). (b)* Orthion subsessile, *nickel hyperaccumulator found in northeastern Chiapas, Mexico (image by S. Alvarez)*

6.3.3. Rinorea multivenosa, *the first zinc hyperaccumulating species discovered in the Amazon basin*

At present, about 20 zinc hyperaccumulating species ($\geq$3,000 ppm) are known worldwide. These species are distributed in 12 genera belonging to nine families, most of which represented the Brassicaceae family. The genus *Rinorea* (Violaceae) is one genus in which several nickel hyperaccumulators and one zinc hyperaccumulator (*Rinorea longiracemosa* identified in Sabah, Malaysia) have been found (Reeves *et al.* 2018). We systematically screened the 278 specimens (57 species) belonging to the genus *Rinorea* collected in the Americas and Madagascar regions and present in the MNHN herbarium.

Our research revealed very high levels of zinc in *Rinorea multivenosa* specimens collected in Brazil. XRF measurements on the four samples of this species indicated leaf zinc levels ranging from 9,300 to 17,800 ppm, well above the threshold for zinc hyperaccumulation. The screening of 33 other species of the genus *Rinorea* in the American zone did not reveal any other zinc-accumulating species.

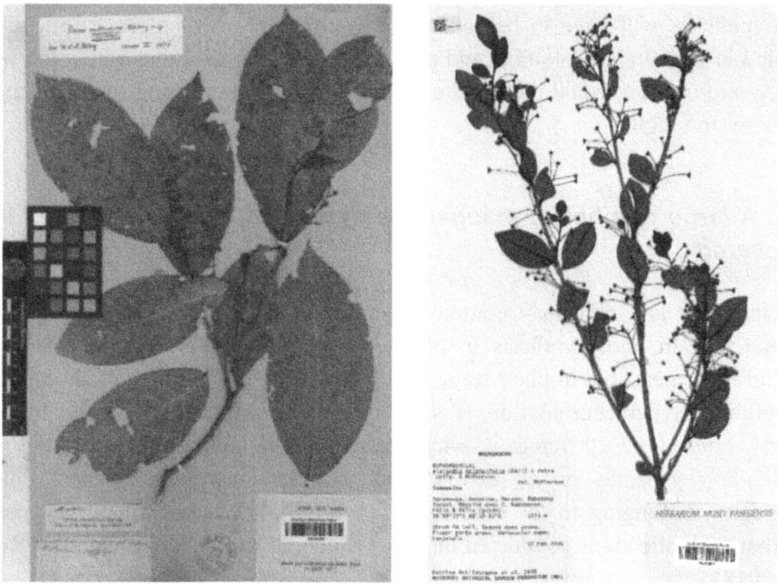

a) b)

Figure 6.5. *(a) Specimen of* Rinorea multivenosa *collected in Brazil by J.W.H. Trail in 1874 (P02141355, MNHN herbarium). (b) Specimen of* Wielandia oblongifolia, *this species from Madagascar is able to accumulate both manganese and cobalt (P05472829, MNHN herbarium)*

Three specimens of *R. multivenosa* present in the MNHN herbarium come from an expedition by Professor James W.H. Trail, a Scottish botanist, carried out in Brazil at the end of the 19th century. These samples were collected in the northwestern region of the Amazon (Amazonas), between the Purus River and the Jutaí River. However, because of the lack of precise geographic and soil information related to these old specimens, it is difficult to know if the hyperaccumulating character of *R. multivenosa* is associated to zinc-rich soils. The last specimen analyzed came from the collection of Auguste F.M. Glaziou, a French landscape engineer who directed the gardens of the Imperial House of Rio de Janeiro from 1869. The label of this specimen indicates that it was collected on October 16, 1874 at the Quinta de S. Christovão (Rio de Janeiro). This specimen, then attributed to the species *Alsodeia macrocarpa*, is thought to have come from a tree grown in the former botanical gardens of the São Critóvão Palace, the residence of the Brazilian emperors in the 19th century, which became the Quinta da Boa Vista and houses the National Museum of Brazil (Glaziou 1905). This last specimen suggests that *R. multivenosa* would be able to hyperaccumulate zinc, even from soils that are not particularly rich with this metal.

This hypothesis will have to be confirmed by the analysis of more recent samples collected in Brazil and Colombia and present in other herbaria throughout the world. *Rinorea multivenosa* would thus be the first zinc hyperaccumulating species identified in the Amazon region.

6.3.4. *A large number of manganese hyperaccumulating species to be discovered*

Manganese is a relatively common element in the earth's crust, and plays an essential role in photosynthesis in plants. Consequently, the average manganese concentration measured in plant leaves varies from 30 to 500 ppm and the threshold for defining hyperaccumulation is 10,000 ppm (1% of dry mass). Until very recently, only about 20 manganese hyperaccumulating species had been described worldwide (Fernando *et al.* 2013). This number has very recently doubled (42 species, belonging to 24 genera and 16 families) because of XRF screening of herbarium collections conducted in New Caledonia and Sabah, Indonesia (Reeves *et al.* 2018).

Species of the Proteaceae family are known to accumulate significant amounts of manganese due to their phosphate acquisition strategy, based on massive excretion of organic acids by their proteoid roots leading to metal solubilization. We systematically screened the species of Proteaceae in the MNHN herbarium from the American sector, which represented 73 species in eight genera. Although no manganese hyperaccumulators had been identified so far on the new continent, we identified nine new species in the genera *Lomatia*, *Roupala* and *Embothrium*, some of which exceeded the 10,000 ppm manganese concentration measured by XRF. The most notable species is *Lomatia dentata*, for which we measured over 40,000 ppm manganese in a specimen collected in the Valdivia region of Chile. Other specimens of *Lomatia ferruginea* and *Embothrium coccineum* collected in the same region also have manganese concentrations exceeding 10,000 ppm. The Valdivia region is known for its temperate rainforest with a remarkable biodiversity.

The island of Madagascar is also an area for which we did not know any hyperaccumulator species, despite the presence of numerous metalliferous outcrops, such as the Ambatovy site exploited for nickel and cobalt. We screened several genera belonging to the order Malpighiales, known to possess numerous hyperaccumulating species. We identified several species with leaf manganese concentrations above 10,000 ppm in the genera *Scolopia* and *Homalium* (Salicaceae), *Wielandia* and *Thecacoris* (Phyllanthaceae). The Malagasy species *Wielandia oblongifolia* is particularly remarkable because it is able to accumulate 11,000 ppm of manganese and 400 ppm of cobalt, also exceeding the hyperaccumulation threshold defined for this metal. Further investigations on the neighboring island of Mayotte revealed a specimen of *Garcinia livingstonei* (Clusiaceae) with 11,500 ppm of manganese. These species certainly represent the first metal hyperaccumulating taxa identified in Madagascar and Mayotte. In general, our results suggest that there are still many manganese hyperaccumulating species to be discovered throughout the world.

6.4. Conclusion

The work presented in this chapter illustrates the great interest that herbarium collections represent for researching metal hyperaccumulating plants. This interest has been revived recently with the development of handheld XRF technology that allows rapid, non-destructive elemental analysis of specimens, and is therefore compatible with the screening of these collections.

Herbaria are an invaluable source of samples for XRF analysis, providing easy access to a very large quantity of specimens collected from all over the world, sometimes in difficult to access areas, and whose identification has been validated by experts. Moreover, unlike DNA, the elements do not suffer from sample preparation methods or time, and almost all specimens present in the herbaria can be analyzed.

Our research at the MNHN herbarium in Paris has led to the identification of new metal hyperaccumulating species in the Neotropics and Madagascar. The discovery of nickel and zinc hyperaccumulating species in Mexico and Brazil, and of several manganese hyperaccumulating species in Chile and Madagascar, represent the first examples of the identification of metal hyperaccumulating plants in these countries. These results open up prospects for the use of these plants in sustainable phytotechnologies, such as agromine and ecocatalysis, to reduce the impact of the mining industry on the environment.

To date, XRF screening in herbaria has focused primarily on plant groups known to harbor metal hyperaccumulating species. This strategy has proven to be very effective in identifying such species in regions of the world that have been little studied to date. However, the discovery of hyperaccumulating species in new groups will probably require the systematic screening of regional herbaria in the vicinity of metalliferous outcrops, or the ability to easily identify specimens from these areas of interest in large herbaria. For this, it seems important to us that in addition to the precise geographical location of the plant specimens, information concerning the nature of the soils be provided. The effort currently being made in various herbaria to digitize the herbarium sheets and to computerize the data associated with the specimens should, in the near future, enable us to efficiently search for samples collected in metalliferous soils in order to increase our chances of identifying new metal-hyperaccumulating species.

6.5. Acknowledgments

The authors would like to thank the teams at the MNHN herbarium, in particular, the scientific managers of the collections of the studied groups, for access to the collections, as well as the National herbarium of Mexico MEXU (Instituto de Biologia, UNAM), and Saul Alvarez for the images of *Orthion subsessile*. The MNHN herbarium XRF screening project is funded by the CNRS Mission for Transversal and Interdisciplinary Initiatives (X-Life Challenge). *Les Herbonautes* is a citizen science program of the national research infrastructure e-ReColNat (ANR-11-INBS-004).

6.6. References

Angiosperm Phylogeny Group (2009). An update of the Angiosperm Phylogeny Group classification for the orders and families of flowering plants: APG III. *Botanical Journal of the Linnean Society*, 161, 105–121.

Borhidi, A. (1996). *Phytogeography and Vegetation Ecology of Cuba.* Akadémiai Kiadó, Budapest.

Brooks, R.R., Lee, J., Jaffré, T. (1974). Some New Zealand and New Caledonian plant accumulators of nickel. *Journal of Ecology*, 62, 493–499.

Brooks, R.R., Lee, J., Reeves, R.D., Jaffré, T. (1977). Detection of nickeliferous rocks by analysis of herbarium specimens of indicator plants. *Journal of Geochemical Exploration*, 7, 49–57.

Do, C., Abubakari, F., Remigio, A.C., Brown, G.K., Casey, L.W., Burtet-Sarramegna, V., Gei, V., Erskine, P.D., van der Ent, A. (2020). A preliminary survey of nickel, manganese and zinc (hyper)accumulation in the flora of Papua New Guinea from herbarium X-ray fluorescence scanning. *Chemoecology*, 30, 1–13.

van der Ent, A., Baker, A.J.M., Reeves, R.D., Chaney, R.L., Anderson, C.W.N., Meech, J.A., Erskine, P.D., Simonnot, M.-O., Vaughan, J., Morel, J.L., Echevarria, G., Fogliani, B., Rongliang, Q., Mulligan, D.R. (2015). Agromining: Farming for metals in the future? *Environ. Sci. Technol.*, 49, 4773–4780.

van der Ent, A., Ocenar, A., Tisserand, R., Echevarria, G., Erskine, P.D. (2019a). XRF herbarium ionomics of the ultramafic flora of Sabah (Malaysia, Borneo Island). *Journal of Geochemical Exploration*, 202, 49–58.

van der Ent, A., Pollard, A.J., Echevarria, G., Erskine, P.D. (2019b). X-ray fluorescence ionomics of herbarium collections. *Scientific Reports*, 9, 4746.

Fernando, D.R., Marshall, A., Baker, A.J.M., Mizuno, T. (2013). Microbeam methodologies as powerful tools in manganese hyperaccumulation research: Present status and future directions. *Frontiers in Plant Science*, 4, 9.

Gei, V., Erskine, P.D., Harris, H.H., Echevarria, G., Mesjasz-Przybyłowicz, J., Barnabas, A.D., Przybyłowicz, W.J., Kopittke, P.M., van der Ent, A. (2018). Tools for the discovery of hyperaccumulator plant species and understanding their ecophysiology. In *Agromining: Farming for Metals,* van der Ent, A., Echevarria, G., Baker, A.J.M., and Morel, J.L. (eds) Springer International Publishing, Cham.

Glaziou, A.F.M. (1905). *Liste des plantes du Brésil central recueillies en 1861–1895.* Société botanique de France, Paris.

Grison, C., Escande, V., Biton, J. (2015). *Ecocatalysis: A New Integrated Approach to Scientific Ecology.* ISTE Press, London, and Elsevier, Oxford.

Jaffré, T. (1980). Étude écologique du peuplement végétal des sols dérivés de roches ultrabasiques en Nouvelle-Calédonie. PhD Thesis, ORSTOM, Paris.

Jaffré, T. and Schmid, M. (1974). Accumulation du nickel par une Rubiacée de Nouvelle-Calédonie, *Psychotria douarrei* (G. Beauvisage) Däniker. *Comptes Rendus de l'Académie des Sciences de Paris (D)*, 278, 1727–1730.

Jaffré, T., Brooks, R.R., Lee, J., Reeves, R.D. (1976). *Sebertia acuminata*: A hyperaccumulator of nickel from New Caledonia. *Science*, 193, 579–580.

Jaffré, T., Kersten, W.J., Brooks, R.R., Reeves, R.D. (1979). Nickel uptake by Flacourtiaceae of New Caledonia. *Proceedings of the Royal Society B: Biological Sciences*, 205, 385–394.

Jaffré, T., Pillon, Y., Thomine, S., Merlot, S. (2013). The metal hyperaccumulators from New Caledonia can broaden our understanding of nickel accumulation in plants. *Frontiers in Plant Science*, 4, 279.

Kersten, W.J., Brooks, R.R., Reeves, R.D., Jaffré, T. (1979). Nickel uptake by New Caledonian species of *Phyllanthus*. *Taxon*, 38, 529–534.

Le Bras, G., Pignal, M., Jeanson, M.L., Muller, S., Aupic, C., Carré, B., Flament, G., Gaudeul, M., Gonçalves, C., Invernón, V.R., Jabbour, F., Lerat, E., Lowry, P.P., Offroy, B., Pimparé, E.P., Poncy, O., Rouhan, G., Haevermans, T. (2017). The French Muséum national d'Histoire naturelle vascular plant herbarium collection dataset. *Sci. Data*, 4, 170016.

Macnair, M.R. (2003). The hyperaccumulation of metals by plants. *Advances in Botanical Research*, 40, 63–105.

McCartha, G.L., Taylor, C.M., van der Ent, A., Echevarria, G., Navarrete Gutiérrez, D.M., Pollard, A.J. (2019). Phylogenetic and geographic distribution of nickel hyperaccumulation in neotropical *Psychotria*. *American Journal of Botany*, 106, 1377–1385.

Minguzzi, C. and Vergnano, O. (1948). Il continuto di nichel nelle ceneri di *Alyssum bertolonii* Desv. *Atti della Società Toscana di Scienze Naturali*, 55, 49–74.

Pillon, Y., Munzinger, J., Amir, H., Lebrun, M. (2010). Ultramafic soils and species sorting in the flora of New Caledonia. *Journal of Ecology*, 98, 1108–1116.

Pitron, G. (2018). *La guerre des métaux rares : la face cachée de la transition énergétique et numérique*. Editions Les Liens qui Libèrent, Paris.

Reeves, R.D., Baker, A.J.M., Jaffré, T., Erskine, P.D., Echevarria, G., van der Ent, A. (2018). A global database for plants that hyperaccumulate metal and metalloid trace elements. *New Phytol.*, 218, 407–411.

Rouhan, G., Chagnoux, S., Dennetière, B., Schäfer, V., Pignal, M. (2014). The herbonauts website: Recruiting the general public to acquire the data from herbarium labels. In *Botanists of the Twenty First Century: Roles, Challenges and Opportunities, Proceedings of UNESCO International Conference*. UNESCO, Paris.

Severne, B.C. and Brooks, R.R. (1972). A nickel-accumulating plant from Western Australia. *Planta*, 103, 91–94.

7

Fossil Crustaceans in the Light of New Technologies

Sylvain CHARBONNIER and Marie-Béatrice FOREL
CR2P – CNRS, Muséum national d'Histoire naturelle, Sorbonne Université, Paris, France

The French National Museum of Natural History (*Muséum national d'Histoire naturelle* (MNHN) also commonly referred to simply as the *Muséum*) is one of the pillars of research on fossil crustaceans, with a collection of more than 20,000 samples. This collection was built up by the pioneers of marine invertebrate paleontology and micropaleontology. It served as a basis for the establishment of the systematics of crustaceans and their evolutionary history. This collection is today valorized by the use of new technologies that allow us to answer questions on the kinship relations between fossil and modern-day crustaceans, on their origin and their diversification, on their geographical distribution and the occupation of ecological niches during geological times. Ultraviolet light, green light, X-ray radiography, computed tomography (CT) scan X-ray micro-tomography and synchrotron X-ray fluorescence imaging are all techniques that reveal the preservation of anatomical details invisible to the naked eye, internal organs, visual devices and larval stages. The scanning electron microscope documents, from ancient samples, new forms of microcrustaceans and then reveals new paleobiogeographical relationships. The new technologies thus enhance the collection of fossil crustaceans, which is constantly growing.

For a color version of all figures in this chapter, see www.iste.co.uk/pellens/natural.zip.

Natural History Collections in the Science of the 21st Century,
coordinated by Roseli PELLENS. © ISTE Ltd 2021.

7.1. Introduction

The MNHN's paleontology collections contain millions of vertebrate, invertebrate, plant and microfossil specimens, many of which remain uninventoried. Among the groups that abound in the drawers, marine arthropod fossils are major constituents of the invertebrate and microfossil collections.

The MNHN's collection of marine arthropod fossils is very old and is composed of several paleontological collections: large historical collections built up by prestigious naturalists, stratigraphic collections, specialized collections concerning coherent taxonomic groups and geographic collections. It contains about 20,000 specimens including 6,000 trilobites, 2,000 decapod crustaceans, 300 Thylacocephala (enigmatic arthropods that disappeared in the Cretaceous) and thousands of ostracod preparations. More than 2,700 type and figured specimens are now recorded.

This collection is mainly used for fundamental research (taxonomy, comparative anatomy, systematics) but also for museography and teaching. It is regularly solicited by international researchers who wish to consult the reference specimens kept in the marine invertebrate library.

Current research focuses on crustaceans from exceptionally well-preserved deposits in the Jurassic of France (La Voulte-sur-Rhône, Canjuers), Germany (Solnhofen, Eichstätt) and the Cretaceous of Lebanon (Hadjoula, Hakel, Sahel Alma). As far as micropaleontology is concerned, current research focuses, for example, on biodiversity and paleobiogeographic relationships during geological times.

7.2. Fossil crustaceans

In nature today, crustaceans constitute one of the most diversified zoological groups with nearly 70,000 described species. Among the crustaceans, the decapods group together about 15,000 present-day species for only 3,500 fossil species described from the Devonian (about 380 million years ago) to the Quaternary. As for the ostracods, it is estimated that there are about 20,000 species today and more than 10,000 have already been described from the Ordovician (about 485 million years) to the Quaternary.

To date, fossil crustaceans have been largely underexploited as objects and models of study in paleontology. Not particularly abundant in the fossil record, they are clearly underestimated from the point of view of their paleobiodiversity. They are also very often absent from paleoecological, paleoenvironmental and especially paleobiogeographic reconstructions. Moreover, the evolutionary history of crustaceans regularly ignores the fossil record. This is closely related to the conditions of fossilization and to the nature of the crustacean shell, which degrades very rapidly after the death of the organisms. Thus, when crustaceans are present in a given assemblage, they are generally fragments of appendages or carapaces whose identification is always tricky for the specialist, which limits attempts at further interpretation.

NOTE.– *History of the collection of fossil crustaceans.*

The collection is the result of the work of the pioneers of crustacean paleontology and dates back to the end of the 18th century when Faujas de Saint Fond built up the first coherent set of specimens brought back to France by the armies of the Revolution. From the beginning of the 19th century, illustrious naturalists from the *Muséum* or close to the *Muséum* founded the basis of the systematics of marine arthropods: Latreille, Brongniart, Desmarest and Milne Edwards. All these people left an international scientific legacy and built up renowned collections.

During the 19th century, the collection was enriched by donations, bequests and purchases. Numerous specimens also arrived following the exploration of distant countries at the time, such as Madagascar and Indochina. The enrichment continued in the 20th century with large exploration missions in the Middle East (Dubertret and Arambourg missions), Madagascar (Colcanap, Collignon and Bésairie missions) and North Africa (Hollard and Drot missions). The micropaleontology collections of the Museum contain remains of organisms that are no larger than 1 mm.

Among these microfossils are many ostracods, which are tiny crustaceans with a carapace made of two calcified and golden-articulated valves. Ostracodology, the discipline that studies ostracods, both living and fossil, has been distinguished in France for over a century. Thus, the *Muséum* keeps founding collections gathered notably by Nicolas Grekoff, Vespasian Apostolescu and Renée Damotte.

Furthermore, one of the major problems of fossil crustaceans is the access to new deposits, which are and remain rare. Nevertheless, exceptionally well-preserved deposits, also called *Konservat-Lagerstätten*, or more simply *Lagerstätten*, are an important source of information and outline the importance of crustaceans in past communities.

In many *Konservat-Lagerstätten*, marine arthropods and especially crustaceans are not all preserved to the same degree: some specimens are often preserved flattened in the sediment (e.g. in lithographic limestones), while others have been morphologically preserved in three dimensions (e.g. within carbonate nodules).

The observation under good conditions of the numerous anatomical details depends largely on the contrast between the fossil and the sediment containing it. In many cases, photography in natural light is not sufficient, especially to reveal small anatomical details of fossil crustaceans. There are then different techniques to improve this contrast such as UV light, green light or X-ray radiography. These techniques give access to different information useful to the paleontologist: anatomy, systematics, phylogeny, paleobiology, paleoecology and paleoenvironment.

7.3. The radiation of fossil crustaceans

7.3.1. *Revealing characters with UV light (yellow fluorescence)*

Many crustacean fossils contain autofluorescent minerals (Haug *et al.* 2009, 2011). In concrete terms, autofluorescence consists of the absorption of light at a given wavelength (specific to each mineral) and its re-emission at a lower wavelength (also specific to each mineral). This phenomenon allows for specific highlighting of fossil tissues when illuminated by an appropriate light (Audo 2014). A large number of fossil crustaceans from lithographic limestones from the Upper Jurassic (e.g. Solnhofen in Germany, Cerin and Canjuers in France) and the Cretaceous (e.g. Hadjoula, Hakel in Lebanon) possess, for example, an autofluorescent carapace.

The specimens, illuminated by a UV light source (e.g. the portable UV lamp used by philatelists), are photographed with a camera equipped with a UV filter. This filter eliminates the UV and collects only the yellow light. UV light photography has two major drawbacks: (1) due to the danger of UV light to the eyes, protective glasses must be worn and (2) the presence of textile dust or white paper on the specimen is problematic, because this dust has a very strong blue fluorescence that distorts the calculation of the exposure time and can mask fossil details.

This technique has been used on a very large number of fossil crustaceans from the MNHN collections that have been the subject of several publications over the past decade: crustaceans from the European Upper Jurassic (Charbonnier *et al.* 2012; Audo *et al.* 2014a, 2014b, 2017) and crustaceans from the Cretaceous of Lebanon (Petit and Charbonnier 2012; Charbonnier *et al.* 2017a). In each of these cases, UV light revealed anatomical characters that were invisible in natural light and yet crucial for the description of the fossil organisms (Figure 7.1). Even better, in the case of the Cretaceous fossils from Lebanon, UV light revealed the oldest representatives of present-day shrimp families: Hippolytidae, Pandalidae, Glyphocrangonidae and Gnathophausiidae (Charbonnier *et al.* 2017a).

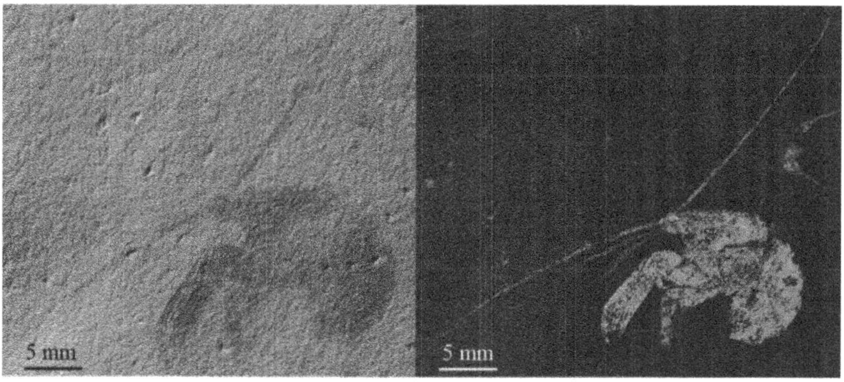

a) b)

Figure 7.1. *Crustacean from the lithographic limestones of the Cretaceous of Lebanon. (a) Specimen in natural light. (b) Specimen in UV light (yellow fluorescence) (© Lilian Cazes – CR2P, MNHN)*

7.3.2. Revealing characters with green light (green–orange fluorescence)

Green–orange fluorescence photography (Haug *et al.* 2009) follows the same principle as UV–yellow fluorescence photography. However, in this case, the light source is a white light source equipped with blue–green filters. The camera lens is equipped with a red filter. The colors of the filters being complementary, all the blue–green light illuminating the fossil is practically eliminated by the red filter, the objective being to allow only the weak orange light emitted by the fossil to pass through.

Compared to UV fluorescence, orange fluorescence has the advantage of not being sensitive to the frequent dusting of paper or white textiles. Many fossils show autofluorescence under green light, including crustaceans from the Jurassic lithographic limestones of Italy, Germany, France and the Cretaceous of Lebanon (Haug *et al.* 2016; Charbonnier *et al.* 2017b).

This technique has been particularly successful in the study of Mesozoic crustacean larvae (Figure 7.2). Through the characteristics revealed, it could be demonstrated that the larval stages of some fossil crustaceans were much more complex than in present-day nature with larvae often exhibiting combinations of characteristics now unknown (Bravi *et al.* 2014; Haug *et al.* 2016). It would appear that the development of some fossil crustaceans was more gradual or “less metamorphic" than in present-day crustaceans. This implies that what we now call "metamorphosis", the result of a long evolutionary process, related to selection pressures on initial developmental stages that are likely related to paleoecological (e.g. predation) and paleoenvironmental (e.g. bathymetry, temperature, salinity) factors.

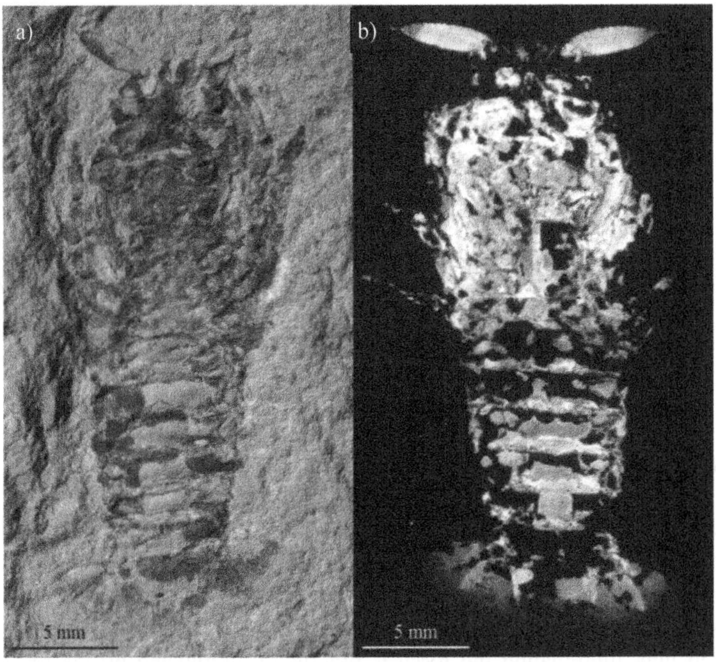

Figure 7.2. *Crustacean larvae from the lithographic limestones of the Cretaceous of Lebanon. (a) Specimen in natural light. (b) Specimen in green light (green–orange fluorescence), image desaturated to reveal anatomical details (© Joachim Haug – SYNTHESYS, MNHN)*

7.3.3. *X-ray radiography*

The advantage of the radiographic study of fossils was discovered shortly after the discovery of X-rays (Lemoine 1996). The principle is simple: the fossil to be studied is placed between a point source of X-rays and a detector (scintillation detector coupled with a CCD sensor). The X-rays are more or less absorbed as they pass through the fossil, which results in a more or less dense shadow on the detector. This technique obviously depends on the difference in absorption between the minerals constituting the fossil and the surrounding sediment.

In the case of fossils, the results with medical type X-ray machines generally do not allow the contours of the fossil in the sediment to be correctly distinguished. Other instruments such as X-ray micro-tomography (e.g. CT-scan and synchrotron) are preferred. This technique is the product of the above-mentioned planar X-ray imaging methods and serial slicing techniques (cutting the specimen into slices to study the microscopic structure and the interior) or polished slides (stripping the specimen to obtain successive cutting planes). Indeed, X-ray tomography allows virtual sections in the fossil to be made.

The specimens are placed on a rotating medium between an X-ray source and a detector. The rotating medium, controlled by computer, allows for a gradual rotation of the specimen. Thus, it is possible to obtain a series of X-rays from different angles of the same fossil. These X-rays are used by the X-ray tomography software to produce a three-dimensional matrix representing the specimen's X-ray absorbance. This matrix is exported as a series of images that will be processed with segmentation software.

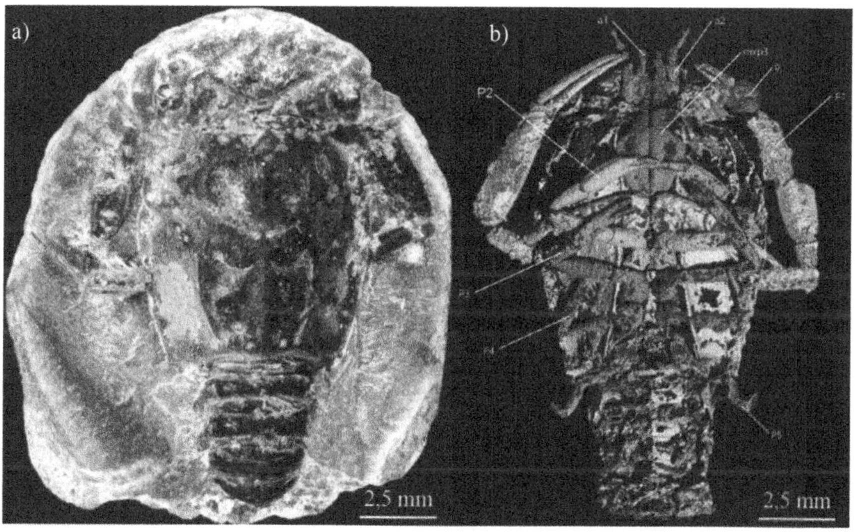

Figure 7.3. *Crustacean preserved in a carbonate nodule from the Jurassic of La Voulte-sur-Rhône. (a) Dorsal surface of the organism. (b) 3D model of the ventral surface showing the appendages (AST-RX microtomography, MNHN) (© Denis Audo – CR2P, MNHN)*

Spectacular results have been obtained on various fossil crustaceans from the MNHN, notably those from the exceptionally well-preserved deposit of La Voulte-sur-Rhône (Middle Jurassic). For example, it was possible to reconstruct the external and internal anatomy of a small crustacean preserved in a nodule in La Voulte (Figure 7.3): X-rays revealed the internal organs and in particular the presence of the ovaries making it possible to demonstrate that it was a female ready to lay eggs (Jauvion *et al.* 2016; Audo *et al.* 2019).

7.4. Exceptional preservation of fossil crustaceans

X-ray microtomography does not always make it possible to distinguish the soft parts from the hard parts, in particular when the fossils are very flattened by sedimentary compaction. Moreover, these crushed fossils also undergo physicochemical modifications during their fossilization, complicating their interpretation. However, these fossils can contain invaluable information, in particular, when the anatomy is well preserved and when soft tissues such as muscles are present. Nevertheless, locating these fossil soft tissues remains particularly difficult because of the limited contrast achieved with light microscopy and the limitations of microtomography.

In 2010, a new approach was developed at the MNHN in order to discriminate on a small scale the different fossilized tissues (muscles, fibers, bones) in fish and crustaceans collected in a Cretaceous deposit in Morocco. This non-destructive method was developed within the framework of the Ipanema platform (partnership between the SOLEIL synchrotron and the MNHN). The method is based on the localization of rare earths. These chemical elements (e.g. yttrium, lanthanides) are known to be contained in trace amounts in fossils, typically from 1 to 1,000 μg/g of material. However, depending on the type of tissue, the quantities of trace elements incorporated during fossilization differ. This preferential fixation makes it possible to discriminate the anatomical parts of a fossil. It is materialized by a significant contrast of the various chemical elements according to the fossil's types of tissues when this is characterized by fast X-ray fluorescence imaging under synchrotron radiation. X-ray fluorescence is a secondary emission of X-rays by an atom bombarded by X-rays. The emission spectrum is characteristic of the chemical elements constituting the sample. Used in imaging mode, it allows these elements to be located.

In our case, the very high intensity of the synchrotron light allows access to trace elements that are unreachable with laboratory X-ray sources. Synchrotron X-ray fluorescence imaging of decapod crustaceans and teleost fish from the Cretaceous of Morocco has been particularly powerful in revealing the anatomy of carapaces and skeletons that have preferentially incorporated yttrium (Figure 7.4) (Gueriau *et al.* 2014).

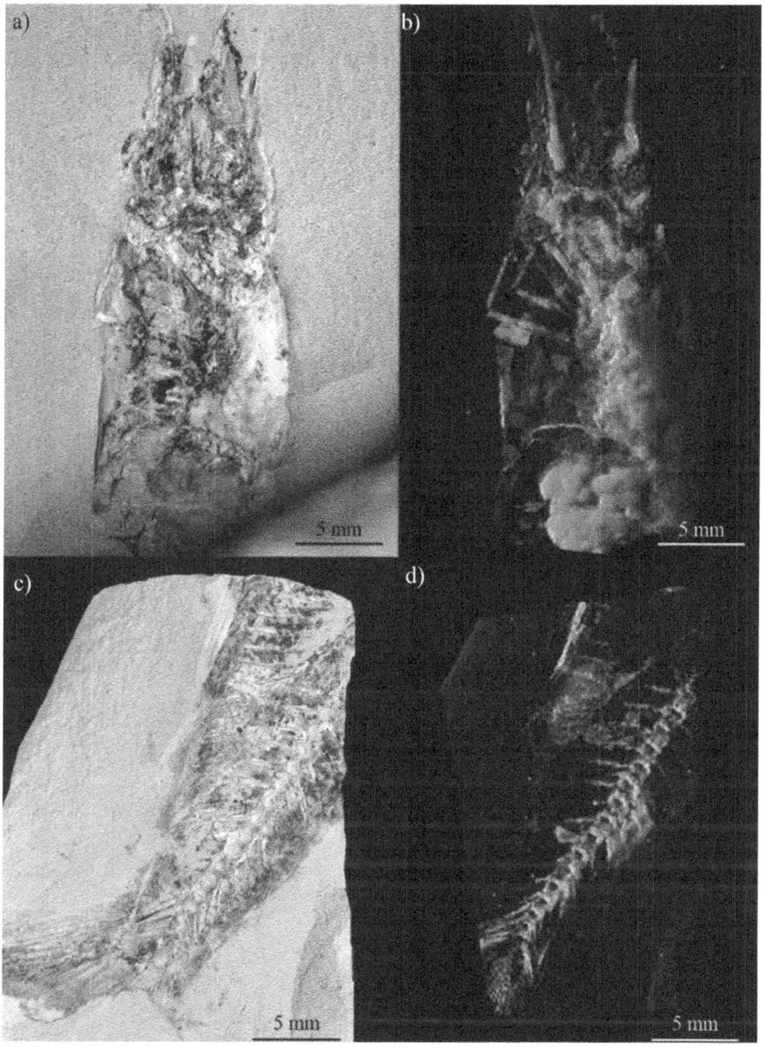

Figure 7.4. *Synchrotron X-ray fluorescence imaging of a decapod crustacean and a teleost fish from the Cretaceous of Morocco. © Pierre Gueriau – CR2P, MNHN*

COMMENT ON FIGURE 7.4.– *Natural light photographs (a and c) and false color images (b and d) of the distributions of iron (blue) and two rare earths: neodymium (red) and yttrium (green), reconstructed from a spectral decomposition. Note the strong elemental contrast between skeletal elements and soft tissues.*

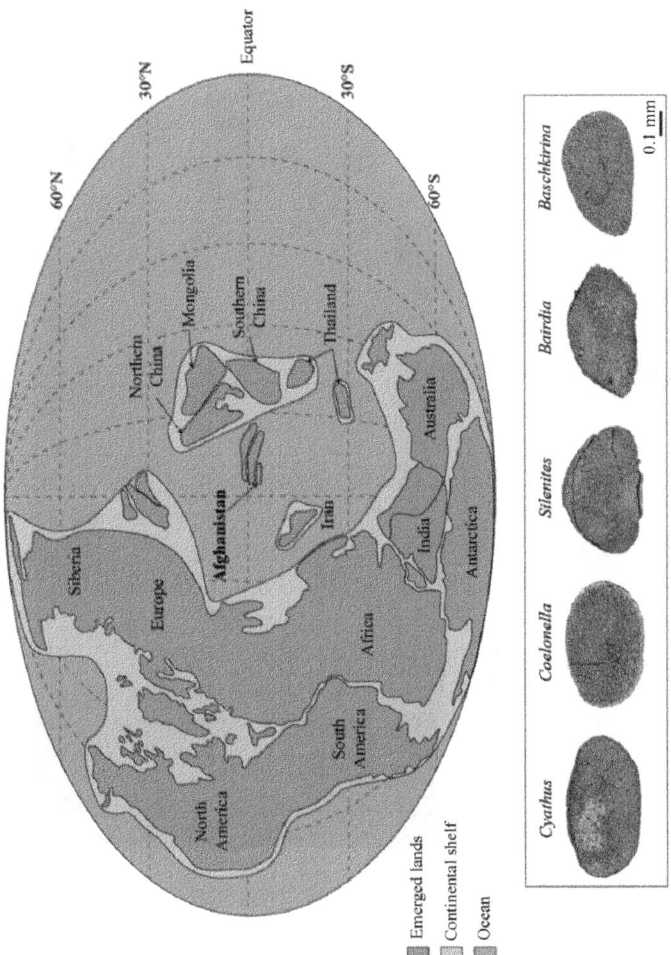

Figure 7.5. *Paleogeographic map of the position of the continents at the end of the Paleozoic modified from Angiolini* et al. *(2013). The inset shows some of the ostracods from Afghanistan described by Forel (2018) (© Marie-Béatrice Forel – CR2P, MNHN)*

7.5. Ostracods and paleogeography at the end of the Paleozoic

The end of the Paleozoic, about 250 million years ago, was marked by great changes with the disappearance of more than 90% of marine species during the largest extinction in the living world (Erwin 1993). At the same time, the continental masses reunited into a single continent, Pangea, and a string of islands became individualized at the paleoequator of the time (Dercourt *et al.* 1993). These islands correspond to the rocks that today outcrop in Iran, Turkey and Afghanistan.

The MNHN micropaleontology reserves contain rocks, thin sections and microfossils collected in Afghanistan. Because of the conflicts and security problems that have plagued the country since the late 1970s, field research is impossible. The samples preserved in the MNHN collections are therefore unique opportunities to study the organisms that populated these islands at the end of the Paleozoic.

Very recently, some of these rocks have been treated for the first time by chemical methods. These methods mainly use acetic acid in which the rocks are immersed in order to slowly disintegrate them and release the microfossils. The microfossils were observed and photographed by scanning electron microscopy, which uses a beam of electrons to produce high-resolution images of the surface of an object. This work has made it possible to describe the first known fossil ostracods from this period in Afghanistan. Unexpectedly, it also revealed a large biogeographic province composed of Afghanistan and Thailand: at the end of the Paleozoic, the ostracods of these regions were very different from those known at the same time in other regions such as China or Europe. These differences show that there were significant barriers to the dispersal of ostracods, such as deep ocean bottoms or particular ocean currents, which are still unknown to this day (Figure 7.5) (Forel 2018).

7.6. References

Angiolini, L., Zanchi, A., Zanchetta, S., Nicora, A., Vezzoli, G. (2013). The Cimmerian geopuzzle: New data from South Pamir. *Terra Nova*, 25, 352–360.

Audo, D. (2014). Les Polychelida, un groupe de crustacés énigmatiques : systématique, histoire évolutive, paléoécologie et paléoenvironnements. PhD Thesis, Muséum national d'Histoire naturelle, Paris.

Audo, D., Charbonnier, S., Schweigert, G., Saint Martin, J.-P. (2014a). New eryonid crustaceans from the Late Jurassic Lagerstätten of Cerin (France), Canjuers (France), Wattendorf (Germany) and Zandt (Germany). *Journal of Systematic Palaeontology*, 12(4), 459–479.

Audo, D., Schweigert, G., Haug, J.T., Haug, C., Saint Martin, J.-P., Charbonnier, S. (2014b). Diversity and palaeoecology of the enigmatic genus *Knebelia* (Eucrustacea, Decapoda, Eryonidae) from Upper Jurassic plattenkalks in southern Germany. *Palaeontology*, 57(2), 397–416.

Audo, D., Schweigert, G., Charbonnier, S., Haug, J.T. (2017). Systematic revision and palaeobiology of *Rosenfeldia triasica* and *Rogeryon oppeli* gen. et comb. nov. (Eucrustacea, Polychelida). *European Journal of Taxonomy*, 367, 1–23.

Audo, D., Robin, N., Luque, J., Krobicki, M., Haug, J.T., Haug, C., Jauvion, C., Charbonnier, S. (2019). Palaeoecology of *Voulteryon parvulus* (Eucrustacea, Polychelida) from the Middle Jurassic La Voulte-sur-Rhône Fossil-Lagerstätte (France). *Scientific Reports*, 9, 5332.

Bravi, S., Garassino, A., Bartiromo, A., Audo, D., Charbonnier, S., Schweigert, G., Thévenard, F., Longobardi, C. (2014). Middle Jurassic Monte Fallano Plattenkalk (Campania, southern Italy): First report on terrestrial plants, decapod crustaceans and fishes. *Neues Jahrbuch für Geologie und Paläontologie*, 272(1), 79–107.

Charbonnier, S., Garassino, A., Pacaud, J.-M., Schweigert, G. (2012). Rediscovery of the type material of *Eryon cuvieri* Desmarest, 1817 (Crustacea, Decapoda, Eryonidae) and nomenclatural consequences. *Geodiversitas*, 34(4), 849–855.

Charbonnier, S., Audo, D., Garassino, A., Schweigert, G., Hyžný, M. (2017a). Fossil Crustacea of Lebanon. *Mémoires du Muséum national d'Histoire naturelle, Paris*, 210, 1–252.

Charbonnier, S., Teruzzi, G., Audo, D., Lasseron, M., Haug, C., Haug, J.T. (2017b). New thylacocephalans from the Cretaceous Lagerstätten of Lebanon. *BSGF-Earth Sciences Bulletin*, 188(3), 1–19.

Dercourt, J., Ricou, L.E., Vrielynck, B. (1993). *Atlas Tethys Palaeoenvironmental Maps.* Gauthier-Villars, Paris.

Erwin, D.H. (1993). *The Great Paleozoic Crisis: Life and Death in the Permian.* Columbia University Press, New York.

Forel, M.-B. (2018). Rediscovering the Lys collection in the Muséum national d'Histoire naturelle (Paris, France): First report on Middle Permian ostracods from occidental Hindu Kush, Afghanistan. *Geodiversitas*, 40, 471–496.

Gueriau, P., Mocuta, C., Dutheil, D.B., Cohen, S.X., Thiaudière, D., The OT1 Consortium, Charbonnier, S., Clément, G., Bertrand, L. (2014). Trace elemental imaging of rare earth elements discriminates tissues at microscale in flat fossils. *PLoS ONE*, 9(1), e86946.

Haug, C., Haug, J.T., Waloszek, D., Maas, A., Frattigiani, R., Liebau, S. (2009). New methods to document fossils from lithographic limestones of southern Germany and Lebanon. *Palaeontologia Electronica*, 12(3), 1–12.

Haug, J.T., Haug, C., Kutschera, V., Mayer, G., Maas, A., Liebau, S., Castellani, C., Wolfram, U., Clarkson, E.N.K., Wasloszek, D. (2011). Autofluorescence imaging, an excellent tool for comparative morphology. *Journal of Microscopy*, 244, 259–272.

Haug, J.T., Audo, D., Charbonnier, S., Palero, F., Petit, G., Saad P.A., Haug, C. (2016). The evolution of a key character, or how to evolve a slipper lobster. *Arthropod Structure & Development*, 45, 97–107.

Jauvion, C., Audo, D., Charbonnier, S., Vannier, J. (2016). Virtual dissection and lifestyle of a 165 million-year-old female polychelidan lobster. *Arthropod Structure & Development*, 45, 122–132.

Lemoine, V. (1996). De l'application des rayons de Röntgen à la Paléontologie. *Comptes Rendus des Séances de l'Académie des Sciences*, 123(19), 764–765.

Petit, G. and Charbonnier, S. (2012). Fossil sponge gemmules, epibionts of *Carpopenaeus garassinoi* n. sp. (Crustacea, Decapoda) from the Sahel Alma Lagerstätte (Late Cretaceous, Lebanon). *Geodiversitas*, 34(2), 359–372.

8

The “Cyanobacteria and Microalgae” Collection in the Time of “-omics”

Sébastien DUPERRON, Charlotte DUVAL, Sahima HAMLAOUI, Katia COMTE, Claude YÉPRÉMIAN and Cécile BERNARD

MCAM – CNRS, Muséum national d'Histoire naturelle, Paris, France

The “cyanobacteria and microalgae” collection of the *Muséum national d'Histoire naturelle* includes, respectively, 870 and 890 living strains maintained in the laboratory. It is regularly enriched, notably within the framework of research programs conducted both in France and abroad. This collection supports fundamental research, particularly in taxonomy and ecophysiology, but also numerous activities for the development of bioactive molecules and expertise in environmental diagnosis. These different activities benefit from the emergence of high-throughput “-omics” approaches, which now offer new possibilities to enable such a collection to respond to the current challenges of both fundamental and targeted research, and of the conservation of biological resources.

8.1. Introduction

Collections are central to the major missions of natural history museums, herbaria and some university centers in most countries. In addition to its historical collections (for example, in botany, anthropology, mineralogy), the

For a color version of all figures in this chapter, see www.iste.co.uk/pellens/natural.zip.

Natural History Collections in the Science of the 21st Century,
coordinated by Roseli PELLENS.

Muséum national d'Histoire naturelle has been involved in the conservation of biological resources since the last century, with the appearance, as early as 1928, of a living collection of freshwater microalgae, then, in 1929, a collection of fungal strains (Figure 8.1) (Lefèvre 1960; MNHN 1954a, 1954b) and finally, since 1997, a collection of cyanobacteria. The existence of these collections participates in the MNHN's naturalistic approach, but in an original form, that of cultures of living cells (see Chapter 9 of this book by Gerbault-Seureau and Dutrillaux). This particularity allows the conservation of organisms' genetic information as well as their cellular and metabolic potential. The MNHN's "cyanobacteria and microalgae" collection is currently the largest French collection of freshwater phytoplankton. Over time, it has evolved in its organization, content and methods to adapt to emerging disciplinary fields and to a diversity of uses. At the interface of fundamental research, this collection benefits from new "-omics" approaches, applicable, in particular, to living material. In the applied field, the exploration of bioactive molecules produced by microorganisms is an innovative use of the collection, and in the field of expertise, the collection of reference strains is essential for ecotoxicological tests and constitutes an important orientation. In addition to its classic role as a reference collection for taxonomy and the description of natural communities, the collection of cyanobacteria and microalgae is opening up to new fields, some examples of which are illustrated in this chapter.

Figure 8.1. *Collection belonging to the* laboratoire de Cryptogamie *(ca. 1930), and catalog of the living collections, herbaria and documents (1953)*

8.2. A living collection supported by research

The "cyanobacteria and microalgae" collection currently contains 1,764 specimens belonging to 97 genera and 322 species divided into 26 orders (Table 8.1) with reference to the current taxonomy[1]. A specimen corresponds to a collection entry number of a strain or cell line of cyanobacteria or microalgae. Each specimen can be present under different conservation conditions: alive (870 cyanobacteria and 890 microalgae), cryopreserved in liquid nitrogen (200 cyanobacteria), or as a DNA extract (550 specimens). The term "microalgae", a widely and commonly used name, has no taxonomic validity and is applied to several distinct taxa without a phylogenetic relationship. Indeed, over the last 20 years, several revisions of the classification of eukaryotes and particularly unicellular eukaryotes have demonstrated the polyphyletic nature of "microalgae" (Adl *et al.* 2019). Cyanobacteria, on the other hand, are monophyletic and their classification is also in a state of change (Oren and Ventura 2017).

Clades	Cyanobacteria	Chlorophyta	Streptophyta	Ochrophyta	Rhodophyta	Euglenozoa	Total
Orders	6	11	3	4	1	1	**26**
Genera	38	40	10	7	1	1	**97**
Species	91	185	27	13	1	5	**322**

Table 8.1. *Distribution of the number of specimens in the "cyanobacteria and microalgae" collection in the main taxonomic groups*

In the "cyanobacteria and microalgae" collection, the nomenclature applied is that accepted by AlgaeBase (Guiry and Guiry 2019). Based on this classification, most of the specimens in the collection belong to Chlorophyta (185 species, Figure 8.2) and to Cyanobacteria (91 species, Figure 8.3). The great majority of the specimens come from Europe, and particularly from metropolitan France; however, more than 100 specimens come from other geographical areas on five other continents.

1 Available at: https://www.ncbi.nlm.nih.gov/Taxonomy.

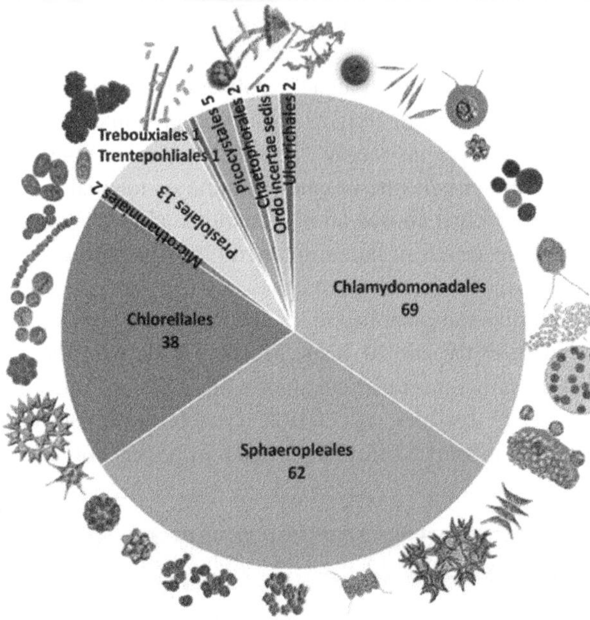

Figure 8.2. *Distribution of the number of microalgal specimens of the taxon Chlorophyta by order*

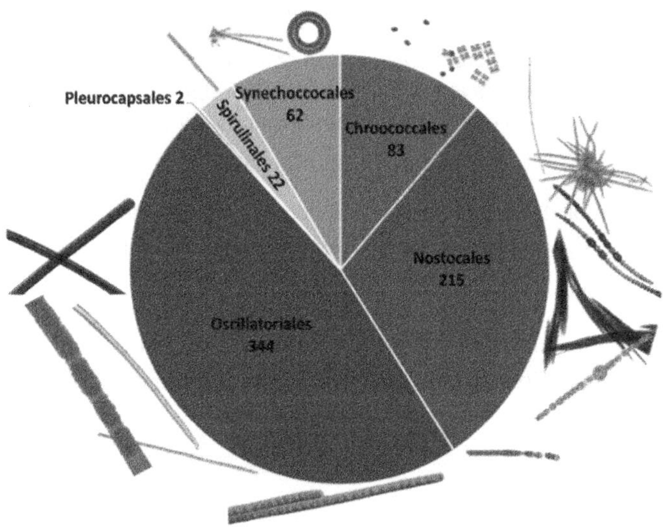

Figure 8.3. *Distribution of the number of Cyanobacteria specimens by order*

The enrichment strategy of the collection follows several axes. The first is to isolate specimens belonging to *taxa* that are poorly represented in national or international collections because of the difficulties in isolating and culturing them, such as the "picoeukaryotes" (<1 μm) recently sampled in Mayotte (Bernard *et al.* 2019). Another axis consists of relying on research programs to enrich the collection by making isolations from poorly known ecosystems (extreme environments, mangroves, etc.) and in areas that are under-represented in the collection (Southeast Asia (Laos) or Africa (Mayotte)), which makes it possible to obtain new reference materials and the description of new species (Figure 8.4).

a)

b)

Figure 8.4. *Two environments from which enrichments of the cyanobacteria collection were made in 2018. (a) Freshwater stream in Mayotte and detail of Oscillatoriales isolated from this environment. (b) Mangroves in Guadeloupe and Oscillatoriales isolated from this environment*

8.3. New uses of the collection in basic research

As its name indicates, a living collection is, well, living! The collection thus preserves the properties and biological potential of the organisms, which can be used as much as necessary in the light of methodological developments that allow an increasingly detailed study, particularly through new high-throughput approaches. Beyond its legacy and reference value, the collection is used in the framework of these original and new approaches, of which three examples are discussed below.

8.3.1. *Polyphasic identification and taxonomy of cyanobacteria and microalgae*

The goal of many natural history collections is to allow the accumulation of reference specimens on the basis of which the identification of new specimens can be performed. In the microbial world, however, seeing is not identifying, and a collection of microorganisms requires the support of complex methodologies, some of which are new, in order to explore the collection and extract relevant information. The description of the collection's species is done by molecular approaches and sequencing, for example, the gene coding for the 16S rRNA that serves as an "identity card" in prokaryotes (Woese *et al.* 1990; Amann *et al.* 1995). Beyond molecular typing, a coupled so-called polyphasic approach is applied to cyanobacteria (Komárek *et al.* 2014). This approach uses microscopy (light and electron) to document the cytological and morphological characteristics of the taxa as well as their molecular characteristics. This coupled approach is progressively gaining ground in the taxonomy of cyanobacteria (Komárek *et al.* 2014). The description of a new prokaryotic species from an environmentally isolated strain requires the deposition of the 16S rRNA sequence in an international public database such as the EMBL-EBI[2]. Any new sequence can thus be compared to all available sequences and contributes to improving our understanding of the relationship between species.

8.3.2. *Contribution to the evolutionary sciences*

The comparison of marker gene sequences with those from other collections and databases allows the construction of phylogenetic trees that trace hypothetical relationships between present-day species and provide information about the evolution of groups (Woese 1987). Through these approaches, the characterization of cyanobacteria in collections can directly contribute to our understanding of their evolution, as well as to the selection of particular strains whose study could provide

2 Available at: https://www.ebi.ac.uk/.

information on the evolution of certain traits of interest. An example is the isolation and study of a Chlorophyta of the genus *Picocystis* and a cyanobacterium of the genus *Arthrospira* in the waters of Lake Dziani Dzaha in Mayotte. This lake, located in a crater of the Petite Terre island, presents alkaline and hypersaline waters, in which oxygen is available only in the first tens of centimeters. Microbialites, concretions formed by the activity of microorganisms, develop on the periphery. Such a lake constitutes, on the geochemical level, an analog of the oceans that could have existed at the Precambrian time. *Picocystis* and *Arthrospira* are dominant (Figure 8.5), and their study can provide information on the role played by phototrophic microorganisms in ancient oceans, and on their evolution (Leboulanger *et al.* 2017; Bernard *et al.* 2019). The factors governing the codominance of these two species could thus be studied from cultures, testing the hypothesis of a complementarity of photosynthetic pigments for the exploitation of the light resource. In addition to re-culturing the strains of interest for the study of their physiology, it is also possible to sequence their complete genome, and thus refine phylogenetic reconstructions through so-called multi-loci approaches based on the comparison of the sequence of a large number of genes.

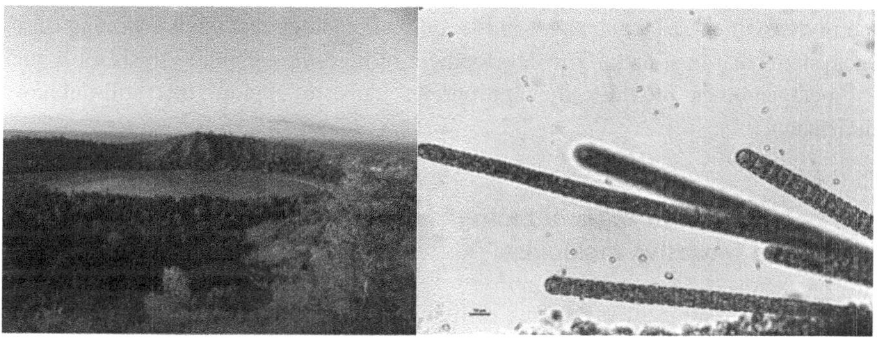

a) b)

Figure 8.5. *(a) Lake Dziani Dzaha in Mayotte. Hypersaline, alkaline environment with a predominantly microbial component. (b) The high biomass consists mainly of* Arthrospira fusiformis *(filamentous cyanobacteria of the Oscillatoriales group) and* Picocystis salinarum *(Picoeucaryote)*

8.3.3. *Contribution to the study of interactions between organisms*

A major paradigm shift in biology in recent years is the realization of the importance of interactions between organisms. Just as humans cannot live without their microbiota, it is increasingly clear that interactions are a necessary condition for the survival of most organisms (Dubilier *et al.* 2015). The isolation of strains as practiced in the "cyanobacteria and microalgae" collection results in non-axenic cultures, that is, along with a cyanobacterium or microalga of interest, the procession of prokaryotes and viruses that live in physical contact with it is isolated, meaning it becomes possible to study. The sequencing of the genome of several strains of the cyanobacterium *Aphanizomenon* present in a collection and originating from a single natural population reveals that many types of bacteria coexist with these cyanobacteria in cultures (in preparation). This associated microbiome, called a "cyanosphere", can contribute to the survival and metabolism of cyanobacteria, as shown by the recently described example of the cyanosphere associated with *Microcoleus vaginatus*. This associated microbiota performs the atmospheric nitrogen fixation that benefits the cyanobacteria (Couradeau *et al.* 2019). This still underestimated role is just beginning to be explored, and it is conceivable that such associations may, for example, produce substances co-constructed by the different partners, similar to sponges that produce bioactive substrates in close collaboration with their symbiotic bacteria (Lopanik 2014). The living collections that maintain, often without knowing it and without characterizing it, the cyanosphere or phycosphere associated with each of the referenced strains can thus become important tools, allowing access to a better understanding of the way these organisms live in their environment, as much as a window on their ecology and the functioning of the ecosystems that host them. The development of high-throughput analysis tools such as metagenomics provides the appropriate tools to explore this still-unknown dimension.

8.4. Enhancing the value of biological resources through the search for innovative bioactive molecules

The conservation of organisms in living collections allows us to explore their biological properties. Among these, the production of bioactive metabolites is a rapidly developing theme. The increase of antibiotic resistance phenomena leads to the search for new antimicrobial substances, and the search for new anti-pain or anti-tumor molecules is a significant challenge. Nature directly provides or inspires such molecules. Their production is facilitated when the organisms that produce them can be maintained in the laboratory, which is not the case for the majority of bacteria that cannot be cultivated (Amann *et al.* 1995). Today, cyanobacteria are identified as a source of new bioactive compounds, due to their ability to produce numerous metabolites with diverse effects (Figure 8.6) (Leão *et al.* 2012; Shah *et al.* 2017; Demay *et al.* 2019).

Thus, the collection of cyanobacteria and microalgae, in collaboration with the research center in chemistry of natural substances of the UMR 7245, MCAM

(*Molécules de communication et adaptation des micro-organismes*, Communication Molecules and Adaptation of Micro-organisms), the MNHN explores the antibacterial properties of cyanobacteria forming benthic biofilms in mangroves of tropical areas. These species are still unknown, and a sampling program has been launched in the mangroves of Guadeloupe and Mayotte (X-Life program, CNRS). Several strains showing selective antibacterial activities against pathogens (human and animal) are being studied to identify the molecules at the origin of these activities, and thus are of pharmacological interest (Duperron *et al.* 2020).

Two other projects also focused on the metabolites produced by cyanobacteria that have been isolated from thermal muds (Demay *et al.* 2020). Isolation, taxonomic identification and sequencing of whole genomes have allowed us to predict, using a bioinformatics approach, the presence of molecules with beneficial properties (e.g. eucapsitrione or kulolide-like peptide, molecules with antimicrobial properties). A biomass production stage coupled with a metabolic approach allowed for the isolation of molecules and carrying out activity tests for their desired properties: anti-inflammatory, antimicrobial and antioxidant.

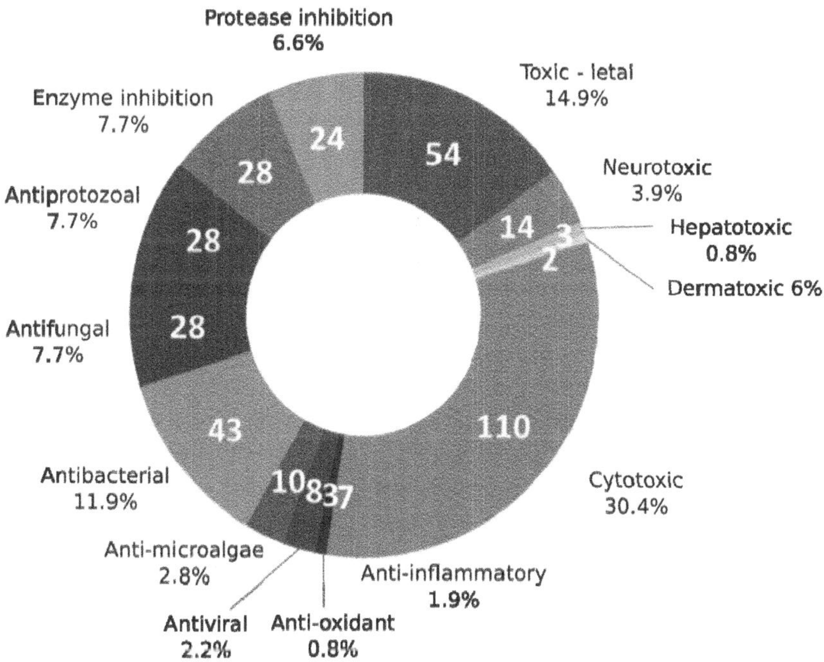

Figure 8.6. *Diagram illustrating the number (in white) and proportion (%) of molecules with various types of bioactive properties in cyanobacteria, compiled from data from a review of 670 papers. Source: adapted from Demay* et al. *(2019)*

Such projects are based on the analysis of the metabolic profiles of the specimens (metabolomics), the measurement of the activities and the structural characterization of the molecules produced by the microorganisms. They require the close association between biologists that are able to produce the important microbial biomass necessary for this exploration, and chemists that are able to evaluate and characterize the substances. As such, the collection constitutes a bank of molecules whose potential remains to be explored.

8.5. Expertise in environmental diagnosis

The "cyanobacteria and microalgae" collection is a reference collection for environmental diagnosis. A number of microorganisms belonging to these groups are indeed likely, under certain conditions, to grow rapidly, forming efflorescences or blooms (Huisman *et al.* 2018). These blooms are originally natural phenomena, but they are increasingly accentuated by direct human activity that increases the eutrophication of waters through the release of nutrients (phosphorus, nitrogen, etc.), or by climate change that warms and/or decreases water flow. These blooms disrupt the functioning of hydrosystems by causing a decrease in biodiversity, alteration of the functioning of the microbial loop, and anoxia, causing the death of teleost fish. In addition to these deleterious effects, in some cases, and, depending on the species, the production of toxins that are likely to present a danger to macro-organisms, including humans, occurs (Bernard and Catherine 2017, pp. 55–82). This phenomenon requires the identification of the microorganisms involved and the evaluation of their potential to produce these toxins. The cyanobacteria collection here plays a key role in the MNHN's mission in terms of expertise, since it includes representatives of the genera grouping the main species at the origin of cyanotoxin-producing blooms (e.g. *Microcystis*, *Planktothrix*, *Phormidium*, *Aphanizomenon*, *Dolichospermum*, *Cylindrospermopsis*). Each reference strain is genotyped and characterized for one type of cyanotoxin (e.g. *Phormidium* PMC 240.0 4, producing anatoxin-*a* (Gugger *et al.* 2005)) in order to be used as a positive control for ecotoxicological tests performed in the laboratory. The collection is regularly enriched with new isolates from water bodies in which the teams are conducting research and expertise programs (e.g. local authorities in Seine-Saint-Denis). The metabolites produced by these strains are characterized by metabolomic and genomic (multi-omic) approaches, which evaluate their capacity to produce the main families of toxins (microcystins, anatoxins, saxitoxins and cylindrospermopsins), and thus to highlight the influence of various taxonomic or physicochemical parameters on these capacities.

Some microalgae strains are also used in standardized tests by ecotoxicology laboratories to which the collection provides strains (e.g. the standard NF EN ISO

8692: 2012, "Water quality – Fresh water algal growth inhibition test with unicellular green algae").

8.6. The living collection of cyanobacteria and microalgae of today and tomorrow

A living collection is flexible. Its uses have evolved and can continue to do so over time by following methodological advances. By building on its fundamentals, it can respond to new problems and emerging approaches by multiplying interfaces with other disciplines. However, a living collection also presents specific challenges. Its maintenance requires a significant amount of manpower, and the long-term maintenance of strains requires a sustained reflection on the modes of conservation. Maintaining living cells inevitably leads to genetic drift of organisms, which is all the more significant in the case of microorganisms with a short generation time. Alternatives, or complementary approaches such as cryopreservation or conservation in a protective atmosphere, can constitute progress. However, experience in the collection shows that it is often necessary to adapt protocols to different taxa, strains or cell lines. The sharing of experiences through national and international networks is an asset in optimizing this work. Similarly, the integration of the cyanobacteria and microalgae collection into the larger collection of living and cryopreserved biological resources (RBCell) of the MNHN[3] makes it possible to optimize the management of certain current tasks.

In the future, collections of microorganisms will have to play other roles. They will certainly constitute a means of safeguarding biological resources, but, like zoos and botanical parks, "microbe zoos" cannot replace the necessary protection of their natural habitats. On the other hand, they can feed reflection on the protection of natural spaces. Indeed, if there is no IUCN list of threatened bacteria, the presence of microorganisms with remarkable properties in certain environments could constitute a selection criterion for their protection. Lake Dziani Dzaha (Figure 8.5), for example, is populated solely by microorganisms, so the presence of endangered species cannot be used to justify its protection. The originality of the exceptional microbial ecosystem that it harbors would, however, fully justify protection measures, all the more urgent since this lake, which was once left aside due to local beliefs, is now seeing its banks cleared for cultivation (Figure 8.7).

3 Available at: https://www.mnhn.fr/fr/collections/ensembles-collections/ressources-biologiques-cellules-vivantes-cryoconserves.

a)

b)

Figure 8.7. *Crater slopes of the Dziani Dzaha in (a) April 2014 and (b) November 2018, showing crop establishment in the interval*

In general, the prospection of atypical and unexploited environments, likely to shelter new endemic or original species, makes it possible to propose new references of the current diversity and to safeguard the resources in the event of habitat disappearance. The "cyanobacteria and microalgae" collection today constitutes an original and remarkable medium for describing the biodiversity of photoautotrophic microorganisms in an unstable environmental context and on a global scale. Rich in terms of its past, the recent evolution of this collection allows it to answer the current stakes of fundamental and finalized research, and of the conservation of biological resources.

8.7. References

Adl, S.M., Bass, D., Lane, C.E., Lukeš, J., Schoch, C.L., Smirnov, A., Agatha, S., Berney, C., Brown, M.W., Burki, F. *et al.* (2019). Revisions to the classification, nomenclature, and diversity of eukaryotes. *J. Eukaryot. Microbiol.*, 66, 4–119.

Amann, R.I., Ludwig, W., Schleifer, K. (1995). Phylogenetic identification and *in situ* detection of individual microbial cells without cultivation. *Microbiol. Rev.*, 59, 143–169.

Bernard, C. and Catherine, A. (2017). Les cyanobactéries toxiques : quels impacts sur le fonctionnement des écoystèmes aquatiques ? In *Écotoxicologie, des communautés au fonctionnement des écosystèmes*, Bernard, C., Mougin, C., Péry, A. (eds). ISTE Editions, London.

Bernard, C., Escalas, A., Villeriot, N., Agogué, H., Hugoni, M., Duval, C., Carré, C., Got, P., Sarazin, G., Jézéquel, D., Leboulanger, C., Grossi, V., Ader, M., Troussellier, M. (2019). Very low phytoplankton diversity in a tropical saline-alkaline lake, with co-dominance of *Arthrospira fusiformis* (Cyanobacteria) and *Picocystis salinarum* (Chlorophyta). *Microb. Ecol.*, 78, 603–617.

Couradeau, E., Giraldo-Silva, A., De Martini, F., Garcia-Pichel, F. (2019). Spatial segregation of the biological soil crust microbiome around its foundational cyanobacterium, *Microcoleus vaginatus*, and the formation of a nitrogen-fixing cyanosphere. *Microbiome*, 7, 55.

Demay, J., Bernard, C., Reinhardt, A., Marie, B. (2019). Natural products from cyanobacteria: Focus on beneficial activities. *Mar. Drugs*, 17(6), E320.

Demay, J., Halary, S., Knittel-Obrecht, A., Villa, P., Duval, C., Hamlaoui, S., Roussel, T., Yéprémian, C., Reinhardt, A., Bernard, C., Marie, B. (2020). Anti-inflammatory, antioxidant, and wound-healing properties of Cyanobacteria from thermal mud of Balaruc-Les-Bains, France: A multi-approach study. *Biomolecules*, 11(1), 28.

Dubilier, N., McFall-Ngai, M., Zhao, L. (2015). Microbiology: Create a global microbiome effort. *Nature*, 526, 631–634.

Duperron, S., Beniddir, M.A., Durand, S., Longeon, A., Duval, C., Gros, O., Bernard, C., Bourguet-Kondracki, M.L. (2020). New benthic Cyanobacteria from Guadeloupe mangroves as producers of antimicrobials. *Mar. Drugs*, 18(1), E16.

Gugger, M., Lenoir, S., Berger, C., Ledreux, A., Druart, J.C., Humbert, J.F., Guette, C., Bernard, C. (2005). First identification in a French river of the benthic cyanoacterium *Phormidium favosum* producing anatoxin-*a* associated with dog neurotoxicosis. *Toxicon*, 45, 919–928.

Guiry, M.D. and Guiry, G.M. (2019). AlgaeBase world-wide electronic publication. Document, National University of Ireland, Galway.

Huisman, J., Codd, G.A., Paerl, H.W., Ibelings, B.W., Verspagen, J.M.H., Visser, P.M. (2018). Cyanobacterial blooms. *Nat. Rev. Microbiol.*, 16, 471–483.

Komárek, J., Kaštovský, J., Mareš, J., Johansen, J.R. (2014). Taxonomic classification of cyanoprokaryotes (cyanobacterial genera), using a polyphasic approach. *Preslia*, 86, 295–335.

Leão, P.N., Engene, N., Antunes, A., Gerwick, W.H., Vasconcelos, V. (2012). The chemical ecology of cyanobacteria. *Nat. Prod. Rep.*, 29, 372–391.

Leboulanger, C., Agogué, H., Bernard, C., Bouvy, M., Carré, C., Cellamare, M., Duval, C., Fouilland, E., Got, P., Intertaglia, L., Lavergne, C., Le Floc'h, E., Roques, C., Sarazin, G. (2017). Microbial diversity and cyanobacterial production in Dziani Dzaha Crater Lake, a unique tropical Thalassohaline environment. *PLoS ONE*, 12, e0168879.

Lefèvre, M. (1960). Exposé des titres et des travaux. Document, CNRS, Paris.

Lopanik, N.B. (2014). Chemical defensive symbioses in the marine environment. *Funct. Ecol.*, 28, 328–340.

MNHN (1954a). La chaire de Cryptogamie du Muséum national d'Histoire naturelle, Organisation et buts. MNHN, Paris.

MNHN (1954b). Catalogue des collections vivantes, Herbiers et Documents : IV, L'ALGOTHEQUE. MNHN, first supplement, Paris.

Oren, A. and Ventura, S. (2017). The current status of cyanobacterial nomenclature under the "prokaryotic" and the "botanical" code. *Antonie Van Leeuwenhoek*, 110, 1257–1269.

Shah, S.A.A., Akhter, N., Auckloo, B.A., Khan, I., Wang, K., Wu, B., Guo, Y.W. (2017). Structural diversity, biological properties and applications of natural products from Cyanobacteria: A review. *Mar. Drugs*, 15(11), 354.

Woese, C.R. (1987). Bacterial evolution. *Microbiol. Rev.*, 51(2), 221–271.

Woese, C.R., Kandler, O., Wheelis, M.L. (1990). Toward a natural system of organisms: Proposal for the domains Archaea, Bacteria and Eucarya. *Proc. Natl. Acad. Sci. USA*, 87, 4576–4579.

9

The Collection of Cryopreserved Cells and Tissues of Vertebrates: Methods and Application

Michèle GERBAULT-SEUREAU and Bernard DUTRILLAUX

ISYEB – CNRS, Muséum national d'Histoire naturelle, Sorbonne Université, École Pratique des Hautes Études, Université des Antilles, Paris, France

We wish to briefly describe here the MNHN's cryopreserved tissue and cell bank (TCCV). Its principle consists of developing cell cultures from animal samples, which are frozen in liquid nitrogen, under conditions enabling their survival. Created in 1975, it has enabled the progressive accumulation of samples, mainly from mammals. The aim of this research collection was to facilitate the multiplication of analyses and in particular, chromosomal study. Thus, after more than 40 years of enrichment, this bank provides living cells of more than 500 species for research, whose identification remains controllable by their karyotype. Beyond chromosomal analysis, live cells provide material for a wide range of analyses, including DNA, RNA, proteins and many other cellular components. Some examples of applications are given. The current massive reduction of biodiversity makes it all the more useful to develop such a bank whose potentialities are still far from being fully exploited.

For a color version of all figures in this chapter, see www.iste.co.uk/pellens/natural.zip.

Natural History Collections in the Science of the 21st Century,
coordinated by Roseli PELLENS.

9.1. Introduction

Before establishing any collection, it is necessary to ask some fundamental questions regarding its purpose, the adequacy and durability of the means used, and whether the collection will remain usable in an uncertain future. The importance of this last question is significant when conserving strictly classified biological material, because any classification is likely to evolve, or change radically over the course of time. Thus, in zoological systematics, any good species of the moment is likely to constitute several tomorrows, or inversely be gathered with another. Similarly, in pathology, each disease may later constitute several distinct syndromes according to the progress of diagnostic means. The registers of collected objects must therefore be flexible and open enough to undergo any modification. Equally important is the technical aspect concerning the method of preservation. We are just emerging from the "all-DNA" period, when the genome was supposed to explain all of biology. The development of high-throughput analysis methods, with DNA chips and the possibility of quantifying RNA, has given renewed interest to the quantification of gene expression, but it is quite obvious that the DNA–RNA–protein continuum will lead to the unquestionable predominance of proteins for the understanding of all biological phenomena. Today, apart from the criterion of hybrid fertility, which is difficult to implement, the number of mutations on a DNA sequence of a few hundred nucleotides (the CO1 mitochondrial gene is particularly informative) and the number of chromosomal rearrangements are the best biological criteria for proposing or rejecting the membership of two individuals to the same species. Let us bet that in the future, this criterion will also be provided by proteins.

As Jean Rostand once disclosed, it is possible to preserve certain animals, batrachians or fish alive by intense cold. This does not apply to homeotherms, but fortunately, cryopreservation of their living tissues or cells is possible. This allows for, after thawing and culture, multiplying and functioning the cells, which become an inexhaustible source of any cellular component, including proteins. This is a poor solution for the conservation of species, but apart from the maintenance of breeding farms, it is the best way to preserve the living characteristics of a species. Especially since the progress made on stem cells and cellular differentiation/dedifferentiation suggest that somatic cells have enormous potential, which may one day allow the reconstruction of entire individuals.

9.2. History of the collection

It was at the end of the 1960s that a decision was made to build a cell bank for research on genetic and chromosomal diseases in particular. The promoter was

Professor Jérôme Lejeune, at the French Institute of Progenesis (Cordeliers Faculty of Medicine at Paris V University). The development was entrusted to Sophie Carpentier – it was still a research project at that time – and Martine Lombard was in charge of the technical aspect until the end of the 1990s. In 1975, working on the chromosomal evolution of primates, one of the authors of this chapter (Bernard Dutrillaux) started to enrich this bank with animal samples, mainly taken from the *Muséum national d'Histoire naturelle* because of an excellent collaboration with professors Guy Chauvier, Francis Petter and Jean-Jacques Petter. Outside the MNHN, Yves Rumpler (professor at the Tananarive Faculty of Medicine, then Strasbourg) and Dr. Jean-Marc Lernould (director of the Mulhouse Zoo) were among the main collaborators.

At the beginning of the 1980s, the collection of animal cells joined with our center (URA 620, then UMR 147), the Curie Institute, where it continued to be enriched alongside the medical samples, notably by Dr. Vitaly Volobouev. The latter joined the MNHN in the mid-1990s, bringing a duplicate of the mammalian cell collection. Finally, in 2003, a small team, composed of Florence Richard, Michèle Gerbault-Seureau, Anne-Marie and Bernard Dutrillaux, joined the MNHN, correlating the rest of the animal cell collection. This collection continues to be enriched according to requests for collaboration and field collections. Today, one of the authors of this chapter (Michèle Gerbault-Seureau) is in charge of it under the responsibility of Professor Bertrand Bed'Hom. This collection (of cryo-preserved vertebrate tissues and cells: TCCV; biological resources of living cells) is entirely computerized with a database management system. For each specimen, a unique MNHN number is assigned and the geographical origin (GPS) and the collector's name are listed. Some examples are shown in Figure 9.1.

Other taxon-specific data, including the karyotype, have been systematically added. The collection, hosted by the ISYEB (*Institut de systématique, évolution, biodiversité,* French Institute of Systematics, Evolution and Biodiversity), remains one of the most important in the world, with some 500 animal species represented. Only the collection of cryopreserved cells at the San Diego Zoo (USA) is comparable.

9.3. Can all living beings be cryopreserved?

Cryopreservation can theoretically be applied to living beings as well as to organs, tissues or cells. However, it does not apply to homeothermic animals, outside their first embryonic stages. The great amplitude of the variation of temperatures, as well as during freezing and thawing, involves a strong cellular mortality, and it is all the more important that the sample is large and that it concerns organized tissues. Cryopreservation is therefore limited to cells, including eggs and sperm (semen), tissues, small biopsies and young embryos. In the current state of science, preservation in anticipation of reviving individuals, humans or other homeotherms, remains a pure utopia exploited by some. For maximum chances of success, it is preferable to go to the cellular level (Figure 9.1).

9.3.1. *Collection, culture and freezing*

The first step is to collect a biopsy of a few hundred mm^3 under good conditions (see Figure 9.2). It is essential, for each step, to proceed as sterilely as possible: removal of skin (after shaving and disinfection), muscle or tendon; placement in a sterile isotonic medium, and storage in a cool place, but not frozen (1). Note that necropsies can also be performed, after a period of time proportional to the animal's size. This can take place more than eight days after the death of a large animal. It is then necessary to remove fibrous tissue with little vascularity, such as tendon.

In the laboratory, the remaining operations are completed under a laminar flow hood. The sample is cut into small fragments (explants) (2), which are placed on the bottom of a 25 cm^2 culture vial (3) and covered with a culture medium (DMEN + 15% fetal calf serum + antibiotics). After a few days, cells start to grow around each explant. When there are enough fibroblasts, the explants are removed and then frozen in a cryo-tube containing a freezing medium (culture medium as above with addition of 15% dimethyl sulfoxide [DMSO]) (Figure 9.2).

The cells remaining in the vial are then detached by the action of trypsin, resuspended with a pipette, and sown again in two 25 cm^2 vials (4).

When the cells have reached confluence, they are detached with trypsin as before, rinsed and transferred to 75 cm^2 vials (5). When they invade these vials (about 5–8 million cells), they are peeled off again and frozen with 1 mL of freezing medium in a cryo-tube (6). The karyotype in the strips will be established from one of these vials (7).

The freezing of the vials (8) must be done in stages: one to several hours in a refrigerator at +4°C, then one to several days in a freezer at –80°C, and finally in liquid nitrogen, until a possible thawing, which follows the reverse protocol. Each cryo-tube is labeled with the name of the species, its MNHN number, the date of freezing and its content (explants or cells). Its unique location in the liquid nitrogen container is recorded in the database.

Figure 9.1. *Some examples of animals whose cells are preserved in our bank*

COMMENT ON FIGURE 9.1.– *Their karyotype, shown next to them, is systematically established at the time of collection, allowing for control of the cells and thus to avoid any identification error later on. (a) Snow leopard* Panthera uncia *Schreber, 1775. (b) Giant panda* Alluropoda melanoleuca *David, 1869. (c) White-faced saki* Pithecia pithecia *Linnaeus , 1766. (d) Spider monkey* Ateles paniscus *Linnaeus, 1758. (e) Barking deer* Muntiacus muntjak *Zimmermann, 1780. (f) Hippopotamus* Hippopotamus amphibius *Linnaeus, 1758. (g) Black crow* Corvus corone *Linnaeus, 1758. (h) Northern treeshrew* Tupaya belangeri *Wagner, 1841.*

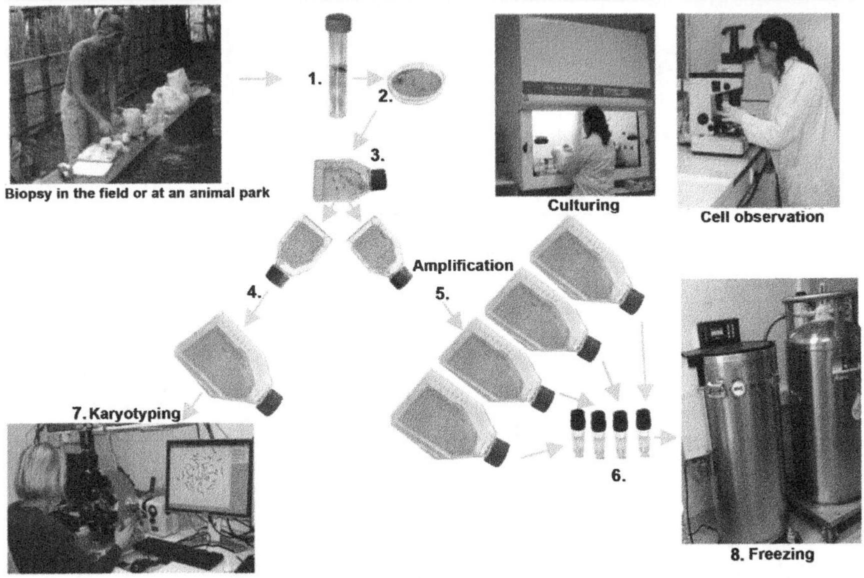

Figure 9.2. *Cell culture and freezing steps. Before any use of the cells, care is taken to make a passage in culture, to preserve the stock. Steps 1–8 are described in detail in section 9.3.1*

9.4. Current applications

With the culture method used, the proliferating cells, and therefore those that are stored, are fibroblasts that are not very well differentiated, forming the support tissue of the organ removed. They have, of course, the same nuclear and mitochondrial genome as any other cell of the removed individual. They can, therefore, be used for genomic sequencing and karyotyping (Figure 9.3). The fact that genomic sequencing, which is now one of the major tools for the study of phylogeny, did not yet exist more than 10 years after the creation of the bank, shows that it is difficult to foresee what applications it will have. In fact, the bank was created mainly for cytogenetics, at a time when technical developments were becoming numerous. The cell bank has therefore made it possible to apply each new chromosomal analysis technique without having to repeat the sampling.

The study of Cercopithecinae is a perfect example of the bank's interest. In the 1970s, the MNHN menagerie had a collection of aging Cercopithecinae, left over from the colonial period. Their condition, aggravated by the poor captivity conditions of the time, required frequent veterinary interventions to treat or euthanize them. Most of the species concerned no longer exist at the MNHN, but because of an agreement with Guy Chauvier, samples were taken and put in a bank. Because of this bank, which contains cells from several dozen Cercopithecinae, doctoral theses and several publications have been produced, first with a strictly cytogenetic approach (Dutrillaux *et al.* 1980, 1982), then more recently, with a molecular approach as well (Figure 9.3) (Cacheux 2016). Cytogenetic work demonstrated, as early as the 1980s, that the evolution of Cercopithecinae took place through a succession of hybridizations (Figure 9.4), a process that was ardently rejected by evolutionists (at the MNHN in particular), until it was rediscovered due to molecular approaches 20 years later. Figure 9.4 shows the phylogenetic tree of the primates, established after the reconstruction of the sequence of chromosomal changes. Each "bubble" represents a period of hybridization, and we see that those of the two groups of Cercopithecinae, terrestrial and arboreal, are the most significant.

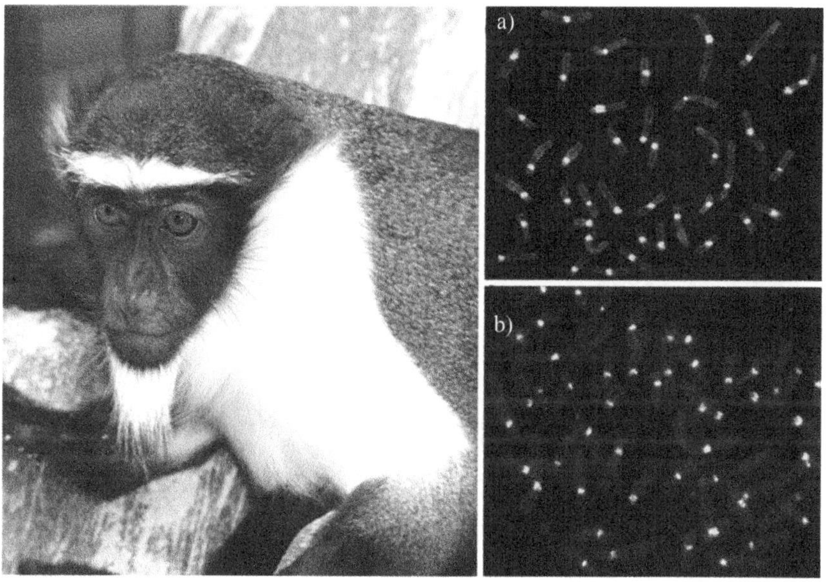

Figure 9.3. *Cercopithecus diana and its chromosomes*

COMMENT ON FIGURE 9.3.– *(a) Chromosomes stained red, with their centromeres in green after* in situ *hybridization of a "pan-centromere" probe and (b) stained blue, with all centromeres in green. Some are also stained red, after* in situ *hybridization of another probe, which shows that they contain a new specific repeated DNA sequence, which was formed during recent evolution. Recall that the chromosomal evolution of Cercopithecinae has been through fissions and the acquisition of new centromeres (see Figure 9.4).*

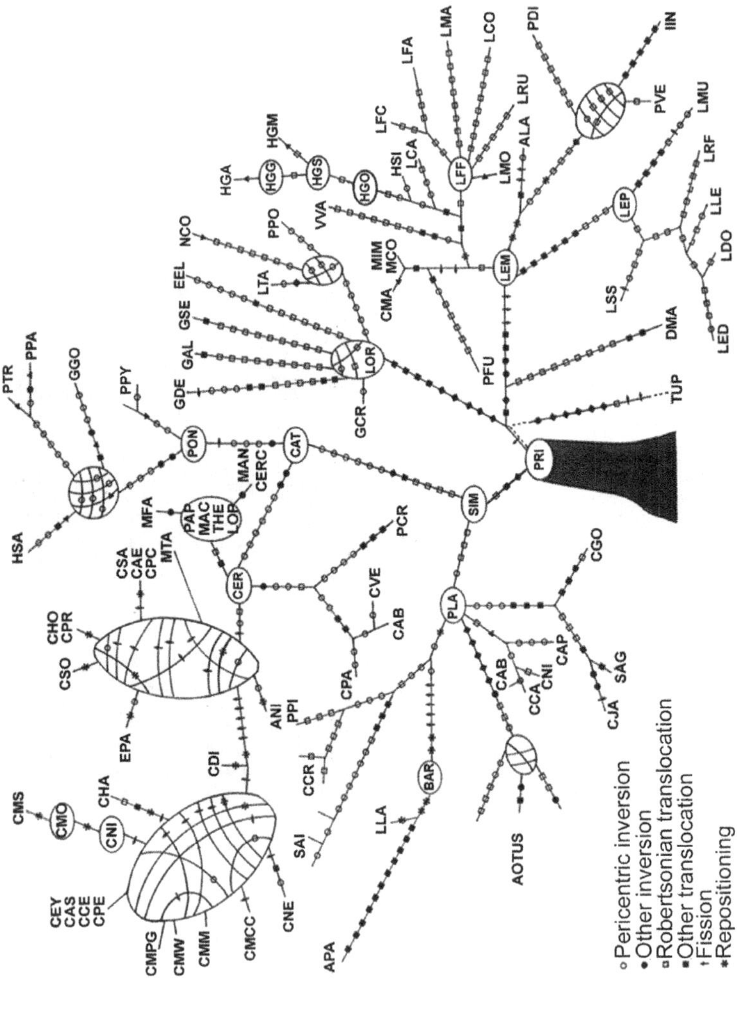

Figure 9.4. *Phylogeny of primates as established by reconstructing the succession of chromosomal rearrangements (modified from Dutrillaux 1998)*

COMMENT ON FIGURE 9.4.– *Each chromosomal rearrangement is represented by a symbol. The three-letter codes represent the living species studied (e.g. HSA =* Homo sapiens*), as well as the ancestral forms whose karyotypes have been reconstructed (PON = Pongidae; CAT = Catarrhini; CER = Cercopithecidae; PLA = Platyrrhini; SIM = Simians; PRI = Primates; LEM = Lemuriformes; LOR = Lorisidae). The bubbles containing lines and symbols indicate the sharing of rearranged chromosomes, showing the origin of the branches leading to the present species. Thus, our species originated from a population of anecestral Pongidae, in which our distant ancestors and those of chimpanzees (PTR and PPA) and gorillas (GGO) hybridized before the separation of each branch.*

In the Cercopithecidae, the intense activity of chromosomal rearrangements in the evolution of the Cercopithecinae is about 10 times higher than in the Papioninae, a close subfamily composed of an equivalent number of species. It was therefore legitimate to ask questions about the mechanisms involved, and work on the composition of centromeres could be resumed 30 years later with new means, qualifying the long-term interest of the bank.

Thus, chromosomal evolution models could be confronted with the molecular modifications encountered in different species (Cacheux 2016). The study of *Laonastes aenigmamus* (Kanyou) is another example of the importance of the cell bank. This rodent, the only representative of a family (Diatomyidae) thought to be extinct, was found only a few years ago in Laos (Jenkins *et al.* 2005). In 2011, a mission funded by the MNHN allowed us to collect several animal specimens from different localities. All of these biopsies were banked and then studied by locality. They were also studied by various cytogenetic methods (Figure 9.5). It appeared that, according to the localities of capture, the karyotypes are so different that one can predict the infertility of certain hybrids with certainty. Thus, even if the phenotypes have not changed or have changed little, a chromosomal reproductive barrier has been created, a strong criterion for the presence of cryptic species. Thus, it is not a relic species that has been discovered, but an entire multispecies genus (Richard *et al.* 2016).

9.5. Current composition of the bank

It should be noted that it has become very difficult to obtain new specimens today, apart from a few species known as laboratory species or species familiar to zoos. Ethical rules have become stricter, insecurity affects many countries, natural areas are disappearing and everyone is trying to preserve their biodiversity, unfortunately more often against the curiosity of researchers than with the aim of slowing down the disappearance of species.

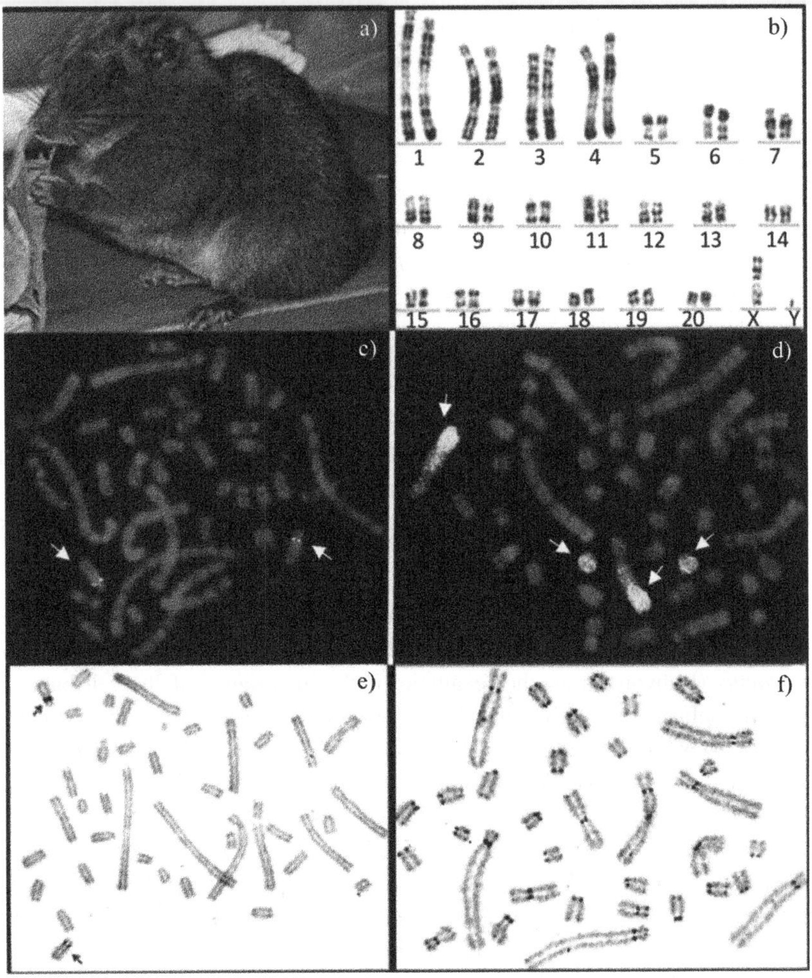

Figure 9.5. Laonastes aenigmamus *(kanyou)*

COMMENT ON FIGURE 9.5.– *(a)* Laonastes aenigmamus*; (b) one of the R-band karyotypes, established after BrdU (5-bromodeoxyuridine) incorporation; (c) metaphase after propidium iodide staining (red) and* in situ *hybridization of a 28S ribosomal DNA probe (green) and (d)* in situ *hybridization of a probe (paint) of human chromosome 5; (e) metaphase after silver staining, showing expression of the nucleolar organizer (same* locus *as in c); (f) metaphase after C-banding = centromere staining.*

Over time, however, the bank has accumulated collections of more than 520 listed species, including about 120 endangered species, classified in Appendix I of CITES, in some 5,500 vials. Some examples are shown in Figure 9.1. Mammals are the most represented, especially primates and rodents (Tables 9.1 and 9.2).

	Mammals	
Order	Number of families	Number of species
Artiodactyla	7	72
Carnivora	12	42
Cetartiodactyla	1	2
Chiroptera	5	23
Dasyuromoprhia	1	1
Dermoptera	1	1
Didelphimorphia	1	1
Erinaceomorpha	1	2
Lagomorpha	2	5
Perissodactyla	2	4
Primates	13	78
Proboscida	1	1
Rodentia	18	154
Scandentia	1	2
Sirenia	1	1
Soricomorpha	1	15

Table 9.1. *Number of families and species by mammalian order currently in the TCCV collection*

9.6. Perspectives

The future of the TCCV bank therefore lies in both the conservation and intelligent use of already conserved samples and the accumulation of samples of new species.

Compared to other collections, cryopreserved tissue banks suffer from some disadvantages, such as the need to obtain living tissue and perform heavy and constraining manipulations. Liquid nitrogen preservation is expensive and requires regular deliveries and handling. Fortunately, these costs can be shared.

	Birds	
Order	Number of families	Number of species
Accipitriformes	1	1
Anseriformes	1	1
Apodiformes	1	1
Charadriiformes	3	3
Columbiformes	1	1
Falconiformes	3	5
Galliformes	2	8
Gruiformes	1	1
Musophagiformes	1	1
Passeriformes	7	10
Piciformes	1	2
Strigiformes	2	2
Struthioniformes	2	2
	Reptiles	
Squamata	3	6

Table 9.2. *Number of families and species by order of birds and reptiles currently in the TCCV collection*

However, the conservation of living cells has certain advantages:

– for cells frozen in liquid nitrogen, time stops because all biochemical functions are stopped;

– by the cell culture that can be derived from it, it represents an almost inexhaustible and immortal source of biological material of great interest for many research projects. The great diversity of species in this collection allows for the initiation of research projects by minimizing the collection missions and the manipulation, even the sacrifice of new animals;

– it is a source of recent DNA;

– it is also a source of RNA that can be exploited, since it is extracted from living cells, but limited to fibroblasts;

– it is a source of proteins and other cellular compounds.

Finally, with the current rapid disappearance of species, it is not forbidden to imagine other potentialities of preserved living cells. In recent years, work on stem cells has made great progress. Their potentialities are immense, and in a more surprising way, it seems that we are able to go back in time, to obtain the dedifferentiation of cells and reach a stem level. Like all "blasts", fibroblasts are not highly differentiated cells. Much progress and development is still to be made, but the dream of reconstituting individuals from somatic cells, such as fibroblasts, is getting closer to reality.

9.7. References

Cacheux, L. (2016). Diversité et histoire évolutive de l'ADN alpha satellite chez les Cercopithèques. PhD Thesis, Muséum national d'Histoire naturelle, Université de Paris V, Paris.

Dutrillaux, B. (1988). Chromosome evolution in Primates. *Folia Primatologica*, 50, 134–135.

Dutrillaux, B., Couturier, J., Chauvier, G. (1980). Chromosomal evolution of 19 species or sub-species of Cercopithecinae. *Annales de Génétique*, 23, 133–143.

Dutrillaux, B., Couturier, J., Muleris, M., Lombard, M., Chauvier, G. (1982). Chromosomal phylogeny of 42 species or subspecies of cercopithecoids (Primates, Catarrhini). *Annales de Génétique*, 25, 96–109.

Jenkins, P.D., Kilpatrick, C.W., Robinson, M.F., Timmins, R.J. (2005). Morphological and molecular investigations of a new family, genus and species of rodent (Mammalia: Rodentia: Histricognatha) from Lao PDR. *Systematics and Biodiversity*, 2, 419–454.

Richard, F., Gerbault-Seureau, M., Douangboupha, B., Keovichit, K., Hugot, J.P., Dutrillaux, B. (2016). Karyotype diversity suggests that *Laonastes aenigmamus* (Laotian rock rat) (Rodentia, Diatomyidae) is a multi-specific genus. *Chromosome Research*, 24, 379–392.

10

Herbaria, the Last Resort for Extinct Plant Species

Serge MULLER[1], Valérie PRIOLET[2], Éric BADEL[3] and Stéphane BUORD[4]

[1] *ISYEB – CNRS, Muséum national d'Histoire naturelle, Sorbonne Université, École Pratique des Hautes Études, Université des Antilles, Paris, France*

[2] *Muséum national d'Histoire naturelle, Paris, France*

[3] *PIAF, INRAE, Université Clermont Auvergne, Clermont-Ferrand, France*

[4] *Conservatoire botanique national de Brest, France*

Herbaria, which gather collections of plants sometimes collected centuries ago, can contain specimens of species that are now extinct in the wild, or even in cultivation. They represent a last resort to restore living plants from seeds or embryonic tissues that are sometimes still viable in these specimens. The rarity of the available seeds, and therefore their heritage value, requires the development of appropriate protocols on closely related species ("affine species") first, before applying them to the targeted extinct species. In this chapter, we present the experimental approach and the first results obtained for two extinct species, *Filago neglecta*, an annual herb that used to occur in the open environments of Western Europe, and *Astiria rosea*, a woody species that was endemic to Mauritius (Mascarene Islands). We conclude by proposing several avenues for further investigation.

For a color version of all figures in this chapter, see www.iste.co.uk/pellens/natural.zip.

Natural History Collections in the Science of the 21st Century,
coordinated by Roseli PELLENS.

10.1. Context and objectives

Herbaria (collections of dried plants) (see Chapter 16) constitute a scientific and cultural heritage of great interest, reflecting the natural history expeditions and the work of botanists of the past centuries that led to the inventory of flora. They preserve specimens of plants collected from distant to recent times (sometimes several centuries ago), some of which may have become extinct in the wild, and sometimes even in cultivation (category EX of the IUCN) since they were collected. For this category, the seeds preserved in herbarium specimens represent a possibility (sometimes the last one) of regenerating plants and reconstituting populations, which can then be reintroduced into the natural environment. Indeed, several cases are reported in the literature, in which seeds collected about 100 or more years ago were successfully germinated and regrown, for example, the 97-year-old seeds of *Astragalus neglectus* (Bowles *et al.* 1993), 129-year-old seeds of *Geranium bohemicum* (Milberg 1994), 130-year-old seeds of *Astragalus contortuplicatus* (Molnar *et al.* 2015) and 158 years-old seeds of *Cassia multijuga* (Becquerel 1934). In Belgium, Godefroid *et al.* (2011) investigated the potential for the "resurrection" of species that had become extinct in that territory, by germinating seeds contained in herbarium samples preserved at the National Botanical Garden of Belgium. Among the 23 tested species, they were able to successfully germinate seeds of *Bupleurum tenuissimum*, from herbarium specimens that were older than 100 years. For most species, however, the simple germination did not give satisfactory results and it was necessary to resort to *in vitro* culture and micro-propagation techniques, making it possible to regenerate a whole plant from living cells. Very interesting results have been obtained using such methods. For example, a team from the *Conservatoire botanique national* (CBN) of Brest was able to regenerate populations of *Cylindrocline lorencei* (*Asteraceae*), an endemic species that had disappeared from Mauritius, through *in vitro* culture of embryos from seeds harvested before the disappearance of the species and preserved at –18°C (Buord 2001; Sarasan *et al.* 2016). A Russian team was even able to obtain plants of the species *Silene stenophylla* (*Caryophyllaceae*) through *in vitro* culture of placental tissues from seeds preserved in the permafrost of Siberia for over 30,000 years (Yashina *et al.* 2012).

The study that we have launched follows initial experiments conducted at the *Muséum national d'Histoire naturelle* (MNHN) by Stéphane Buord and Valérie Priolet nearly 20 years ago, on 15 species that were considered extinct, using seeds from the MNHN herbarium (Buord 2001). These first experiments did not result in the development of seedlings. However, recent advances in knowledge and technology (Buord 2012; Cosac *et al.* 2016), both for the evaluation of seed viability (using radiographic techniques) and in *vitro* culture methods (by applying biotechnological techniques to this problem), prompted us to resume the approach and extend it to other species that had not been tested in the late 1990s.

10.2. Proposed approach and protocol

The MNHN herbarium (see Chapter 6) is integrated into the French national research infrastructure e-ReColNat (ANR-11-INBS-0004), a network of natural history collections that links French natural history institutions to promote their inventory and valorization[1]. Among their collections, these herbaria hold specimens of species that are completely extinct in the wild and in cultivation, possibly containing seeds that have the potential to regenerate plants after germination.

The first phase of the project was the acquisition of a better knowledge of extinct species and the presence of specimens of these species in herbaria. The steps that followed involved assessing the viability of seeds and carrying out *in vitro* culture experiments for these species, starting with non-extinct and phylogenetically related species, called "affine species". The implemented approach was as follows:

– consultation of the IUCN[2] red lists of threatened species and identification of species considered extinct ("EX" species); classification by systematic groups (families) and geographical groups (island endemic species, large biogeographical sectors, biodiversity hotspots, etc.);

– search for specimens corresponding to these species in the MNHN Herbarium and the e-ReColNat collections; identification of the number of collections, their age, their geographical distribution, the authors, the modalities of these collections, the presence or absence and the (approximate) number of seeds, as well as their state of conservation;

– additional literature review on the pre-selected species in order to complete the knowledge on their distribution, systematics, biology and ecology;

– state-of-the-art on similar experiments carried out on rare taxa, on the conservation of collections, techniques used in *in vitro* culture, etc.;

– selection of "target species" with the highest stakes and the best regeneration potential (abundance of herbarium specimens, not too old, good conservation status and potentially high longevity of seeds);

– extraction of a portion (10 or 20%) of these seeds (with precise tracing) and evaluation of their viability (by non-destructive methods);

---

1 Available at: https://www.recolnat.org/.

2 Available at: http://www.iucnredlist.org/.

– identification of “affine” species of the target species, that is, species that are phylogenetically related and if possible, ecologically close, but not extinct and possibly common enough, to allow experimentation with *in vitro* culture methods. Indeed, as the target species usually have very limited numbers of seeds preserved in the herbarium, it is essential not to “waste” them and to use them only for good purposes;

– viability tests and development of *in vitro* culture protocols for these affine species;

– *in vitro* culture in axenic conditions of viable embryos or cell masses extracted from seeds of the target species and development of the appropriate culture medium;

– if the results are favorable, after multiplication of the plants, acclimatization in culture in the botanical garden;

– reintroduction into suitable habitats in the wild.

10.3. First results

10.3.1. *Selection of target species and identification of affine species*

The first step was to establish a state of knowledge on the lists of extinct plant species. The latest version (2018) of the global red list of extinct plants mentions 118 species in the EX category (i.e. species totally extinct, even in cultivation) worldwide. The red list of the vascular flora of Metropolitan France (UICN France *et al.* 2012) identifies two “EX” species (*F. neglecta* (Soy.-Will.) DC and *Viola cryana* Gillot), that of Reunion Island (2010) three “EX” species (*Mucuna pallida* Cordem, *Fernelia pedunculata* C.F. Gaertn. and *Claoxylon grandifolium* (Poir.) Müll.Arg.), that of French Polynesia endemic plant species six “EX” species (*Achyranthes mangarevica* Suess., *Fitchia mangarevensis* F.Br., *Myoporum rimatarense* F.Br., *Pavonia domatiifera* J. Florence, *Pisonia rapaensis* J. Florence, *Psychotria adamsonii* Fosberg), while there are no species classified as “EX” in the red lists of plants from Guadeloupe (2013), Martinique (2013) and Mayotte (2014). Red lists have not yet been published for French Guiana, Saint-Pierre and Miquelon, Wallis and Futuna and New Caledonia (work in progress by the Endemia association).

Among these extinct species in France (mainland and overseas), the one that is by far the best represented in the Paris Herbarium collection is *F. neglecta* (Soy.-Will.) DC (family *Asteraceae*), with 93 herbarium parts (Figure 10.1), while *Viola cryana is* represented by only 14 parts and the other taxa by four parts at most.

This species of *Filago* was discovered in Badonvillers (54) in 1834 and described by the Lorraine-native botanist Soyer-Willemet (1835). This species was the subject of a relatively recent detailed herbarium survey to identify the exising collections (Seznec 2004). The last known collection of this species was made on September 28, 1977 in Vioménil by G. Bosc and R. Auriault (Alphand 1994). We were able to receive seeds from this last harvest, from the Botanical Conservatory of National Pyrénées et de Midi-Pyrénées, which is why this taxon was retained as one of the target species for the study. The chosen affine species were *Filago arvensis* L., *F. pyramidata* L. and *Gnaphalium luteoalbum* L. species that are well represented in the MNHN's Herbarium and seed collection.

Figure 10.1. *Herbarium part of the collection of* Filago neglecta *by P. Le Brun in 1922 in Vioménil (88) (P00318820, MNHN Herbarium)*

The second chosen target species was *Astiria rosea* Lindl., an endemic species that disappeared from Mauritius (Mascarene Islands, Indian Ocean), of which the Paris Herbarium holds eleven parts (Figure 10.2).

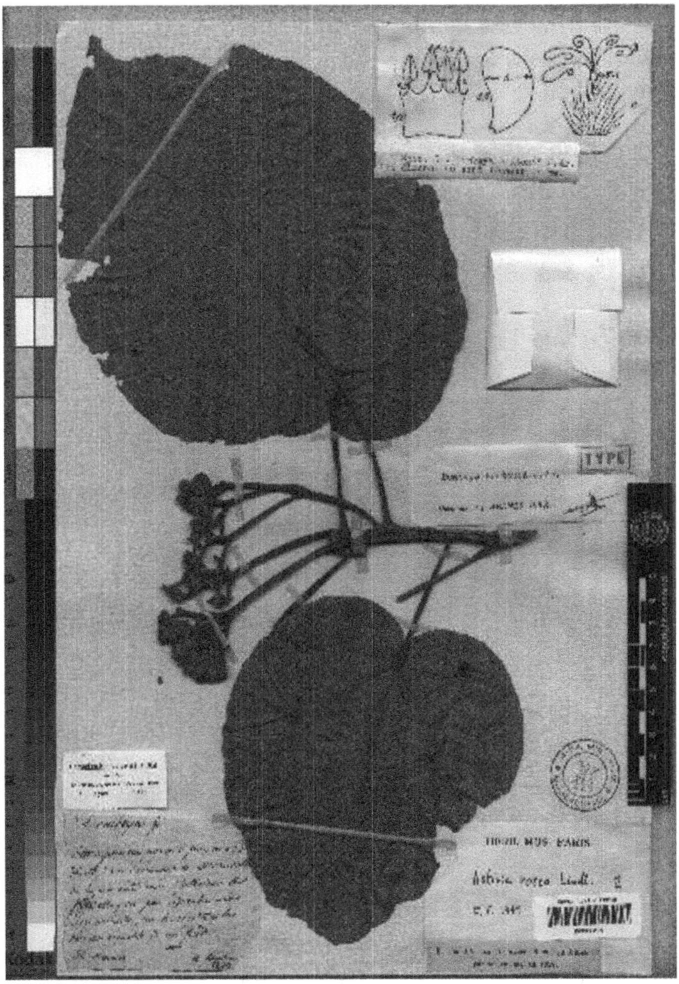

Figure 10.2. *Herbarium specimen of* Astiria *rosea, collected in 1830 by L. Bouton in Mauritius (P00093406, MNHN Herbarium)*

It is a dioecious species of the *Malvaceae* subfamily (*Dombeyoideae* subfamily), corresponding to a monospecific genus. It has not been seen in the field since around 1860, even if Lesouef still hoped to find it in the 1980s (Lesouef 1988). The species was described by Lindley in 1844 and represented in the *Botanical Register* from a plant brought back from Mauritius and cultivated in the greenhouses of the Duke of Northumberland, where it had flowered in 1844, but unfortunately withered afterwards. A related species, also corresponding to a monospecific genus, is *Ruizia cordata* Cav., a dioecious species endemic to the island of Reunion, which also nearly disappeared (Friedmann 1977). Rescue, propagation and population reinforcement of this species was carried out during this same period by the *CBN de Brest* (Lesouef 1988; Folgoat 2011), but the number of remaining "stump" individuals is very small, in the range of three or six trees (Pichillou *et al.* 2013; Augros *et al.* 2015). The affine species of these two taxa are members of the genus *Dombeya* from the Mascarene Archipelago, well studied by Le Péchon and well represented in the Paris Herbarium and the *CBN de Brest*'s seed collections (Arenes 1959; Friedmann 1987; Le Péchon *et al.* 2009).

10.3.2. *Assessment of the viability of available seeds*

10.3.2.1. *X-ray microtomography*

The second step was to assess the "viability" (ability to produce seedlings) of available seeds, using X-ray microtomography to see the cellular structures inside of the seeds without any invasive and destructive cutting (Cochard *et al.* 2015). This initial exploratory work was only performed on *Dombeyoideae* (3–4 mm seeds). Seeds of various species of this subfamily, from cultures in the greenhouses of the *CBN de Brest* and then preserved at – 18°C, *Dombeya acutangula* Cav. (seeds collected in 1992), *D. mauritiana* Friedmann (collected in 2001), *R. cordata* Cav. (harvested in 1996) and *Trochetia boutoniana* Friedmann (harvested in 2015) (Figure 10.3), were tested by one of the authors of this chapter, E. Badel (INRAE, Clermont-Ferrand), who performed 3D X-ray imaging scans at the Psiché beamline of the SOLEIL synchrotron in Gif-sur-Yvette (91).

The 3D volumes resulting from these scans enabled the visualization of the entire internal structure of the seeds. Each species was scanned in a dry state followed by a second scan after 24 hours of soaking under imaging conditions aimed at limiting the X-ray irradiation dose.

The results showed a very good 2D and 3D restitution (resolution up to 0.64 μm) (Figure 10.4) of the intracellular structures for the first three species, which can be considered as an indicator of germination potentiality, whereas the seeds were empty (dry and collapsed) for *T. boutoniana*.

This imaging technique, however, while allowing for detailed observations of the quality of the various tissues composing the internal structure in a non-destructive way, remains a qualitative observation that does not offer proof that the seed has a real germination capacity. Only culture experiments can attest to this.

An *in vitro* germination test was performed for each of the seeds irradiated during the tomographic scan. While germination was observed within the same time frame as the non-irradiated controls for *R. cordata* and *D. acutangula*, only the germinations obtained from the seeds that were scanned at low resolution developed seedlings. It should be noted that from the first weeks of cultivation, the growth of *R. cordata* was much slower than the control. Rooting was limited to the formation of a single, brittle root. Only two leaves were developed, and incompletely. The seedling did not develop beyond this stage. *D. acutangula*, on the other hand, did not survive beyond 21 days of culture. At this stage of this first and only experiment, the death of this seedling could not be directly attributed to its exposition to radiation, with the mortality of young seedlings during transplanting being a frequently observed phenomenon.

Figure 10.3. Trochetia boutoniana *in culture in the CBN de Brest's greenhouses (photo by Charlotte Dissez/CBN de Brest)*

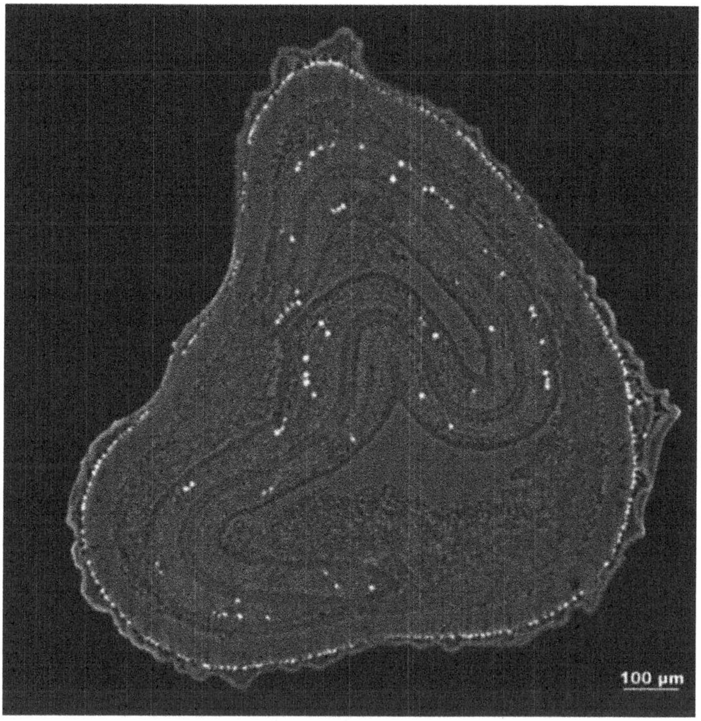

Figure 10.4. *Internal structure of a* Dombeya mauritiana *seed (2D section, resolution 0.64 µm) observed by X-ray microtomography (photo by E. Badel/PIAF Clermont-Ferrand)*

The exploratory character of this approach to assess seed viability led us to use standard working protocols for both X-ray imaging and *in vitro* germination tests of the irradiated seeds. It would be interesting to continue the experiments by using other working protocols, aiming in particular at further reducing the radiation doses absorbed by the seeds during the microtomography scans. It also seems essential to follow the long-term evolution of the seedlings obtained after the germination of irradiated seeds, in order to note the possible effects of radiation on the genotype and phenotype.

10.3.2.2. *Germination test: tetrazolium red test*

Other methods that are simpler to implement provide information on seed viability and germination ability. This is the case for germination tests on wet filter paper and the tetrazolium red test (TZT).

The germination tests on filter paper enable the assessment of the germination capacity of the seeds, but the absence of germination does not necessarily imply that the seed is dead. The TZT test provides additional information by coloring the viable parts of the seed red, with the non-viable parts remaining colorless. While the staining is technically easy to perform, its interpretation is sometimes difficult. These tests have the major disadvantage of being destructive for the seeds, so they have been used on affine species.

The important representation of *G. luteoalbum*, affine species of *F. neglecta*, in the MNHN's seed bank allowed us to carry out these two types of tests.

Several batches of seeds harvested in 2017, 2007, 1994 and 1987 were placed on wet filter paper in Petri dishes. The results showed a germination capacity maintained after 30 years of cold storage, with the percentage of germination at 21 days of culture varying from 4 to 11% depending on the batches for the 1987 harvest (80% for the 1994 harvest and 74% for the 2017 harvest). This result was a positive indicator of the possible viability of *F. neglecta* seeds stored for 40 years. It should be noted here that the seeds of herbarium preserved plants are not harvested and stored for cultivation purposes, and that their stage of maturity at harvest and storage conditions may alter their ability to germinate and regenerate seedlings. The same tests were performed on herbarium specimens of *F. arvensis* (harvest 1981), *F. pyramidata* (harvest 1968) and *F. neglecta* (harvest 1977). As no germination was obtained, we decided to further test the viability of the seeds by using the TZT. The control coloration of the genus was observed on the 2017 harvest of *G. luteoalbum* from the MNHN seed bank. Only *Gnaphalium luteo-album* (1987) was tested for "aged" affine species. While 98% of the 2017 seeds were fully colored red, only 32% of the 1987 harvest were colored, 54% were totally colorless and 13% partially colored, demonstrating metabolic activity in reduced areas of the seed, but an inability to germinate under traditional growing conditions.

Similarly, the control coloration of seeds belonging to the *Dombeyoideae* group was observed on *D. acutangula* (1992 harvest) and *R. cordata* (1996 harvest) (Figures 10.5(a) and (c)). In comparison, the various *R. cordata* embryos collected in the herbarium (plate no. P06644141, harvested around 1800) remained colorless (Figure 10.5(b)), in contrast to the *D. acutangula* embryo (plate no. P06645607, harvested in 1872), which showed staining on the embryonic axis (Figure 10.5(d)).

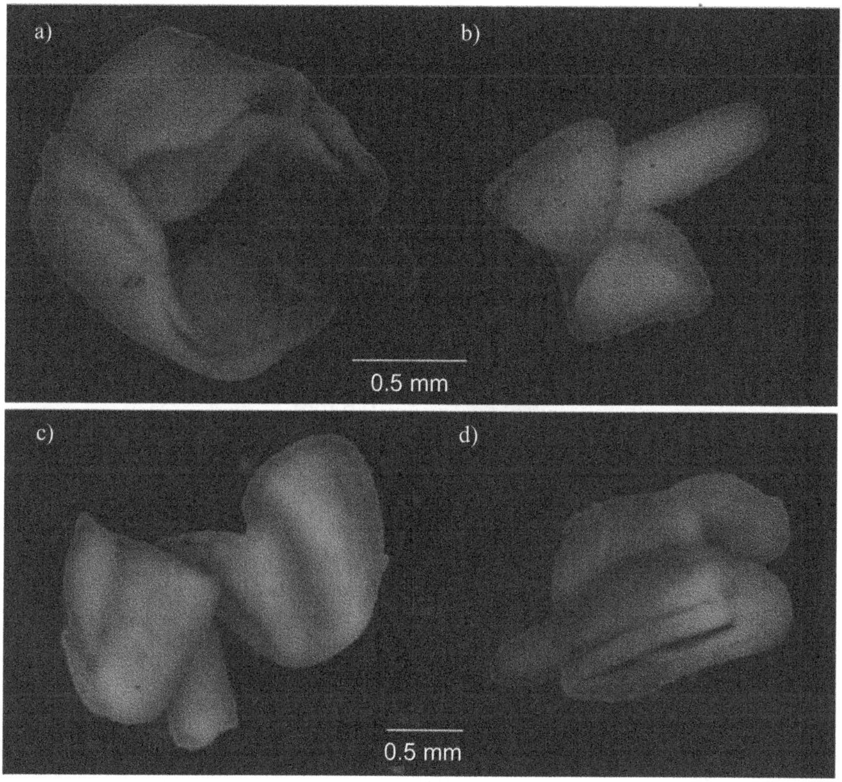

Figure 10.5. *Tetrazolium red viability test (TZT): (a)* Ruizia cordata *(1996) control; (b)* R. cordata *P06644141 (undated harvest, circa 1800, from L.M.A du Petit-Thouars on Reunion Island); (c)* Dombeya acutangula *(1992) control; (d)* D. acutangula *P06645607 (collected in 1872 on Reunion Island) (photos by Valérie Priolet, DGD MJZ, MNHN)*

10.3.3. *Cultivation experiments on affine species of the target species*

The third step consisted of conducting growth experiments on affine species in order to define the cultivation protocol best suited to the regeneration of each target species.

Based on the expression of the cellular totipotency of plants, *in vitro* cultures allow us, in theory, to regenerate a whole plant from any plant fragment (Margara 1989); they thus offer a possibility of regeneration from all or part of the seed.

NOTE.– Each seed is examined before handling, viability testing or cultivation. After observation with a binocular magnifying glass, seeds that are altered or do not conform to the control are discarded and returned to the herbarium plates (Figure 10.6).

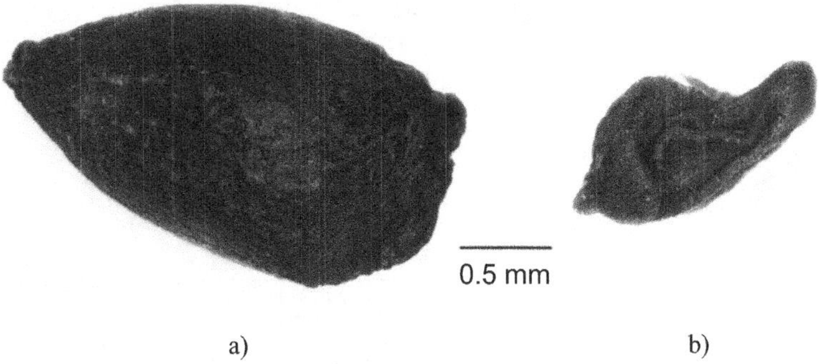

Figure 10.6. *Seeds of* D. acutangula*: (a) control seed-harvest CBN Brest 1992; (b) altered seed (specimen P06645605, MNHN Herbarium) (photo by Valérie Priolet, DGD MJZ, MNHN)*

With regard to the results obtained during the evaluation of the viability of seeds, the technical protocol was organized around three axes: (1) the germination of the whole seed, (2) the germination and the development of the isolated embryo, (3) the regeneration by organogenesis or embryogenesis, starting from the various components of the embryo, embryonic axis and cotyledons, or cellular massifs of embryonic origin.

Preliminary work allowed us to define, for each of the three axes, the most suitable growing conditions for each of the target species: environmental conditions (temperature, luminosity), seed scarification, transplanting sequence and acclimatization of the resulting seedlings.

10.3.3.1. *Germination of whole seeds*

Without going into the technical details of the different protocols, the mineral base chosen for *in vitro* germination and growth of the *Dombeyoideae* group was the Woody Plant Medium (Lloyd and McCown 1980).

The Gamborg B5 base (Gamborg *et al.* 1968) was preferred for the *Gnaphalium-Filago* group because of its lower mineral load and nitrogen concentration than the traditional base of Murashige and Skoog (1962). The action of gibberellic acid, a growth regulator with a stimulating effect on seed germination, was tested on this group by adding GA3 at concentrations between 1 and 30 mg·L^{-1}, without any observed effect.

10.3.3.2. *Germination and development of the isolated embryo*

The germination and development of single embryos was achieved using dual media, one in the center of the dish and the other poured into the outer ring. The media diffused into each other over time, allowing the needs of the embryo to be met during the different phases of germination and development (Tegan *et al.* 2011).

Although the results were good for both groups with respect to the percentage of germination obtained, it should be noted that there was a significant vitrification of plants in the *Gnaphalium-Filago* group and a delay in growth for those in the *Dombeyoideae* group. This led us to consider adapting this technique by modifying the concentration of certain elements, such as nitrogen or sucrose, whose presence at concentrations higher than 60 g·L^{-1} appeared to inhibit germination (*Filago-Gnaphalium* group) or limit the development of seedlings (*Dombeyoideae* group). The mannitol–sucrose combination proved to be more favorable in the various experiments.

10.3.3.3. *Regeneration by organogenesis or embryogenesis*

The use of zygotic embryos to provide micropropagation in a number of species via organogenesis (organ formation) or somatic embryogenesis (embryo formation from non-sexual cells) has been frequently reported in the literature (Elhiti and Stasolla 2011). Consisting of juvenile tissues capable of dedifferentiating and then regenerating new tissues and organs, embryos represent a prime material for this type of multiplication (Figures 10.7(a)–(c)).

This potentiality is being explored on the affine species of the two studied groups. Various protocols using the isolated embryo, as well as its components, embryonic axis, cotyledons, cell masses of embryonic origin aim to reach the formation of an organogenic and embryogenic callus, the first step of the process. This induction phase is sought with different concentrations of growth regulators, used alone or in combination.

Initial results showed that the *Gnaphalium-Filago* group had a high sensitivity to growth regulators, with the group responding at low concentrations. For this group, as for the *Dombeyoideae* group, many green and friable non-organogenic and non-embryogenic calluses were rapidly formed from excised cotyledons.

For *Dombeyoideae*, it appears that the isolated embryonic axis was able to develop a seedling directly after excision of the cotyledons, if the terminal meristem was preserved. No callus was formed if the whole embryo was put in culture, the presence of the embryonic axis seemingly inhibiting the process. It could therefore be interesting to conduct further experiments using fragments of the embryonic axis.

Figure 10.7. Gnaphalium luteoalbum. *(a) Evolution of the germination of an embryo, (b) seedling in rooting phase and (c) vitrified seedling (photos by Valérie Priolet, DGD MJZ, MNHN)*

10.4. Discussion and conclusion

The Herbarium of the *Muséum national d'Histoire naturelle* represents an incomparable wealth of data on flora, among the most important in the world. Some of these herbarium specimens represent the last witnesses of species that are now extinct. This program proposes to explore the potential for regeneration of these extinct species from their seeds: a novel challenge in terms of conservation of plant heritage and an innovative experimental approach.

Inspired by previous experiments carried out at the MNHN, the objective here is to demonstrate the benefit of developing new scientific and technical approaches to solve problems linked to biodiversity conservation. It is essentially a question of adapting techniques and tools from the biotechnology field, in particular *in vitro* cultivation, which are relatively common in agricultural and horticultural production, to the field of plant conservation, where they sometimes represent the only possible and necessary recourse for the safeguard of certain species.

These trials also illustrate the will to build new bridges between these two fields of botany: scientific knowledge (fundamental or applied as in the case of biotechnology), on the one hand, and the conservation of flora, on the other hand, two fields that are still too often ignored.

The first investigations described here have shown the feasibility and relevance of the proposed approach of using ancient material (herbarium samples) and modern bio-technologies (X-ray microtomography, *in vitro* culture) for the purpose of conservation and restoration of biodiversity.

The irreplaceable character of the concerned samples, testimonies of a heritage that has now disappeared, may have seemed contradictory to a possible experimental use at first sight. It was therefore necessary to find a methodological workaround. The identification of species affine to the target species (the extinct species) and the prior development of germination techniques on these affine species, before applying them to the target species appeared to be the most appropriate solution to limit the impact of the experiments on the very precious material represented by the seeds of the target species.

The next step of the program will be to continue the experiments on the affine species before extending these viability tests, as well as the *in vitro* culture experiments on the seeds of the two target species.

While it is difficult to presume the future lessons to be learned from the work yet to come, it will probably provide useful new information concerning the possible toxic effects of the various preservative treatments (mercury salts, methyl bromide, etc.), and the sometimes surprising and above all little-known protective capacities of the seminal teguments.

The success of these operations should not obscure the fact that they can only generate a very small number of genetically different individuals. But a small living population, even with poor genetic diversity, is infinitely preferable to a definitive extinction, and provides ample justification for the proposed approach.

Many other extinct or near-extinct species could be apt candidates for future regeneration trials. This is why it is also planned to extend the project to other extinct species of France (especially overseas), as well as to other hotspots of plant biodiversity where dozens of other taxa would be concerned. For example, among many others, the regeneration of new individuals of *Sophora toromiro* Skottsb., an extremely rare shrub from Easter Island and emblematic of the island's culture, could be attempted from seeds collected *in situ* in 1934–1935 during the Metraux–Lavachery expedition. The objective here would be to expand the extremely reduced genetic pool of this species currently maintained in cultivation in various botanical gardens worldwide. Indeed, a few individuals from seeds collected in the wild in 1955–1956 by Thor Heyerdahl on the last living specimen are at the single origin of all of the subjects cultivated in Europe (Maunder 1997).

It is undeniable that the establishment of a true policy of conservation management of natural areas is infinitely preferable and more satisfying than the development *in extremis* of *ex situ* rescue procedures for a few taxa. Nevertheless, these innovative experiments are highly symbolic and promising. They demonstrate that, despite an unfavorable initial context and a particularly compromised future, the will to implement truly adequate means of rescue and conservation could lead to success.

10.5. Acknowledgments

The authors would like to thank all of the people who offered help and advice throughout this study: C. Ramambason (Master's student at the University of Aix-Marseille), Y. Pauthier, S. Lang, V. Invernon, M. Pignal, R. Garrouste (MNHN), T. Le Pêchon (Jardin botanique de Meise, Belgium), J. Lherminier (INRA), as well as G. Largier and J. Cambadélis (CBN Pyrénées and Midi-Pyrénées) for providing seeds from the last harvest of *F. neglecta* made in 1977, the SOLEIL synchrotron for providing the synchrotron radiation facilities, Andrew King for his help with the Psiché beamline, and the Sorbonne Universities for their contribution to the funding of this project as part of the 2017 call for exploratory projects of OPUS Observatoire des patrimoines.

10.6. References

Alphand, J. (1994). Note sur *Logfia neglecta* (Soyer-Will.) Holub. *Le Monde des Plantes*, 449, 28–30.

Arenes, J. (1959). Les *Dombeya* des îles des Mascareignes. *Mémoire de l'Institut Scientifique de Madagascar*, 9, 189–216.

Augros, S., Hoareau, D., Paillusseau, J., Lombard, C., Thueux, P., Fontaine, C., Hivert, J. (2015). Découverte d'un nouvel individu de Bois de senteur blanc, *Ruizia cordata* Cav., dans le Nord de l'île de La Réunion et bilan des connaissances sur l'état de conservation de l'espèce (Malvales, Malvaceae, Dombeyoideae). *Cahiers scientifiques de l'océan Indien occidental*, 6, 1–22.

Becquerel, P. (1934). La longévité des graines macrobiotiques. *Comptes Rendus Hebdomadaires des Séances de l'Académie des Sciences*, 199, 1662–1664.

Bowles, M.L., Betz, R.F., DeMauro, M.M. (1993). Propagation of rare plants from historic seed collections: Implications for species restoration and herbarium management. *Restoration Ecology*, 1, 101–106.

Buord, S. (2001). Exploration des potentialités de régénération de taxons disparus à partir de semences conservées en herbier. In *Études écologique et biosystématique de la flore littorale remarquable de l'arc dunaire Gâvres-Quiberon, futur grand site national : intérêts d'une approche pluridisciplinaire et des biotechnologies pour une gestion integrée de la biodiversité*, Buord, S. and Couderc, M.H. (eds). Muséum national d'Histoire naturelle, Paris.

Buord, S. (2012). Réintroduction d'espèces végétales endémiques à l'île Maurice. Une première dans le domaine de la conservation. *Le Courrier de la Nature*, 267, 22–29.

Cochard, H., Delzon, S., Badel, E. (2015). X-ray microtomography (micro-CT): A reference technology for high-resolution quantification of xylem embolism in trees. *Plant, Cell & Environment*, 38(1), 201–206.

Cosac, A.C., Teodorescu, G., Buord, S., Isac, V., Nicola, C. (2016). Exploration of regeneration potential of extinct plants starting from old seeds, by in vitro technology. *The Annals of "Valahia" Universiy of Targoviste*, 10(1), 68–71.

Elhiti, M. and Stasolla, C. (2011). The use of zygotic embryos as explants for in vitro propagation: An overview. In *Plant Embryo Culture: Methods and Protocols*, Thorpe, T.A. and Yeung, E.C. (eds). Humana Press, New York.

Folgoat, N. (2011). Bilan de l'opération de sauvetage d'une espèce végétale endémique de La Réunion, Bois chanteur, Bois de senteur blanc, *Ruizia cordata* Cav. Mission report, Société réunionnaise pour la protection de l'environnement, Sainte-Clotilde.

Friedmann, F. (1977). Le bois de senteur, arbre très rare de La Réunion. *Info-Nature – Bulletin de Liaison de la SREPEN*, 15, 58–62.

Friedmann, F. (1987). 53. Sterculiacées. In *Flore des Mascareignes : La Réunion, Maurice, Rodrigues. 51. Malvacées à 62. Oxalidacées*, Bosser, J., Cadet, T., Guého, J., Marais, W. (eds). The Sugar Industry Research Institute, île Maurice, Office de la recherche scientifique et technique d'outre-mer, Paris, and The Royal Botanic Gardens, Kew.

Gamborg, O.L., Miller, R.A., Ojima, K. (1968). Nutrient requirement of suspensions cultures of soybean root cells. *Experimental Cell Research*, 50, 151.

Godefroid, S., van de Vyver, A., Stoffelen, P., Robbrecht, E., Vanderborght, T. (2011). Testing the viability of seeds from old herbarium specimens for conservation purposes. *Taxon*, 60, 565–569.

Le Bras, G., Pignal, M., Jeanson, M., Muller, S., Aupic, C., Carré, B., Flament, G., Gaudeul, M., Gonçalves, C., Jabbour, F., Lerat, E., Lowry, P., Offroy, B., Pérez, E., Poncy, O., Rodrigues Invernon, V., Rouhan, G., Haevermans, T. (2017). The French Muséum national d'Histoire naturelle vascular plant herbarium collection dataset. *Scientific Data*, 4, 170016.

Le Péchon, T., Cao, N., Dubuisson, J.-Y., Gigord, L.D. (2009). Systematics of Dombeyoidae (Malvaceae) in the Mascarene Archipelago (Indian Ocean) inferred from morphology and new insights on the evolution of breeding systems. *Taxon*, 58(1), 1–13.

Lesouef, J.Y. (1988). The rescue of *Ruizia cordata* and the possible extinction of *Astiria rosea. Botanic Gardens Conservation News*, 1, 35–39.

Lloyd, G. and McCown, B. (1980). Commercially-feasible micropropagation of mountain laurel, Kalmia latifolia, by use of shoot-tip culture. *Combined Proceedings – International Plant Propagators' Society*, 30, 421–427.

Margara, J. (1989). *Bases de la multiplication végétative : les méristèmes et l'organogenèse.* INRA, Paris.

Maunder, M. (1997). Conservation of the extinct Toromiro tree, *Sophora toromiro. Curtis's Botanical Magazine*, 14(4), 226–231.

Murashige, T. and Skoog, F. (1962). A revised medium for rapid growth and bio assays with tobacco tissue cultures. *Physiologia Plantarum*, 15, 473–497.

Milberg, P. (1994). Germination of up to 129-years old, dry-stored seeds of *Geranium bohemicum* (*Geraniaceae*). *Nord. J. Bot.*, 14, 27–29.

Molnar, V.A., Sonkoly, J., Lovas-Kiss, A., Fekete, R., Takacs, A., Somlyay, L., Török, P. (2015). Seed of the threatened annual legume, *Astragalus contortuplicatus*, can survive over 130 years of dry storage. *Preslia*, 87, 319–328.

Pichillou, S., Lavergne, C., Gigord, L. (2013). Le bois de senteur blanc, *Ruizia cordata* Cav. Plan national d'actions 2012–2016 : outils d'aide à la conservation des espèces végétales menacées d'extinction, version 2013 (updated 19 April 2013). Conservatoire Botanique National de Mascarin, Saint-Leu, Réunion.

Sarasan, V., Buord, S., Pellicer, J., Sanchez, M., Cowan, R.S., Lesouëf, J.Y. (2016). Approaches to develop a road map for the long-term conservation of an island endemic genus *Cylindrocline*. *Acta Physiol. Plant.*, 38, 1–10.

Seznec, G. (2004). Enquête dans les herbiers à la recherche d'une plante disparue : la cotonnière négligée [*Logfia neglecta* (Soy.-Will.) Holub]. In *Les herbiers : un outil d'avenir. Tradition et modernité*, Pierrel, R. and Reduron, J.P. (eds). AFCEV, Paris.

Soyer-Willemet, H.F. (1835). *Gnaphalium neglectum*, nouvelle espèce du groupe des Filaginées avec des observations sur les autres espèces françaises de ce groupe. *Mémoires de la Société Royale des Sciences, Lettres et Arts de Nancy*, 46–54.

Tegan, M., Yeung, H., Yeung, E.C. (2011). Zygotic embryo culture: An overview. In *Plant Embryo Culture: Methods and Protocols*, Thorpe, T.A. and Yeung, E.C. (eds). Humana Press, New York.

UICN France, FCBN, MNHN (2012). La Liste rouge des espèces menacées en France – flore vasculaire de France métropolitaine : premiers résultats pour 1000 espèces, sous-espèces et variétés. Electronic document, UICN France, FCBN, MNHN, Paris.

Yashina, S., Gubin, S., Maksimovich, S., Yashina, A., Gakhova, E., Gilichinsky, D. (2012). Regeneration of whole fertile plants from 30,000-y-old fruit tissue buried in Siberian permafrost. *PNAS*, 109(10), 4008–4013.

11

Ocean Cores, Climate Archives

Eva MORENO[1] and Annachiara BARTOLINI[2]

[1] *LOCEAN, CNRS, IRD, Muséum national d'Histoire naturelle, Paris, France*

[2] *CR2P – CNRS, Muséum national d'Histoire naturelle, Sorbonne Université, Paris, France*

The study of ocean sediment cores has made it possible to document past climate variability beyond the instrumental period, that is, since direct measurements of climate variables such as water or atmospheric temperature are available. With advances in geotechnical and geophysical technology, cores taken from oceanographic vessels are of better quality and longer duration, allowing more sediment to be obtained and from further back in time.

In paleoclimatology, or the study of past climates, scientists use indirect methods, called "proxies", to reconstruct past climate conditions. A proxy is a physical, chemical or biological parameter that is sensitive to temperature, salinity or any other climatic variable that is obtained from total sediment, but especially from foraminiferal tests. These indirect methods always have a greater or lesser uncertainty. A proxy can be influenced by parameters other than the climate variable it is supposed to represent, which can make its interpretation difficult. As geochemical analyses progress, new proxies are developed. While quantitative estimates of some parameters, such as water temperature, are well established, robust proxies are not yet available for estimating other variables, such as salinity and pH, which are essential for describing the state of the oceans.

For a color version of all figures in this chapter, see www.iste.co.uk/pellens/natural.zip.

Natural History Collections in the Science of the 21st Century, coordinated by Roseli PELLENS.

11.1. Introduction

At a time when there is increasing concern about the impact of humans on the climate, it is important to know what the climate was like in the past, in order to know how global warming might affect us in the future. By studying climate archives, such as marine sediments or ice cores, we are able to reconstruct the temperature and amount of CO_2 in the past. Unfortunately, it is not possible to obtain direct records of these parameters in sediments, so it is necessary to use indirect indicators or what we usually call proxies.

Paleoclimatology developed mainly from the 1950s onwards; thanks to the first oceanographic explorations and especially to the development of isotopic chemistry, mass spectrometry allowed to the creation of one of the first climate proxies, namely $\delta^{18}O$, based on the study of oxygen isotopes. ^{14}C dating has made it possible to date marine sediments back about 50,000 years.

As science progresses, new proxies are developed, which allow us to better understand the state of the Earth during periods that are colder and warmer than today.

In this chapter, we will attempt to cover first a brief history of the MNHN oceanic collection (section 11.2), second, the development of coring techniques (section 11.3) and, finally, the types of proxies currently used to reconstruct sea temperature, salinity and CO_2 concentration, three parameters that are very sensitive to climate change (section 11.5).

11.2. The MNHN's oceanic collection

The oceanic collection consists of about a thousand marine sediment cores representing a total length of about 7 km, coming mainly from the Indian Ocean, but also from the Red Sea, the Atlantic Ocean, etc. The oceanic collection is very recent compared to other MNHN collections: it was initiated in the early 1970s with the development of new tools allowing deep coring. In France, other core repositories have also been created: IFREMER (*Institut français de recherche pour l'exploitation de la mer*) in Brest, LSCE (*Laboratoire des sciences du climat et de l'environnement*) in Gif-sur-Yvette, EPOC (*Environnements et paléoenvironnements océaniques et continentaux*) in Bordeaux and CEREGE (*Centre européen de recherche et d'enseignement des géosciences de l'environnement*) in Aix-en-Provence are the most important. The oldest is the IFREMER core repository, whose first cores date back to 1965. However, the MNHN is the only center in France with a mission of management and conservation of samples that includes an associated database open to the entire scientific community.

The MNHN's oceanic collection is closely linked to the French oceanographic ship *Marion Dufresne,* since a large majority of the samples were collected during scientific expeditions on this ship. The first *Marion Dufresne*, built in 1972, was replaced by the *Marion Dufresne 2* in 1995. It is one of the largest ships in the global fleet (120 m long). We cannot forget the expeditions carried out on other French ships which also contributed to the growth of the collection: *L'Atalante*, *Suroit* and *Pourquoi Pas*. However, the *Marion Dufresne* remains the flagship vessel because it was specially designed for coring operations.

Very close links have been established between the oceanic collection and the micropaleontology collection since microfossils (foraminifers, coccoliths, radiolarians, diatoms, etc.) are one of the main components of marine sediments. These can represent up to 90% of the total sediment, such as the cores taken during the campaign "MD64 SOMIRMAS" in 1990 in the northern Mascarene Basin in the Indian Ocean. These are regions that are far from land input sources.

In addition, the MNHN micropaleontology collection houses Alcide d'Orbigny's prestigious foraminifer collection, the founder of micropaleontology. Several species described by d'Orbigny are widely used in paleoclimatic studies, such as *Globigerinoides ruber* (d'Orbigny 1839, pp. 82–83), *Globigerina bulloides* (d'Orbigny 1826), *Orbulina universa* (d'Orbigny 1839, pp. 82–83) and *Neogloboquadrina dutertrei* (d'Orbigny 1839, pp. 82–83), to name a few.

11.3. Development of core drilling techniques

There are several coring techniques, but most of the cores in the oceanic collection are from "calypso" type piston corers, derived from the Kullenberg corer, invented in 1947 (Figure 11.1). The corer consists of a steel tube inside which is another PVC tube to receive the sediment column. The system is topped by a ballast that can reach several tons and a lever arm that is under tension thanks to a counterweight suspended at its end. When it was invented, this type of corer was able to provide real "cores" of up to 10 m of sediment and provided a major boost to paleoceanographic research (Estival 2003). Of course, the technology has advanced and continues to do so. This development has made it possible to collect longer and longer cores, to increase the diameter of the corer tube and to improve the recovery rate (the length of sediment in relation to tube length). In the collection we have the first core from the *Marion Dufresne*, MD73-0001, taken during the "OSIRIS 1" cruise off Madagascar in 1973, measuring 9.9 m in length and 54 mm in diameter. At that point, it was not possible to core more than 20 m. Moreover, the recovery rate was low (30%–60%) and the cores collected at great depth were often very "pistoned", that is, the sedimentary strata were totally mixed.

Figure 11.1. *Calypso corer on board the* Marion Dufresne *during the "MONOPOL" campaign in 2012 in the Bay of Bengal. Source: © Eva Moreno*

In the 1990s, engineers from the IPEV (*Institut Polaire Paul Émile Victor*) designed a new giant corer called CALYPSO. Taken on-board the *Marion Dufresne*, it made it possible to take cores up to 50 m long and 110 mm in diameter. However, in some cases, the sedimentary structures were "stretched" in the first few meters of the cores. Sequences recovered from long CALYPSO cores could appear up to 1.5–2 m longer than the same sequence recovered by conventional piston cores at the same site (Thouveny *et al.* 2000).

The *Marion Dufresne* was fully modernized in 2015, allowing for further improvements to the giant CALYPSO corer and on-board equipment. These improvements have minimized sediment deformation and increased the recovery rate. Thus, on March 5, 2019, a team of French researchers from UMR EPOC (*Environnements et Paléoenvironnements Océaniques et Continentaux*) at the University of Bordeaux succeeded in collecting a 70 m long core, covering the last million years, north of the Crozet Archipelago in the Southern Ocean. This is the longest core sample ever taken.

The longest core we have in the oceanic collection is MD01-2412: it is 58.10 m long and was taken in 2001 off Japan in the Western Pacific on the *Marion Dufresne* at 1,225 m depth. It covers the last 115,000 years, which reflects an exceptional sedimentation rate and has allowed the detailed study of rapid climate changes on scales ranging from a few hundred to a few thousand years (Sakamoto *et al.* 2006).

The deepest core is CAS16-42PC: it was collected at a depth of 6,898 m off the Lesser Antilles in the Atlantic Ocean in 2016 on the *Pourquoi pas*. It measures 28.2 m for a tube length of 31.2 m, a recovery rate of 90%. This type of core would have been impossible to sample in the 1970s when the oceanic collection was created.

11.4. Ocean cores: archives of past climate variability

The first detailed information on atmospheric temperature dates back to the invention of the first thermometers in the mid-17th century. Ocean surface temperature records, which are much more informative about climate than those of the continent, are even more recent. During the first oceanographic campaigns, oceanographers measured the temperature in a bucket that they brought aboard the ship. Later, measurements were made directly from the ship, including the inverting thermometer, an invention of the French oceanographer Georges Aimé (Geistdoerfer 2015). In 1872, the first major world oceanographic campaign was conducted on the oceanographic ship the *HMS Challenger*. The expedition collected numerous temperature and salinity data. It was also the first campaign to explore the ocean floor. John Murray (1841–1914) studied the sedimentary deposits of the ocean floor and classified them into three categories: globigerina, diatom and pteropod muds, and deep-sea red mud (Geistdoerfer 2015). Today, oceanographers use bathysondes, sensors that acquire a continuous profile of salinity, temperature and pressure measurements from the surface to great depths. These measurements can be made from ships and meteorological buoys that are either fixed or drifting. More recently, since the early 1970s, the radiometer of a meteorological satellite has been used to determine and map sea surface temperatures.

We therefore have no direct information on climate variations before the so-called "instrumental" period. In order to know the climate prior to this period, we use "climate archives": ocean and/or lake cores, ice cores, coral and tree rings. Depending on the climate archive studied, and also on the location of the site, these climate reconstructions can cover all time scales, from annual variations to those that occurred over millions of years. In the case of ocean cores (Figure 11.2), we are theoretically able to study climatic variations backward from the present to the oldest sediment covering the oceanic crust, which dates from the Middle Jurassic, or about 170 Ma (Bartolini and Larson 2001).

Figure 11.2. *Core MD92-1017 taken in 1992 at a depth of 572 m on board the French ship* Marion Dufresne. *It is 15.40 m long. The photo shows intervals between 35 and 90 cm, dated between 10,000 and 25,000 years ago. Source: © MNHN*

However, the older the sediment, the less information we have: indeed, there are often cores with hiatuses (interruption in sedimentation) or from areas with low sedimentation rates. In addition, such ancient sediments may be poorly preserved, making it difficult to estimate climate parameters. The oldest sediments in the oceanic collection date from the Maastrichtian period, between 66 and 72 Ma, and were collected during the various campaigns in the Indian Ocean: DRAKAR, SOMIRMAS and SAFARI 1 in the Mozambique Channel and the Kerguelen Plateau.

11.5. Climate proxies

A climate proxy can be defined as an indirect estimate of environmental parameters made from the measurement of a physical, chemical or biological parameter of a geological record. In the case of marine sediment cores, a proxy can be measured on the total sediment. However, very often, the part that has the most advantage in paleoclimatology is the biogenic fraction, that is, the *tests* (shells) of calcareous microorganisms such as foraminifers. Most climate proxies have been developed on foraminiferal *tests* composed of $CaCO_3$, both planktic and benthic. In order to study them, the sediment is sieved with sieves with mesh sizes ranging from 63 μm to 2 mm. In general, for geochemical studies, tests with a size between 250 μm and 315 μm are used.

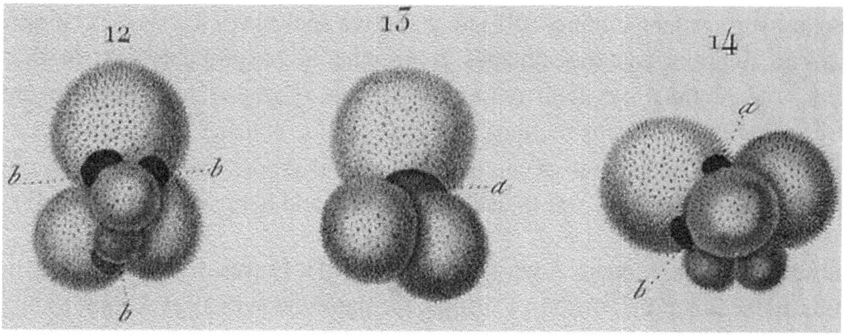

Figure 11.3. *Original drawings by d'Orbigny of the planktic foraminifer* Globigerinoides ruber *(d'Orbigny 1839, pp. 82–83). Source: RECOLNAT (ANR-11-INBS-0004)*

COMMENT ON FIGURE 11.3. – *Illustrations idealized by the author of the species* name-bearing *types of* G. ruber *preserved in the MNHN's d'Orbigny foraminifer collection (MNHN.F.FO302, MNHN.F.FO246) and which summarize the typical characteristics of the species. Drawing 12: dorsal view of the trochospiral shell characterized by the presence of supplementary sutural apertures (b); drawing 13: umbilical view characterized by the presence of three subspherical chambers at the last turn of the spire and the large arch-shaped primary umbilical aperture (a); drawing 14: profile view of the trochospiral shell with primary aperture (a) and supplementary aperture (b).*

Planktic foraminifers (Figure 11.3) provide information on the first few hundred meters of surface waters, while benthic foraminifers provide information on bottom waters at the water/sediment interface, as well as on pore waters in the sediments from the shallow seas to the deep ocean basins. Carbonate muds (rich in foraminifers) occur between 60° N and 60° S near the ridges and down to 4.5–5 km depth. The deeper the water column, the less carbonates there are. Beyond this depth, called the carbonate compensation depth (CCD), which is determined by temperature, pressure and the amount of CO_2 dissolved in the water, all the calcium carbonate brought from the surface is dissolved.

11.5.1. *Temperature proxies*

11.5.1.1. *Analyses of planktic foraminiferal assemblages*

In the surface ocean, the distribution between different species of planktic foraminifers depends primarily on the temperature of their environment (Kucera 2007). Of the many techniques for obtaining paleo-temperatures of surface ocean waters (sea surface temperature, SST), quantitative analysis of the species composition of planktic foraminiferal assemblages is proving to be probably one of the most accurate, after comparison with numerical models (Caley *et al.* 2014). In order to achieve quantitative reconstructions of paleo-SSTs, transfer functions have been further developed, taking into account foraminiferal groups associated with climatic zones: tropical, subtropical and transitional (Imbrie and Kipp 1971, pp. 71–181).

Subsequently, the Modern Analog Technique (MAT) was developed (Prell 1985): unlike transfer functions, the MAT allows for the same statistical weight to be given to both rare and common species when reconstructing SSTs. To use this method, it is necessary to measure the similarity (or distance) between the foraminifer assemblage of the sample studied and the modern assemblages of a database composed by the study of sediments at the top of recent cores (<4,000 years). These core tops are associated with instrumentally measured current surface water temperatures. Assuming that the link between foraminiferal assemblages and SSTs has not varied or has varied very little on the scale of tens to hundreds of thousands of years, it is possible, in a statistical way, to reconstruct past temperature variations through this comparison.

However, the effectiveness of this method depends on the quality of the database. Today, the database used for this technique consists of more than 40 years of data. These data come from foraminifers collected in the water column (plankton nets, sediment traps) and in surface sediments (core tops) from all oceans (Haddam *et al.* 2016). These new generations of databases thus allow for high accuracy in surface temperature reconstructions in the recent past (Neogene).

11.5.1.2. *Stable isotopes of oxygen*

The $\delta^{18}O$ (expressed as ‰) of foraminiferal calcite is a quantification of the ratio of the stable oxygen isotopes ^{16}O and ^{18}O in the sample relative to a standard (PDB, Pee Dee Belemnite), according to the equation:

$$\delta^{18}O_{calcite} = ((^{18}O/^{16}O)_{calcite} - (^{18}O/^{16}O)_{standard}))/(^{18}O/^{16}O)_{standard} \quad [11.1]$$

Although several paleo-temperature equations have been developed for planktic and benthic foraminifers, they are all based on the original temperature equation proposed by Epstein *et al.* (1953):

$$T = 16.5 - 4.3\,(\delta^{18}O^{\,calcite} - \delta^{18}O_{water)} + 0.14\,(\delta^{18}O_{calcite} - \delta^{18}O_{water)})^2 \quad [11.2]$$

where T is temperature (°C), $\delta^{18}O_{water}$ is the oxygen isotopic composition of the seawater in which the organism lived and $\delta^{18}O_{calcite}$ is the oxygen isotopic value of the foraminiferal test calcite. Initially, the change in $\delta^{18}O$ of foraminifers was attributed solely to the temperature of the seawater (Emiliani 1955). In reality, the $\delta^{18}O$ of foraminifers depends on other factors: ice volume, salinity, controlled by river water input, and evaporation–precipitation balance. Advances in understanding and calibration have been made more recently to account for biological effects, including those related to symbiont photosynthesis levels and the effects of carbonate ion variation in seawater (Katz *et al.* 2010). Different species in the same sample may show a shift in $\delta^{18}O$ of up to 1.5‰, a shift attributable to metabolic variations in species-specific isotopic fractionation (the "biological" effects). Large intra-species shifts may also imply different ecological niches (e.g. different calcification depths) (Schiebel and Hemleben 2017).

NOTE.– *Dating methods.*

In order to perform paleoclimatological studies based on sediment cores, it is essential to have a dating method that can transform the depth scale of the sedimentary layers into age. Carbon-14 is used to date the last tens of thousands of years, up to about 50,000 years. This dating is generally done on foraminifer tests.

Beyond 50,000 years, the construction of age scales is based on orbital tuning. It is based on the principle that climatic variations, and in particular the glacial–interglacial oscillations of the Quaternary, are based on cyclical variations in the seasonal and latitudinal distribution of energy from the Sun (insolation). These depend on the variations over time of the position of the Earth in relation to the Sun described by the Milankovitch parameters: eccentricity, obliquity and precession of the equinoxes with periods of 100,000 years, 41,000 years and between 19 and 23,000 years, respectively. The principle of orbital calibration is to correlate a continuous record, considered as a global climate indicator (e.g. a $\delta^{18}O$ curve obtained from foraminifer tests of a core as a function of depth), to a dated reference curve reflecting astronomical forcing (Martinson *et al.* 1987). Various reference chronostratigraphies have been developed using the concept of orbital setting. For example, the LR04 stack curve combines more than 57 benthic $\delta^{18}O$ curves (Lisiecki and Raymo 2005) and covers the last 5.3 million years (Pleistocene–Pliocene).

Intraspecific differences may be related to temperature differences and seasonal variations in hydrological conditions (particularly in mid- and high-latitudes or monsoon-influenced regions). Differences in $\delta^{18}O$ between species in the same sample can then reconstitute the structure of surface waters and seasonal variations in temperature. In addition, during the lifecycle of planktic foraminifers, vertical migration through the water column with a temperature gradient between surface and depth can result in differences in $\delta^{18}O$ between test chambers, depending on their calcification depth. Some species of planktic foraminifers also secrete a layer of calcite near the end of their lifecycle (gametogenetic crust), prior to gamete release, which may have a very different isotopic composition from the rest of the test. New high spatial resolution techniques, such as SIMS (secondary ion mass spectrometry), enable analysis of these $\delta^{18}O$ variations according to the different microstructures of the test.

The $\delta^{18}O$ record of benthic foraminiferal epifaunal tests (living at the water/sediment interface or in the first few centimeters of the surface sediment) is a very good proxy for ice volume and temperature variations of bottom water masses in the past. To untangle the temperature signal from the ice volume signal recorded in the $\delta^{18}O$ of foraminiferal calcite, the current approach is to measure the $\delta^{18}O$ and the Mg/Ca ratio in the same samples (see section 11.5.1) (Katz *et al.* 2010).

11.5.1.3. *Paleo-thermometer based on the Mg/Ca ratio*

The influence of temperature on magnesium (Mg) incorporation in the planktic foraminifer tests was determined using three different calibration methods (Katz *et al.* 2010): (1) analysis of tests from culture experiments where parameters controlling test growth rate are known (e.g. light, temperature, salinity and pH) and tested independently; (2) analysis of live planktic foraminifer tests collected from sediment traps in the water column of sites characterized by significant seasonality in surface water temperature and (3) analysis of fossil planktic foraminifer from sediments of recent core tops.

Magnesium incorporation requires species-specific calibrations. These calibrations suggest an exponential influence of Mg/Ca content in planktic foraminiferal tests with respect to temperature in the form:

$$\text{Mg/Ca (mmol.mol}^{-1}\text{)} = \text{B}e^{\text{AT}} \qquad [11.3]$$

where T is the calcification temperature (K) and B and A are calibration constants. These constants differ between species and depend on the calcification depth of different species (Anand *et al.* 2003), as does the $\delta^{18}O$.

A better understanding of the relationship between Mg/Ca in foraminiferal tests and temperature, as well as the multiple secondary influences on Mg incorporation into foraminiferal calcite, is therefore fundamental to a more robust reconstruction of surface water temperature. A recent calibration of Mg/Ca in tests of the planktic foraminifer *G. ruber*, based on a global-scale sediment trap sample, indicates that not only temperature, but also salinity and carbonate chemistry (pH and $[CO_3^{2-}]$) all have a significant influence on the Mg/Ca ratio of *G. ruber*, making the use of Mg/Ca as a paleo-thermometer more complex (Gray *et al.* 2018).

11.5.2. *Proxies of salinity*

Salinity is one of the key parameters in paleoceanography because, together with temperature, it plays a major role in the density of seawater, which drives ocean circulation, both on a global and regional scale. Despite its importance, there is still no proxy capable of unambiguously reconstructing changes in paleosalinity. Paleosalinity reconstruction can be estimated from combined measurements of stable oxygen isotopes ($\delta^{18}O$) and the Mg/Ca ratio of planktic foraminifers (Elderfield and Ganssen 2000). Oxygen $\delta^{18}O$ in foraminifer tests reflects both the temperature and composition of the ambient seawater, whereas Mg/Ca ratios are primarily controlled by ocean temperature. The uncertainty that exists in the calibration of each of these parameters leads to significant error propagation that makes the results unreliable in most cases.

Therefore, many studies are underway to try to find a new proxy that can measure paleosalinity relatively accurately and independent of temperature. One of the most promising is the measurement of the content of sodium (Na) incorporated in the calcite of foraminifers (Na/Ca ratio). Several experiments have shown a significant positive correlation between seawater salinity and Na/Ca ratio. These experiments were performed on lab cultures of the intertidal benthic foraminifer *Ammonia tepida* (Cushman 1926) and the planktic species *G. ruber* (d'Orbigny 1839, pp. 82–83) pink chromotype and *Trilobatus sacculifer* (Brady 1877), see (Wit *et al.* 2013), (Allen *et al.* 2016) and (Bertlich *et al.* 2018), respectively. There are few studies from the natural environment, but those that have been published show that the amount of sodium incorporated into calcite from foraminifer tests increases with salinity. They involve live planktic foraminifer *G. ruber* (white) and *T. sacculifer* collected alive from surface waters of the Red Sea (Mezger *et al.* 2016). The results of the culture experiments were corroborated by Na/Ca measurements on *T. sacculifer* from Caribbean and Gulf of Guinea surface sediments (Bertlich *et al.* 2018).

All these results suggest that the Na/Ca ratio of foraminifers could potentially provide a new proxy for salinity. However, many additional studies must be carried out. All of the studies discussed here show great variability between measurements made within each area of a test and between different tests within the same sample. At present, the Na/Ca ratio is not used for paleoclimatic studies and reconstruction of the temporal evolution of seawater salinity cannot be satisfactorily documented.

11.5.3. *Paleo-pH proxies and carbonate ion concentration*

More robust reconstructions of pH variations (which depend on CO_2 concentration variation) are crucial to improve understanding of the influence of this greenhouse gas on the Earth's climate (Foster and Rae 2016).

Changes in pH and carbonate ion concentration $[CO_3^{2-}]$ were estimated from size-weighted changes in test weight of planktic foraminifers (Moy *et al.* 2009), with mixed results (Kucera 2007). This method assumes that test weight varies as a function of (1) test weight while the organism was alive, which would be determined solely by surface water $[CO_3^{2-}]$; and (2) post-depositional carbonate loss, which would result exclusively from dissolution, in relation to bottom water $[CO_3^{2-}]$ concentration. However, other factors may influence the weight of fossil tests, which may complicate interpretation solely in terms of $[CO_3^{2-}]$ concentration. Dissolution that may occur during initial deposition, at the water/sediment interface, cannot be distinguished from dissolution in the pore water after burial. Furthermore, the initial thickness of the test wall also depends on other environmental parameters during growth, such as temperature and nutrient availability, making it difficult to establish a direct and unambiguous link between test weight analysis and variations in $[CO_3^{2-}]$.

The ratio between fragments and complete tests of planktic foraminifers is the most commonly used proxy for assessing the intensity of calcite dissolution. It is based on visual identification and counting, usually over 300 particles, of the relative abundance of fragments and intact tests. It has a number of advantages: it is fast and simple to determine. However, each researcher must develop their own criteria for discriminating between fragments and whole tests, making the index highly subjective and difficult to reproduce (Kucera 2007).

11.5.3.1. *Stable boron isotopes ($\delta^{11}B$) and B/Ca ratio*

There are geochemical proxies for pH that are currently in full development, such as boron isotopes ($\delta^{11}B$), derived from the ratio of ^{11}B to ^{10}B isotopes and the B/Ca ratio.

$\delta^{11}B$ is based on the equilibrium between the two dominant boron forms in seawater: boric acid $B(OH)_3$ and borate $B(OH)_4^-$:

$$B(OH)_3 + H_2O \leftrightarrow H^+ + B(OH)_4^- \qquad [11.4]$$

At low pH (<7), almost all of the boron is present as $B(OH)_3$ and at high pH (>10) it is present as $B(OH)_4^-$. The $B(OH)_3$ form is enriched in the heavier isotope, ^{11}B relative to $B(OH)_4^-$. Hemming and Hanson (1992) hypothesized that only

$B(OH)_4^-$ is incorporated into marine carbonates. Thus, the pH of the seawater in which $CaCO_3$ has precipitated is inferred from the $\delta^{11}B$ of $B(OH)_4$. These results have been confirmed by numerous empirical calibration studies based on experimental approaches to growing and precipitating inorganic carbonates in a wide range of seawater pH (Foster and Rae 2016).

The $\delta^{11}B$/pH relationship was calibrated for several planktic foraminifer species through culture experiments, and the impact of photosynthetic activity of symbionts on pH and $\delta^{11}B$ recorded in the test was also quantified under varying light conditions (Foster and Rae 2016).

Analysis of the B/Ca ratio in foraminiferal tests to reconstruct paleo-pH is an emerging approach that may allow for much faster and more numerous analyses than $\delta^{11}B$ analysis (Yu and Elderfield 2007). As with the $\delta^{11}B$ proxy, B/Ca is based on the balance between boric acid and borate in seawater. As previously stated, boron is preferentially incorporated into marine carbonates as borate. Since borate availability is pH dependent, this results in an influence of the pH on the amounts of borate incorporated within carbonates (Hemming and Hanson 1992) and the B/Ca ratio in foraminiferal tests should increase at higher pH. However, surface sediment studies indicate that B/Ca in planktic foraminiferal calcite is also influenced by temperature (Yu *et al.* 2007; Foster 2008). Temperature estimates by Mg/Ca can be used to correct for its influence and extract the seawater pH changes recorded by B/Ca in planktic foraminifers (Yu *et al.* 2007).

In contrast, for benthic foraminifers, the effect of temperature is negligible and the B/Ca ratio shows a linear correlation with the concentration of $[CO_3^{2-}]$ in seawater, suggesting that the B/Ca ratio is a good proxy for assessing past $[CO_3^{2-}]$ (Yu and Elderfield 2007).

11.6. Analytical techniques

The development of new geochemical proxies and the improvement of analytical techniques applied to calcareous foraminifers have multiplied in recent decades, and have improved our ability to reconstruct a wide range of climatic and oceanographic parameters.

The $^{18}O/^{16}O$ ratio in foraminiferal calcite has been measured by mass spectrometry in conjunction with $^{13}C/^{12}C$ for decades. The first isotopic curves date from 1955 (Emiliani 1955). However, technological innovations have reduced the amount of material needed for the analysis and have increased its accuracy. Thus, the $\delta^{18}O$ curve remains the reference proxy in paleoclimatology.

Two classical techniques can be used to determine a wide range of inorganic elements in small samples with extreme reliability and accuracy: optical emission spectrometry (ICP-OES) and inductively coupled plasma mass spectrometry (ICP-MS). Both techniques are used to measure the Mg/Ca ratio of calcite and estimate the temperature of seawater.

A less common but very useful approach is the use of laser ablation coupled with ICP-MS (LA-ICP-MS). This method allows the detection and quantification of chemical elements with minute concentrations (only a few ppm) at several points on the surface of foraminiferal tests. Measurements can be performed with an ablated hole diameter of a few micrometers. Spatial variations in elemental ratios can also be detected by electron probe microanalysis (EPMA) or nanoSIMS (nanometric secondary ion mass spectrometry). SIMS is used to analyze isotopic variations of $\delta^{18}O$ and other elements within the test.

Finally, boron isotopes in calcite can be measured by multi-collector ICP-MS (MC-ICPMS) and by TIMS (thermal ionization mass spectrometry), in negative or positive ions. Because the quantities required are small (100 ng boron), the technique allows for boron isotope analyses from the dissolution of a small number of foraminifers (only 0.1 mg) (Foster 2008).

11.7. Conclusion

Technological advances in the last century have led to the development of the science of paleoclimatology, thanks to which we know more and more about the evolution of the Earth's climate since its formation. On the one hand, these advances have allowed us to improve coring techniques and thus obtain better quality sediment cores. On the other hand, alongside more classical temperature proxies (stable oxygen isotopes, Mg/Ca), the emerging development of new salinity (Na/Ca) and pH ($\delta^{11}B$) proxies, parameters that are sensitive to climate, is crucial for tracing past climate variations from marine sediments. All these proxies are based mainly on geochemical measurements of foraminiferal tests.

Although knowledge of foraminiferal biology and geochemistry has made significant progress in recent years, much work remains to be done to make these new proxies more robust.

11.8. References

Allen, K.A., Hönisch, B., Eggins, S.M., Haynes, L.L., Rosenthal, Y., Yu, J. (2016). Trace element proxies for surface ocean conditions: A synthesis of culture calibrations with planktic foraminifera. *Geochimica et Cosmochimica Acta*, 193, 197–221.

Anand, P., Elderfield, H., Conte, M.H. (2003). Calibration of Mg/Ca thermometry in planktonic foraminifera from a sediment trap time-series. *Paleoceanography*, 18, 1050.

Bartolini, A. and Larson, R.L. (2001). Pacific microplate and the Pangea supercontinent in the Early to Middle Jurassic. *Geology*, 29, 735–738.

Bertlich, J., Nürnberg, D., Hathorne, E.C., de Nooijer, L.J., Mezger, E.M., Kienast, M., Nordhausen, S., Reichart, G.J., Schönfeld, J., Bijma, J. (2018). Salinity control on Na incorporation into calcite tests of the planktonic foraminifera *Trilobatus sacculifer* – evidence from culture experiments and surface sediments. *Biogeosciences*, 15, 5991–6018.

Brady, H.B. (1877). Supplementary note on the foraminifera of the Chalk (?) of the New Britain group. *Geological Magazine*, 4(12), 534–536.

Caley, T., Roche, D.M., Waelbroeck, C., Michel, E. (2014). Constraining the Last Glacial Maximum climate by data-model (iLOVECLIM) comparison using oxygen stable isotopes. *Climate of the Past Discussion*, 10, 105–148.

Cushman, J.A. (1926). Recent foraminifera from Porto Rico. *Publications of the Carnegie Institution of Washington*, 342, 73–84.

d'Orbigny, A. (1826). Tableau méthodique de la classe des Céphalopodes. *Annales des Sciences Naturelles*, 7, 96–169, 245–314 [Online]. Available at: http://biodiversitylibrary.org/page/5753959.

d'Orbigny, A. (1839). Foraminifères. In *Histoire physique, politique et naturelle de l'Ile de Cuba*, de la Sagra, R. (ed.). A. Bertrand, Paris.

Elderfield, H. and Ganssen, G. (2000). Past temperature and $\delta^{18}O$ of surface ocean waters inferred from foraminiferal Mg/Ca ratios. *Nature*, 405, 442.

Emiliani, C. (1955). Pleistocene temperatures. *The Journal of Geology*, 63, 538–578.

Epstein, S., Buchsbaum, R., Lowenstam, H.A., Urey, H.C. (1953). Revised carbonate-water isotopic temperature scale. *Geological Society of America Bulletin*, 64, 1315–1326.

Estival, B. (2003). *Un siècle de navires scientifiques français*. Éditions du Gerfaut, Chartres.

Foster, G.L. (2008). Seawater pH, pCO_2 and $[CO_3^{2-}]$ variations in the Caribbean Sea over the last 130 kyr: A boron isotope and B/Ca study of planktic foraminifera. *Earth and Planetary Science Letters*, 271, 254–266.

Foster, G.L. and Rae, J.W.B. (2016). Reconstructing ocean pH with boron isotopes in foraminifera. *Annu. Rev. Earth Planet. Sci.*, 44, 207–237.

Foster, G.L., Pogge von Strandmann, P.A.E., Rae, J.W.B. (2010). Boron and magnesium isotopic composition of seawater. *Geochemistry, Geophysics, Geosystems*, 11(8), Q08015.

Geistdoerfer, P. (2015). *Histoire de l'océanographie : de la surface aux abysses*. Nouveau monde, Paris.

Gray, W.R., Weldeab, S., Lea, D.W., Rosenthal, Y., Gruber, N., Donner, B., Fischer, G. (2018). The effects of temperature, salinity, and the carbonate system on Mg/Ca in *Globigerinoides ruber* (white): A global sediment trap calibration. *Earth and Planetary Science Letters*, 482, 607–620.

Hachour, H., Szoniecky, S., Abouad, S. (2014). Semio-cognitive spaces: The frontier of recommender systems. In *Recommender Systems*, Kembellec, G., Chartron, G., Saleh, I. (eds). ISTE Ltd, London and John Wiley & Sons, New York.

Haddam, N.A., Michel, E., Siani, G., Cortese, G., Bostock, H., Duprat, J.M., Isguder, G. (2016). Improving past sea surface temperature reconstructions from the Southern Hemisphere oceans using planktonic foraminiferal census data. *Paleoceanography*, 3, 822–837.

Hemming, N.G. and Hanson, G.N. (1992). Boron isotopic composition and concentration in modern marine carbonates. *Geochimica et Cosmochimica Acta*, 56, 537–543.

Imbrie, J. and Kipp, N.G. (1971). A new micropaleontological method for quantitative paleoclimatology: Application to a late Pleistocene Caribbean core. In *The Late Cenozoic Glacial Ages*, Turekian, K.K. (ed.). Yale University Press, New Haven, CT.

Kakihana, H., Kotaka, M., Satoh, S., Nomura, M., Okamoto, M. (1977). Fundamental studies on the ion-exchange of boron isotopes. *Bulletin of the Chemical Society of Japan*, 50, 158–163.

Katz, M.E., Cramer, B.S., Franzese, A., Hönisch, B., Miller, K.G., Rosenthal, Y., Wright, J. (2010). Traditional and emerging geochemical proxies in Foraminifera. *Journal of Foraminiferal Research*, 40(2), 165–192.

Kucera, M. (2007). Planktonic Foraminifera as tracers of past oceanic environments. *Developments in Marine Geology*, 1, 213–262.

Lisiecki, L.E. and Raymo, M.E. (2005). A Pliocene-Pleistocene stack of 57 globally distributed benthic $\delta^{18}O$ records. *Paleoceanography*, 20, PA1003.

Martinson, D.G., Pisias, N.G., Hays, J.D., Imbrie, J., Moore, T.C., Shackleton, N.J. (1987). Age dating and the orbital theory of the ice ages: Development of a high-resolution 0 to 300,000-year chronostratigraphy. *Quaternary Research*, 27(1), 1–29.

Mezger, E.M., de Nooijer, L.J., Boer, W., Brummer, G.J.A., Reichart, G.J. (2016). Salinity controls on Na incorporation in Red Sea planktonic foraminifera. *Paleoceanography*, 31, 1562–1582.

Moy, A.D., Howard, W.R., Bray, S.G., Trull, T.W. (2009). Reduced calcification in modern Southern Ocean planktonic foraminifera. *Nature Geoscience*, 2, 276–280.

Prell, W.L. (1985). The stability of low-latitude sea-surface temperatures: An evaluation of the CLIMAP reconstruction with emphasis on the positive SST anomalies. Final report, Brown University, Providence, RI.

Sakamoto, T., Ikehara, M., Uchida, M., Aoki, K., Shibata, Y., Kanamatsu, T., Harada, N., Iijima, K., Katsuki, K., Asahi, H., Takahashi, K., Sakai, H., Kawahata, H. (2006). Millennial-scale variations of sea-ice expansion in the southwestern part of the Okhotsk Sea during the past 120 kyr: Age model and ice-rafted debris in IMAGES Core MD01-2412. *Global and Planetary Change*, 53, 58–77.

Schiebel, R. and Hemleben, C. (2017). *Planktic Foraminifers in the Modern Ocean*. Springer-Verlag, Berlin Heidelberg.

Thouveny, N., Moreno, E., Delanghe, D., Candon, L., Lancelot, Y., Shackleton, N.J. (2000). Rock magnetic detection of distal ice-rafted debries: Clue for the identification of Heinrich layers on the Portuguese margin. *Earth and Planetary Science Letters*, 180, 61–75.

Wit, J.C., de Nooijer, L.J., Wolthers, M., Reichart, G.J. (2013). A novel salinity proxy based on Na incorporation into foraminiferal calcite. *Biogeosciences*, 10, 6375–6387.

Yu, J. and Elderfield, H. (2007). Benthic foraminiferal B/Ca ratios reflect deep water carbonate saturation state. *Earth and Planetary Science Letters*, 258, 73–86.

Yu, J., Elderfield, H., Hönisch, B. (2007). B/Ca in planktonic foraminifera as a proxy for surface seawater pH. *Paleoceanography*, 22(2), PA2202.

12

Clarifying the Radiocarbon Calibration Curve for Ancient Egypt: The Wager of Herbaria

Anita QUILES[1], Vanessa R. INVERNÓN[2], Lucile BECK[3], Emmanuelle DELQUE-KOLIC[3], Myriam GAUDEUL[2], Serge MULLER[2] and Germinal ROUHAN[2]

[1] *IFAO, Cairo, Egypt*

[2] *ISYEB – CNRS, Muséum national d'Histoire naturelle, Sorbonne Université, École Pratique des Hautes Études, Université des Antilles, Paris, France*

[3] *LMC14 – LSCE/IPSL, CEA – CNRS – UVSQ, Université Paris-Saclay, Gif-sur-Yvette, France*

Taken as a chronometer, ^{14}C is a radioisotope particularly used in archeology to date ancient objects. In Egypt, the "peculiarity of the Nile", with the regular floods that occurred on a recurrent basis before the construction of the Aswan dams, could have been at the origin of seasonal effects that caused the ^{14}C content of plants to fluctuate on regional scales, leading to deviations from the reference ^{14}C calibration curve. Thus, through the study of plants collected before the construction of dams and preserved in herbaria, our project is interested in reproducing the ^{14}C content in the atmosphere in Egypt during the 18th and 19th centuries. This is in order to evaluate possible deviations from the international IntCal calibration curve. The dating issue concerns archeological objects and is crucial for Egypt, but beyond Egyptian history, it is also a question of reproducing the past times of all the neighboring

For a color version of all figures in this chapter, see www.iste.co.uk/pellens/natural.zip.

Natural History Collections in the Science of the 21st Century, coordinated by Roseli PELLENS.

civilizations of the eastern Mediterranean basin of the 3rd and 2nd millennia, whose chronologies are mainly fixed by synchronisms with Egypt.

12.1. Introduction

Herbaria are collections of plants pressed and dried to be preserved and studied; the plants are therefore usually fixed on cardboard materials to enable repeated observations and perennial conservation. The first herbaria were constructed in Italy in the 16th century, and their creation has continued ever since, with various objectives. They are of fundamental interest for the knowledge, description and classification of plant species (taxonomy and systematics), but can also be tools of great interest for other disciplines and applications. The indications of the date and locality of collection associated with each herbarium specimen make them scientific objects, precious and often unique witnesses to the environmental conditions contemporary with their collection (see Chapter 16 by Muller *et al.*).

In Egypt, the construction of the Aswan dams in the 1960s has completely changed the landscape. Thus, the environment in which the plants develop today has nothing to do with the periods during which the flooding of the Nile managed the cycle of the seasons. This ancient "Nile peculiarity" could have been at the origin of seasonal effects that caused the ^{14}C content of plants to fluctuate on regional scales, leading to deviations from the reference calibration curve, IntCal, which is now internationally shared (see the note below on the principle of the radiocarbon dating method). The herbaria, witnesses to these past environments which have now disappeared, allow us to specify this radiocarbon calibration curve for Egypt since they contain well-dated specimens collected before the construction of the dam. The refinement of this curve for the Egyptian territory represents a critical issue for the understanding of a historical period of several millennia, as dense and complex as that of ancient Egypt.

This multidisciplinary project was carried out in close collaboration between the *Institut français d'archéologie orientale du Caire* (Ifao), the *Muséum national d'Histoire naturelle de Paris* (MNHN) and the *Laboratoire de mesure du carbone 14* at the Université Paris-Saclay (LMC14). It aimed at reproducing the ^{14}C content in the atmosphere in Egypt during the 18th and 19th centuries, in order to evaluate possible deviations from the IntCal calibration curve, and to identify the magnitude of these ^{14}C variations according to the geographical regions of Egypt.

12.2. Carbon-14 (^{14}C) dating and Egyptian chronology

12.2.1. *The challenge of calibration*

The applicability of the IntCal13 curve over the entire planet is debated. In his original development of the dating method (see note below), W. Libby had assumed that the atmospheric ^{14}C content was constant over time and could therefore be determined from a contemporary measurement (Arnold and Libby 1949), and then applied for all ancient periods.

NOTE.– *Principle of the radiocarbon dating method.*

Carbon-14 (^{14}C) is a radioactive isotope of carbon, which is produced mainly in the stratosphere by the collision of cosmic protons with nitrogen and oxygen molecules. The ^{14}C formed diffuses and oxidizes into atmospheric carbon dioxide $^{14}CO_2$, which is assimilated by plants through photosynthesis. Every living organism has a ^{14}C content equal to that of the environment in which it lives, corrected to the biological fractionation. When the organism dies, exchanges with its environment cease and its ^{14}C content decreases over time. Evaluating the natural residual ^{14}C content of a dead organism makes it possible to calculate the time that has elapsed since its death by comparing the $^{14}C/^{12}C$ ratio of the dead organism with that of organisms of known ages.

Since the 1950s and the formulation of this dating principle by Willard Libby, Nobel Prize winner in Chemistry (Libby 1960), the radiocarbon (^{14}C) dating method has become the most widely applied dating method in archeology. However, it cannot be considered a "direct" absolute dating method, because it is impossible to measure the minute amount of radiogenic $^{14}N^*$ produced by the decay of ^{14}C, in the presence of a much larger amount of stable ^{14}N on the Earth (Szabo et al. 1997). Thus, ^{14}C dates, expressed in BP – a concept specific to radiocarbon and meaning Before Present (the present being fixed at 1950 of the common era (CE)) – are not "true" ages but "radiocarbon" ages. They are converted into "calendar ages" by means of a so-called "calibration" curve which traces the evolution of the ^{14}C content of the atmosphere over time (Figure 12.1). Developed by a group of international researchers, the IntCal curve (Reimer et al. 2013) is now a reference and covers the entire period of applicability of the method, that is the last 50,000 years. The "calibrated" age intervals are expressed in calBCE (calibrated Before Common Era) or calCE (calibrated Common Era).

De Vries (1958) quickly challenged this hypothesis, which led to the development of a calibration curve (Suess 1970) constructed from materials collected from different locations on the planet (wood rings, corals, stalagmites, lake sediments, etc.) and otherwise dated by other analytical methods such as dendrochronology or decay in uranium families (now IntCal13; Reimer *et al.* 2013).

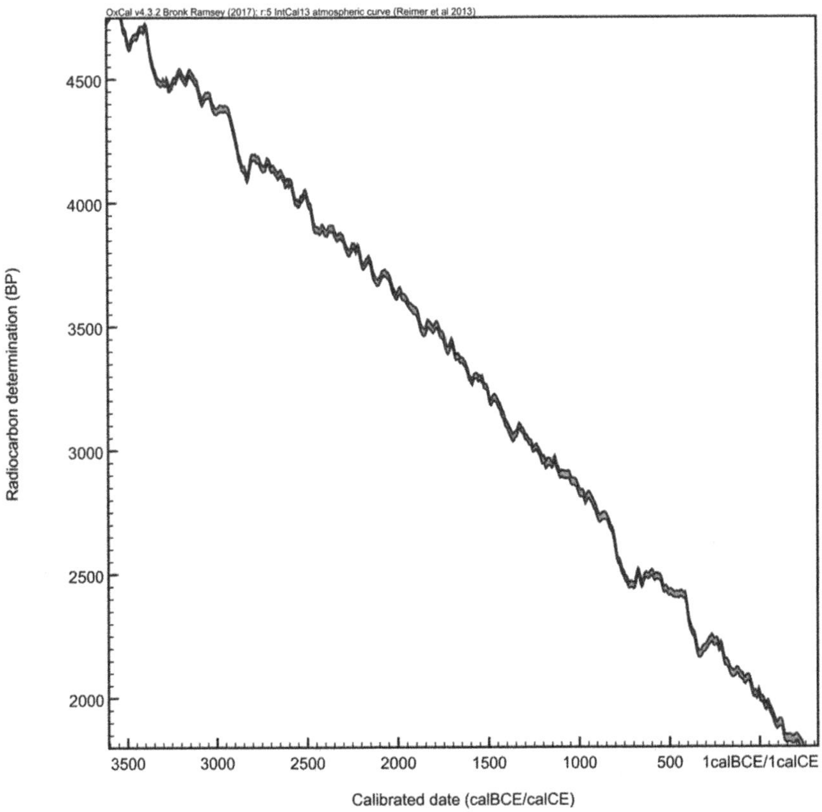

Figure 12.1. *Example of a portion of the* ^{14}C *IntCal13 calibration curve (Reimer* et al. *2013), spanning from 3500 BCE to 1 BCE/CE (ordinates plot radiocarbon ages expressed in BP, abscissa calendar ages expressed in calBCE/calCE)*[1]

Applicable to any organic material less than 50,000 years old (straw, plants, rope, wood, charcoal, shells, etc.), radiocarbon dating now has precise measuring instruments, accelerator mass spectrometers (AMS), and requires only a few milligrams of material. The challenges of its application vary according to the periods invested. At the prehistoric time scale, this method is considered accurate because it offers the possibility of reconstructing chronological scenarios of several millennia with an accuracy of only a few centuries (Quiles *et al.* 2016). Used until

1 Available at: https://c14.arch.ox.ac.uk/oxcal/OxCal.html.

the end of the Middle Ages, regarding the scale of historical times, its application is more difficult as soon as a precision lower than a century is sought. It is then necessary to identify all the methodological pitfalls and all the sources of error in the analysis chain, linked as much to the probabilistic nature of the physical phenomenon at stake (radioactivity), as to the measuring instruments used, and to the artifacts of the methods. In particular, the calibration of ages is an immediate issue that definitively conditions the response of the ^{14}C method to singular historical questions.

12.2.2. *Chronology of ancient Egypt: contribution of ^{14}C and historic debates*

The suggested dates to establish the chronology of Pharaonic Egypt vary greatly from one author to another, especially for the ancient periods. Although Egyptologists have organized themselves around great periods of Egyptian history, with a "prehistory" that ranges from about 20,000 years ago until the birth of writing around 3200 BCE, then a succession of great empires (Ancient, Middle and New) separated by three intermediate periods before Egypt definitively passed under foreign domination with the conquest by Alexander the Great in 331 BCE. To date, there is no consensus on how to reconstruct a detailed chronology of the history of ancient Egypt, reign by reign (Shaw 2000; Hornung *et al.* 2006). Yet the challenge remains, because beyond Egyptian history, it is also a question of reconstructing the past times of all the neighboring civilizations of the eastern Mediterranean basin of the 3rd and 2nd millennia, whose chronologies are mainly determined by synchronisms with Egypt.

As soon as the ^{14}C dating method was developed, the first analyses were carried out on Egyptian objects (Berger 1970; Derricourt 1971), highlighting significant discrepancies between historical and radiocarbon chronologies (Hassan and Robinson 1987). This was also shown by a large study conducted between 1984 and 1995 on monuments of the Old Kingdom and the Middle Kingdom, with a gap of 200–300 years for the 4th dynasty (Bonani *et al.* 2001)[2]. The first ^{14}C chronological modeling covering the entire Pharaonic period[3] dates from 2010 (Bronk Ramsey *et al.* 2010; Shortland and Bronk Ramsey 2013), completed by Dee *et al.* (2013) for the Predynastic period. Conducted exclusively on samples from museum collections, these chronological models also mark significant discrepancies with expected

2 Pyramids Radiocarbon Dating Project.

3 This chronology goes from the Old Kingdom to the 21st dynasty, that is to say from about 2800 BCE to ~1000 BCE.

historical data for the earliest periods. In the current state of research, no chronology is definitively accepted.

Thus, reproducing an absolute chronology of ancient Egypt is a strong and current challenge. In order to address this challenge, the archeometry department of the *Institut français d'archéologie orientale du Caire* (Ifao) has initiated a research program[4] that aims, through a global approach mixing the fields of archeometry, ancient history and archeology (Quiles *et al.* 2013), to model a chronology of ancient Egypt based mainly on the ^{14}C dating of archeological artifacts associated with closed excavation contexts. This work concerns both objects from museum collections and artifacts found in context during archeological excavations currently being conducted throughout the country. Among the pitfalls of methods that can lead to uncertainties on ages, the question of the fine calibration of datings made on samples coming from Egypt is dominant; this constitutes one of the program's major axes of study.

12.3. Specificities of the Egyptian landscape and the objective of the project

While Egypt is a *gift of the Nile*, as Herodotus defined it, the geography and environment of Egypt are atypical because of the Nile, which provides water to an otherwise desert territory. Four types of landscape exist side by side: the Nile valley with its green landscapes, the delta with its swamps in the north, the desert with its oases and the Red Sea coast, defining four unique atmospheric and climatic conditions (Figure 12.2). Until the construction of the Aswan Dam, the annual floods caused significant and specific seasonal variations according to the periods and their magnitudes, and the growing season of the plants was therefore necessarily conditioned by these floods.

This environmental feature of the Nile prior to the construction of the Aswan Dam in the 1960s was the source of possible seasonal effects that caused the ^{14}C content of plants to fluctuate on regional and/or temporal scales, leading to possible deviations from the universal IntCal calibration curve. The work of Dee *et al.* (2010) highlighted an initial shift of (19 ± 5 ^{14}C years) from the expected IntCal ages, based on a study of plants collected throughout Egypt between 1700 and 1900 CE, now held at the University of Oxford Herbaria and the Natural History Museum in London. Manning *et al.* (2014) showed through a dendro-chronological study conducted on a sequence of cedar wood from a funereal boat attributed to the reign of Sesostris III

4 This program has recently been selected by the ANR "Modélisation chronologique de l'Ancien Empire égyptien", MERYT, ANR-19-CE27-0010.

(Middle Kingdom, ~1800 BCE) a possible 22-year gap for the period 2200–1900 BCE. This could be due to an exaggeration of seasonal variations in ^{14}C absorption, due to a climatic change identified 4,200 years ago. Finally, a recently published study highlights a shift of about 19 years in the Southern Levant region around the late Bronze Age/early Iron Age (1200–600 BCE), which would have varied in ^{14}C content over time (Manning *et al.* 2018).

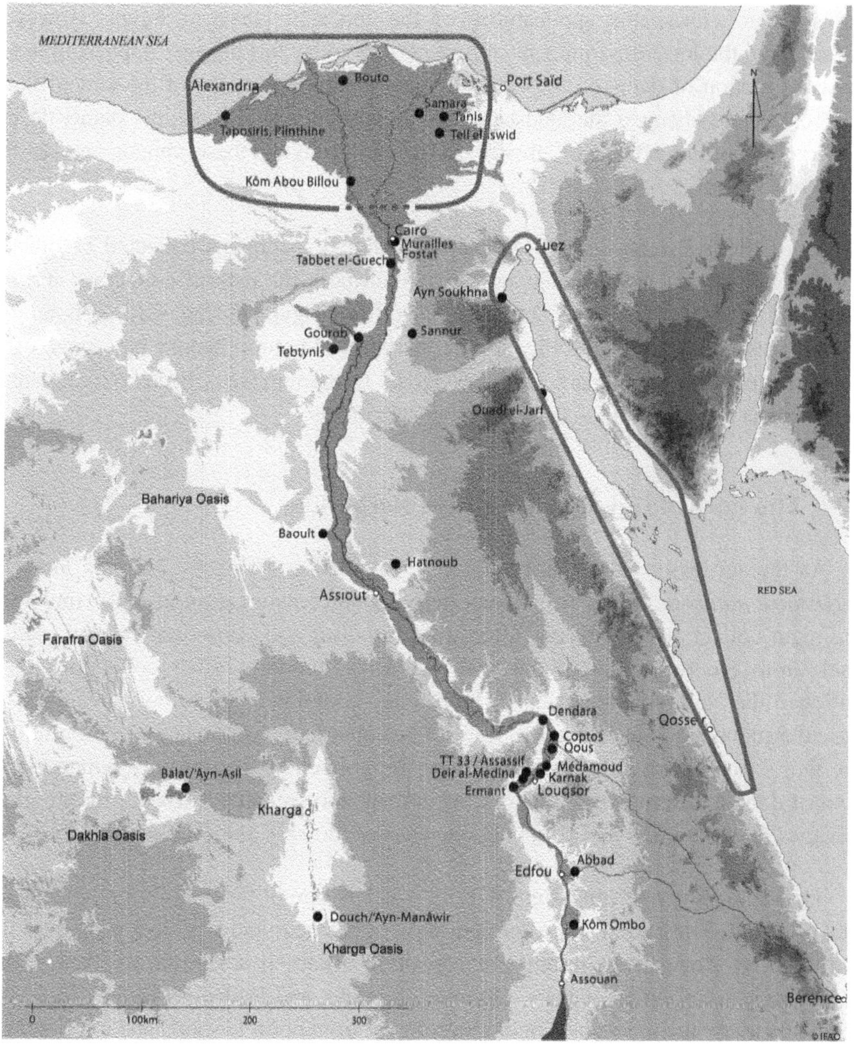

Figure 12.2. *Egypt and the four landscape types. © Ifao*

COMMENT ON FIGURE 12.2.– *The Nile valley and its green landscapes (shaded), the delta and its swamps to the north (framed in red), the Red Sea coast (framed in green) and the desert and its oases (most of the rest of the territory); the reliefs appear in shades of yellow (low altitudes) to brown (high altitudes). The current Ifao excavation missions are marked by a black dot with the locality name.*

To date Egyptian artifacts with precision using the ^{14}C dating method, it is necessary to ensure the applicability of the IntCal13 curve in Egypt. However, reproducing the local environment of past periods today has become impossible since the construction of dams on the Nile, because the disappearance of the annual flood has completely changed the landscape. As a consequence, present-day plants do not undergo the same effects of growth shifts and studying these cannot answer the questions we have.

It is therefore necessary to search for botanical specimens collected before the construction of the Aswan dams and for which both the date of collection and the place are known. Specimens, brought back from Egypt during various botanical expeditions and preserved in the national herbaria of natural history museums, offer a unique field of study. In particular, the MNHN Herbarium (see Chapter 6 by Invernón *et al.*) preserves several thousand Egyptian specimens.

12.4. The flora of Egypt in the MNHN Herbarium

As the successor to the Jardin du Roy, founded in 1635, the *Muséum national d'Histoire naturelle de Paris* has kept herbaria since its creation in 1793, including the main collections of Joseph Pitton de Tournefort and Sébastien Vaillant, dating back from the beginning of the 18th century. Numerous other collections have enriched the National Herbarium, which today contains eight million specimens of vascular plants, mosses, algae and fungi. The constitution of the largest herbarium in the world has historically benefited from the collections of not only botanists but also of doctors, missionaries, sailors, naturalists and soldiers, often during the great voyages of exploration and colonial conquests of the 18th and 19th centuries.

Thus, the Egyptian campaign of 1798–1801 is one of those military expeditions that constituted a windfall for a "commission of sciences and arts" that supported the army led by Bonaparte. Among the 150 or so artists or scientists involved in this commission were three botanists, Montbret, Nectoux and Raffeneau-Delile. The latter was particularly active in traveling through Egypt from north to south, and in bringing a number of plant specimens into the herbarium. Based on collection and observations, 1,030 species from the country are listed in the monumental collective work *Description de l'Égypte, ou recueil des observations et des recherches qui ont été faites en Égypte pendant l'expédition de l'armée française*; some of the plants new to science were described for the first time by Delile (Figure 12.3) and are among the 2,145 species of native or naturalized plants now known from Egypt (Boulos 1999, 2009).

In addition to the Raffeneau-Delile collection, which is the most significant and the one that provides the oldest floristic data for Egypt, the MNHN Herbarium preserves some other important herbaria for knowledge of the country's flora in the 19th century: for example, the collections of C.G. Ehrenberg (1820s), P.M.R. Aucher-Eloy, J.N. Bové and G.W. Schimper (1830s), E. Boissier (1840s), J.-L. Kralik (1850s–1880s), P. Ascherson (1870s), A. Letourneux and G.A. Schweinfurth (1880s).

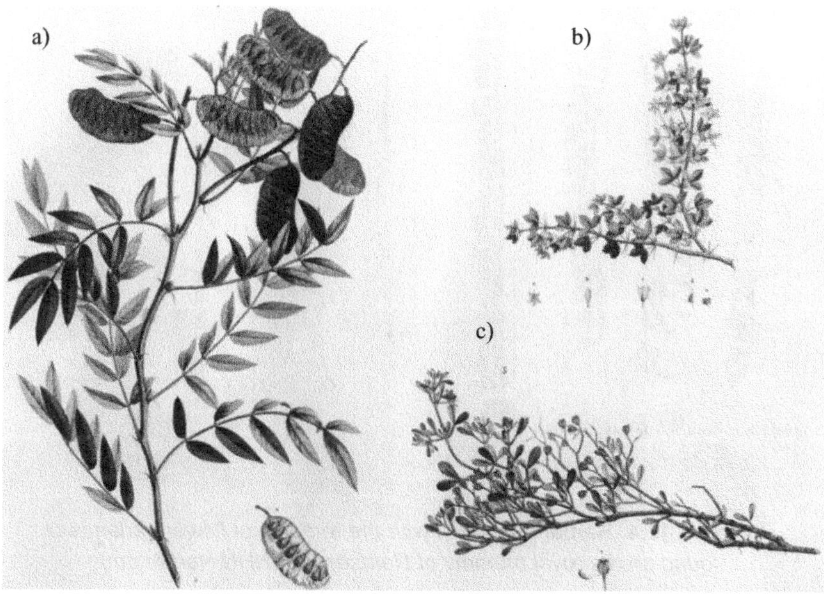

Figure 12.3. *Illustration plate of three species new to science, described by Raffeneau-Delile in the collective work* Description de l'Égypte*: a)* Cassia acutifolia *Delile (=* Senna alexandrina *Mill), b)* Fagonia mollis *Delile, c)* Zygophyllum decumbens *Delile*

However, perhaps the most famous collections of dried plants at the MNHN Herbarium are the remains of flower garlands found on the royal mummy of Rameses II: these garlands were attached to herbarium boards in the 19th century, but the plants are of course among the oldest in the world (Figure 12.4). It is known that these flowers are not the original flowers in the pharaoh's tomb, but those that were placed in the 21st dynasty (ca. 1090–950 BCE) to replace them. The value of these objects, which are more than 3,000 years old, is not only historical, but also scientific, since all the plants making up the garlands have been identified (Schweinfurth 1883) and thus document the flora of ancient Egypt, with some species that are now extinct in the country, such as *Mimusops laurifolia* (Forssk.) Friis (Morat 2004).

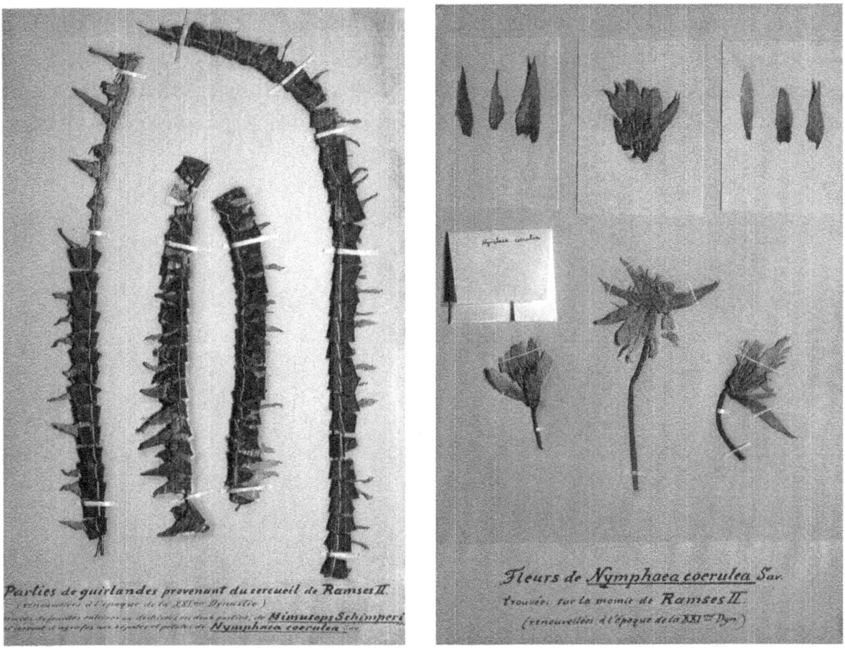

Figure 12.4. *Herbarium plates with the remains of flower garlands found on the royal mummy of Ramses II (MNHN Herbarium)*

12.5. Analytical and statistical challenges

This project aims to identify possible shifts in the IntCal13 calibration curve in Egypt. To do so, it is necessary to evaluate the residual ^{14}C content of botanical specimens preserved in the MNHN Herbarium, and whose year and place of collection in Egypt are specified. The challenge is both to identify appropriate specimens to build coherent datasets that can interrogate the different fields of investigation of the project and to model systematic and statistical differences between the ^{14}C content of Egyptian plants and those of other organisms (mainly trees) integrated in the IntCal calibration curve, for a specific year.

12.5.1. *Selection of herbarium specimens*

12.5.1.1. *Constraints imposed by the collections*

The SONNERAT database of the MNHN Herbarium (see Chapter 6 by Invernón *et al.*) lists 1,760 herbarium plates of plants collected in Egypt. Since only about 15% of the herbarium specimens are computerized with regard to "country" data, the number of herbarium plates of plants from Egypt in the collection is actually much higher, even if their location requires a specimen-to-specimen search in the midst of all the plants from Africa. Thus, the MNHN Herbarium contains an exceptional wealth of material for our study, but one difficulty is to select a sufficient number of samples that simultaneously meet the following three criteria: (1) precise data for the year and locality of harvest (ancient harvests often carry much less precise harvest information than those of today), (2) a verified species determination and (3) species representative of different environmental environments or territories (valley, desert, delta, oasis, Red Sea, etc.)

12.5.1.2. *Constraints imposed by the ^{14}C dating method*

From a methodological point of view, samples collected in Egypt after 1850 are excluded because they are potentially impacted by an atmospheric ^{14}C dilution effect due to CO_2 releases derived from $^{14}CO_2$-depleted fossil fuels, called the Suess effect (Suess 1955). Similarly, materials contaminated with glue or other binding materials are not selectable. It is well known, moreover, that plants in the herbaria of our study period were often poisoned to prevent insect infestation, i.e. treated with toxic solutions of mercury chloride ($HgCl_2$) dissolved from organic solutions whose mixtures are not clearly known (DeWolf 1968). Samples treated in this way may therefore have been contaminated with organic residues after evaporation of the solvent. In most cases, the treatment of a specimen is documented and contentious cases are systematically removed.

12.5.2. *Preliminary results of ^{14}C dating*

Measurements of the residual ^{14}C content of the sampled specimens were carried out at the *Laboratoire de mesure du carbone 14*, CEA Saclay, using the AMS method on the Pelletron 3 MV- Artemis installation.

The first study focused on a selection of samples from the collections made by Raffeneau-Delile during the Egyptian expedition between 1798 and 1801 (Figure 12.5). A total of 27 specimens representing 24 species belonging to 12 different botanical families were collected and measured. This initial sampling met two criteria, an accurate and early date of harvest and plants representative of the diversity of the Egyptian landscape. Thus, the great majority of the plants sampled were collected in 1798 or 1799 during the Egyptian campaigns, covering a large geographical area: surroundings of Cairo, Isthmus of Suez, Nile delta, desert (Arabian, suez, natrou), Nile valley (Luqsor) and Nubia.

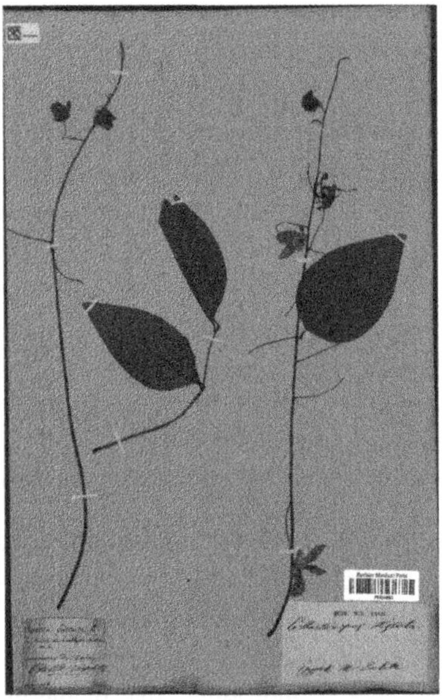

Figure 12.5. Cassia fistulosa *L., collected by Raffeneau-Delile in 1798, near Cairo (P03344563, MNHN Herbarium)*

Samples of 2–3 mg of material were taken following standard protocols using sterile scalpels, and underwent the preparation steps described by Dumoulin *et al.* (2017): they were subjected to a succession of treatments with acidic and basic solutions to remove any contamination derived from carbonates and humic materials, rinsed with ultrapure water, dried and then burned in the presence of copper oxide at 835°C for 5 hours so as to produce approximately 1 mg of carbon in the form of carbon dioxide (CO_2). The carbon dioxide was finally reduced by hydrogen at 600°C in the presence of iron powder to form graphite, compacted into aluminum targets for measurement by AMS.

The ^{14}C content of the samples were obtained by measuring the intensities of the ^{14}C, ^{13}C and ^{12}C beams formed in the AMS and then compared with the ^{14}C content of the international oxalic acid standard (OxII, SRM 4990C) measured in the same sequence as the samples. These concentration ratios were then corrected for the ^{13}C

isotopic fractionation of the sample measured on the installation, and the ^{14}C background due to the chemical preparation steps. The ^{14}C ages of the samples were calculated using the radioactive decay law that relates the ^{14}C concentration ratio of the sample to the time since death (Mook and van der Plicht 1999).

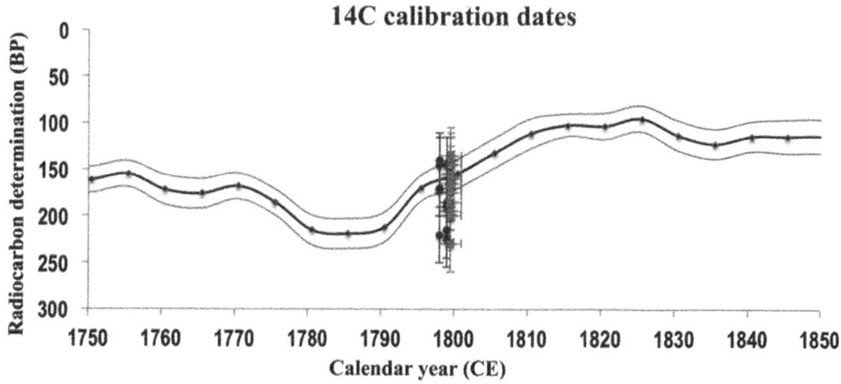

a)

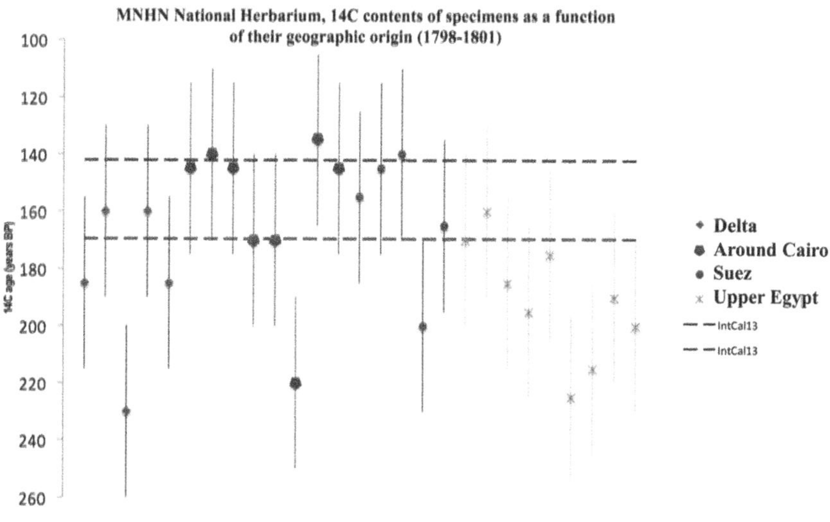

b)

Figure 12.6. *a) Comparison of ^{14}C contents measured on herbarium specimens kept at the MNHN Herbarium and collected in 1798 and 1799 (in purple and red), with the expected values on IntCal13 (in black); b) evolution of ^{14}C contents of specimens according to their geographic origin*

The ^{14}C values measured for each sample were grouped by harvest year, with the majority clustering around 1798 and 1799, for comparison with the expected IntCal values. These results were consistent, yet relatively scattered, potentially indicating a slight shift of about 20 ^{14}C years on average from the values recorded in IntCal for the years 1798 and 1799 (Figure 12.6(a)). Nevertheless, when these results were reorganized according to the geographic origin of the specimens and not the year of harvest (Figure 12.6(b)), it appeared that the majority of the results that were scattered to the expected IntCal values came from a particular geographic sector, in this case the banks of the Nile. However, the number of analyses is not yet sufficient to draw definitive conclusions.

Based on these first observations, a second sampling was carried out at the end of 2018 in the MNHN Herbarium. We sought above all to produce a substantial set of statistical data on a maximum of annual plants (guaranteeing an absorption of atmospheric carbon throughout the year) well identified to the species and also as an archeological object (site of harvest, year, context, botanist). In this sense, 20 samples are currently being analyzed and will quickly complete these first observations.

12.6. Conclusion

The success of this project depends above all on our ability to carry out an optimal sampling to allow a reasoned statistical study. In the medium term, we plan to build a set of more than a hundred analyses, on specimens covering homogeneously the regional sectors and the periods invested. It will be a question, on the one hand, of constructing statistical analytical sets for a few precise years and investing in all the territories, as we have begun to do for 1798 and 1799, and on the other hand, of evaluating the fluctuations of the ^{14}C content on a short time scale, for a given annual species, and by geographical sector. Wherever possible, the quality and extent of sampling should be replicated from one geographic area to another to ensure that equivalent datasets are compared. It is clear that the collections contain more plants from the Nile valley than from the oasis areas. These parameters must be integrated upstream of the collection and analysis of specimens, in order to avoid inducing sampling bias. This requirement in the constitution of the series of study specimens constitutes a major challenge for this project.

A priori, we do not know to what extent any gaps will be identified, nor what their magnitudes will be. The statistical stakes are high. For the ancient Egyptian period, the uncertainties associated with ^{14}C dating are on average of the order of 25 ^{14}C years (1σ). It is therefore very likely that the discrepancies modeled in IntCal are of the same order of magnitude, as shown by the first results. Analyzing homogeneous sample sets that are sufficient to achieve reasonable error margins is essential to reduce the errors associated with these modeled discrepancies.

Finally, this project aims to conclude with precision on the existence of deviations from the IntCal curve for the calibration of ^{14}C dates made in Egypt. This work nevertheless implies testing the relevance of a strong underlying hypothesis: can we transcribe the possible variations observed in the 17th and 19th centuries to the ancient periods (up to 3000 BCE)? Because of the dependence and regularity of the annual flooding of the Nile during the last millennia before the construction of the dam, the hypothesis of considering that the variations observed at the end of the 17th century and beginning of the 19th century are valid over the long term is not aberrant, but must be confirmed. To do this, we plan to work in parallel on Graeco-Roman *papyri* whose year of writing is known, and to compare the ^{14}C contents of plants and shrubs that were used in the manufacture of objects found on site, associated with the same historical periods (same reigns) but having grown in different environments.

12.7. References

Arnold, J.D. and Libby, W.F (1949). Age determinations by radiocarbon content: Checks with samples of known age. *Science*, 110, 678–680.

Berger, R. (1970). Ancient Egyptian radiocarbon chronology. *Philosophical Transactions of the Royal Society A*, 269, 23–36.

Bonani, G., Hass, H., Hawass, Z., Lehner, M., Nakhla, S., Nolan, J., Wenke, R., Wölfli, W. (2001). Radiocarbon dates of old and middle kingdom monuments in Egypt. *Radiocarbon*, 43(3), 1297–1320.

Boulos, L. (1999). *The Flora of Egypt*, Volumes 1–4. Al Hadra Publishing, Cairo.

Boulos, L. (2009). *Flora of Egypt. Check List.* Al Hadra Publishing, Cairo.

Bronk Ramsey, C., Dee, M.W., Rowland, J.M., Higham, T.F.G., Harris, S.A. (2010). Radiocarbon-based chronology for dynastic Egypt. *Science*, 328, 1554–1557.

De Vries, H. (1958). Variation in concentration of radiocarbon with time and location on Earth. *Proceedings of the Koninklijke Nederlandse Akademie van Wettenschapen*, 61, 94–102.

Dee, M.W., Brock, F., Harris, S.A., Bronk Ramsey, C., Shortland, A.J., Higham, T.F.G., Rowland, J.M. (2010). Investigating the likelihood of a reservoir offset in the radiocarbon record for ancient Egypt. *Journal of Archaeological Science*, 37, 687–693.

Dee, M., Wengrow, D., Shortland, A., Stevenson, A., Brock, F., Linus, G.F., Christopher, B.R. (2013). An absolute chronology for early Egypt using radiocarbon dating and Bayesian statistical modelling. *Proceedings of the Royal Society A*, 469, 20130395.

Derricourt, R.M. (1971). Radiocarbon chronology for Egypt and North Africa. *Journal of Near Eastern Studies*, 30(4), 271–292.

DeWolf Jr., G.P. (1968). Notes on making a Herbarium. *Arnoldia*, 28, 69–111.

Dumoulin, J.-P., Comby-Zerbino, C., Delqué-Količ, E., Moreau, C., Caffy, I., Hain, S., Perron, M., Thellier, B., Setti, V., Berthier, B., Beck, L. (2017). Status report on sample preparation protocols developed at the LMC14 Laboratory, Saclay, France: From sample collection to ^{14}C AMS measurement. *Radiocarbon*, 59, 713–726.

Hassan, F.A. and Robinson, S.W. (1987). High-precision radiocarbon chronometry of ancient Egypt, and comparisons with Nubia, Palestine and Mesopotamia. *Antiquity*, 61(231), 119–135.

Hornung, E., Krauss, R., Warburton, D. (eds) (2006). *Ancient Egyptian Chronology: Handbook of Oriental Studies*. Brill, Leiden, and Boston, MA.

Libby, W.F. (1960). Radiocarbon dating, Nobel Lecture [Online]. Available at: https://www.nobelprize.org/prizes/chemistry/1960/libby/lecture/.

Manning, S.W., Dee, M.W., Wild, E.M., Bronk Ramsey, C., Bandy, K., Creasman, P.P., Griggs, C.B., Pearson, C.L., Shortland, A.J., Steier, P. (2014). High-precision dendro-^{14}C dating of two cedar wood sequences from First Intermediate Period and Middle Kingdom Egypt and a small regional climate-related ^{14}C divergence. *Journal of Archaeological Science*, 46, 401–416.

Manning, S.W., Griggs, C., Lorentzen, B., Bronk Ramsey, C, Chivall, D., Timothy Jull, A.J., Lange, T.E. (2018). Fluctuating radiocarbon offsets observed in the southern Levant and implications for archaeological chronology debates. *PNAS*, 115(24), 6141–6146.

Mook, W.G. and van der Plicht, J. (1999). Reporting ^{14}C activities and concentrations. *Radiocarbon*, 41(3), 227–239.

Morat, P. (2004). Les guirlandes de Ramsès II. In *L'Herbier du monde : cinq siècles d'aventures et de passions botaniques au Muséum d'histoire naturelle*, Morat, P., Aymonin, G.G., Jolinon, J.-C. (eds). L'Iconoclaste, Paris.

Quiles, A., Aubourg, E., Berthier, B., Delque-Količ, E., Pierrat-Bonnefois, G., Dee, M.W., Audreu-Lanoë, G., Bronk Ramsey, C., Moreau, C. (2013). Bayesian modelling of an absolute chronology for Egypt's 18th Dynasty by astrophysical and radiocarbon methods. *Journal of Archaeological Science*, 40, 423–432.

Quiles, A., Valladas, H., Bocherens, H., Delqué-Količ, E., Kaltnecker, E., van der Plicht, J., Delannoy, J.J., Feruglio, V., Fritz, C., Monney, J., Philippe, M., Tosello, G., Clottes, J., Geneste, J.-M. (2016). A high-precision chronological model for the decorated Upper Paleolithic cave of Chauvet-Pont d'Arc, Ardèche, France. *PNAS*, 113(17), 4670–4675.

Reimer, P.J., Bard, E., Bayliss, A., Beck, J.W., Blackwell, P.G., Bronk Ramsey, C., Buck, C.E., Cheng, H., Edwards, R.L., Friedrich, M., Grootes, P.M., Guilderson, T.P., Haflidason, H., Hajdas, I., Hatté, C., Heaton, T.J., Hoffmann, D.L., Hogg, A.G., Hughen, K.A., Kaiser, K.F., Kromer, B., Manning, S.W., Niu, M., Reimer, R.W., Richards, D.A., Scott, E.M., Southon, J.R., Staff, R.A., Turney, C.S.M., van der Plicht, J. (2013). IntCal13 and Marine13 radiocarbon age calibration curves 0–50,000 years cal BP. *Radiocarbon*, 55(4), 1869–1887.

Schweinfurth, G. (1883). The flora of ancient Egypt. *Nature*, 28, 109–114.

Shaw, I. (ed.) (2000). *The Oxford History of Ancient Egypt*. Oxford University Press, Oxford.

Shortland, A.J. and Bronk Ramsey, C. (eds) (2013). *Radiocarbon and the Chronologies of Ancient Egypt*. Oxbow Books, Oxford/Oakwill.

Suess, H.E. (1955). Radiocarbon concentration in modern wood. *Science*, 122, 415–416.

Suess, H.E. (1970). *Bristlecone Pine Calibration of the Radiocarbon Time Scale from 5400 B.C. to the Present, Radiocarbon Variations and Absolute Chronology (Olsson, I. U., Ed.), 303–12, and Plates I and II*. Wiley, New York.

Szabo, J., Carmi, I., Segal, D., Mintz, E. (1997). An attempt at absolute ^{14}C dating. *Radiocarbon*, 40(1), 77–83.

13

Herbaria, a Window into the Evolutionary History of Crop Pathogens

Lionel GAGNEVIN[1], Adrien RIEUX[2], Jean-Michel LETT[2], Philippe ROUMAGNAC[1], Boris SZUREK[1], Paola CAMPOS[2,3], Claudia BAIDER[4], Myriam GAUDEUL[3] and Nathalie BECKER[3]

[1] *PHIM, CIRAD – IRD – INRAE, Institut Agro, Université de Montpellier, France*

[2] *PVBMT, CIRAD, INRAE, Université de la Réunion, Saint-Pierre, France*

[3] *ISYEB – CNRS, Muséum national d'Histoire naturelle, Sorbonne Université, École Pratique des Hautes Études, Université des Antilles, Paris, France*

[4] *The Mauritius Herbarium, Ministry of Agro-Industry and Food Security, Port-Louis, Mauritius*

Botanists have generally neglected their cultivated varieties, as beneath their notice.

Charles Darwin

Pathogenic micro-organisms of crops can be responsible for epidemics with a high socio-economic impact. Herbaria in natural history collections are sources of information for studying these epidemics: they cover the last 200 years, including the last wave of major agricultural changes, and complement the data from modern samples. Some ancient samples thus constitute "direct evidence" of the presence of a disease on a host plant, at a given date and locality, which can provide information on the origin

For a color version of all figures in this chapter, see www.iste.co.uk/pellens/natural.zip.

Natural History Collections in the Science of the 21st Century, coordinated by Roseli PELLENS.

of the pathogen and its invasion routes. Current methodological developments (high-throughput sequencing and molecular epidemiology), applied to these historic samples, allow us not only to date divergence times with increased precision, but also to elucidate the microbial evolutionary mechanisms associated with epidemiological dynamics.

The exploration of a herbarium takes into account the date and place of sample collection, and the existence of visible symptoms. The analysis of exploitable sequences from the sample, following the extraction of nucleic acids (DNA or siRNA), is subject to specific constraints. Degradation processes, on the one hand, and evolutionary processes, on the other hand, need to be characterized. Our research is based on models of economic importance in the Indian Ocean or sub-Saharan Africa (*Xanthomonas citri* pv. *citri*, for citrus; geminivirus, for cassava and sweet potato) and on the complementarity of herbaria worldwide. These results underline the importance of herbarium collections for a better understanding of the emergence and evolutionary history, and for a better management of bacterial and viral plant pathogens.

13.1. Epidemics, emergences and re-emergences

Food security is an important societal issue, and losses caused by crop diseases and pests can have dramatic consequences for populations and economies, with cumulative losses that can exceed 40% of crops (Anderson *et al.* 2004; Savary *et al.* 2019). Crop diseases are caused by pathogens[1] that are most often microbial and can be responsible for recurrent epidemics that farmers struggle to control using quarantine measures, pesticides, resistant varieties or adapted cultural practices (Figure 13.1). In this context, the excessive use of pesticides represents a threat to biodiversity, environmental quality and health. A disease may also have recently emerged, that is it may have appeared on a plant or in a region where it had never been described, or it may have re-emerged in a region from which it had previously disappeared. Cases of emergence are difficult because we are in general particularly helpless to quickly offer control solutions (Almeida *et al.* 2019), especially since it is difficult to understand the mechanisms that allowed the emergence to occur. These may be extrinsic (change in climate or cultural conditions, introduction of inoculum, emergence of a vector, etc.) or intrinsic factors to the micro-organism (adaptation to climatic conditions, change of host, acquisition of virulence factors through mutation or exchange of genetic material, resistance breakdown, etc.) (Anderson *et al.* 2004).

1 Viruses, bacteria, fungi, oomycetes and nematodes responsible for diseases (pests are rather superior organisms, like arthropods that cause damage by feeding).

Figure 13.1. *Cassava vascular bacterial blight damage in Colombia, caused by the bacterium* Xanthomonas phaseoli *pv.* manihotis *(photo: C. Trujillo)*

COMMENT ON FIGURE 13.1.– *This bacterium enters at the leaf level and spreads systemically by invading the vessels of the plant where it blocks water and nutrient transport, leading to widespread wilting. It spreads aerially (combination of rain and wind) or through cuttings.*

13.2. Development of agriculture, domestication of cultivated plants and their diseases

Theories on the domestication of plants for agriculture generally hypothesize that it enabled the initial emergence of specialized pathogens causing diseases with a significant socio-economic impact. While microbial pathogens have existed for a long time (Achtman 2016), their emergence and the appearance of epidemics are thought to be contemporary with the Neolithic revolution[2] (Mira *et al.* 2006). This would be due to the fact that plant domestication results in a decrease in the genetic diversity of plants, a genetic cost, an increase in the density of genotypes, rather favorable cultivation conditions, an increase in the exchange of plant material or simply passive transport (through irrigation, for example), and proximity to plants

2 A period in human history during which agriculture and livestock farming were developed, as well as the first organized societies, about 10,000 years ago.

from a different area of origin. These mechanisms constitute a set of parameters that could favor non-generalist, highly virulent micro-organisms with asexual reproduction, and also inoculum transmission mechanisms, horizontal transfers of genetic material and even hybridization (Stukenbrock and McDonald 2008; Stukenbrock and Bataillon 2012). While it was during the Neolithic period that agriculture developed with the (usually local) domestication of many plants and their associated micro-organisms (Larson *et al.* 2014; Zeder 2017), other more recent periods have seen significant changes in agriculture, such as the 18th and 19th centuries with the development of colonial cash crop agriculture, frequent intercontinental exchanges of plant material, the beginnings of conscious varietal selection or the development of grafting techniques. In the 20th century, before and during the green revolution, selection (against diseases or for agronomic or food characteristics) became a major tool accompanied not only by important agronomic changes (chemical fertilizers, synthetic phytosanitary products, seed industry, mechanization, increase in monocultured areas), but also by climatic changes and an increase in exchanges between different regions of the world. All these characteristics either promoted epidemics in absolute terms (e.g. by extending a micro-organism's area of presence), or exerted selection pressures (directional or not) on micro-organisms, causing the classical control strategies to collapse (boom and bust cycles)[3], or giving the opportunity for new host-pathogen combinations to occur (host jumps or host shifts). It is common to speak of co-evolution between plants and their parasites, but the evolution of cultivated plants is in fact artificial, human and rapid, while micro-organisms have an even greater adaptive potential (McDonald and Linde 2002).

While the Neolithic agricultural revolution can be explored by archeology (and today by genetic approaches), the work of the exploratory natural histories undertaken from the 18th century onward has made it possible to catalog planetary plant diversity with, among other things, the objective of moving toward better agriculture. The herbaria they created are testimonies to the plant and microbial diversity at a given time, regarding a given plant, in a given region, and which until recently were the only exploitable data of these dead samples. Later, the intensification of the frequency and magnitude of epidemics, such as the potato famine in the mid-19th century or the first epidemics of bacterial canker of citrus around 1915 in the United States, guided the development of plant pathology and for the first time the establishment of streamlined collections of symptomatic samples (Figure 13.2) (Hasse 1915; Peltier and Frederich 1924).

3 This expression describes the process of expansion of a monoculture ("boom") with, for example, resistance to a pathogen, followed by the collapse of this resistance ("bust") due to its breakdown by the pathogen or the arrival of new pathogen populations not susceptible to this resistance.

13.3. Molecular biology and genomics as a tool for studying phytopathogenic micro-organisms

Today, the study of crop pathogens and the diseases they cause benefits greatly from the contributions of molecular biology and population genetics (Stukenbrock and McDonald 2008; Goss 2015). For nearly 30 years, researchers have developed methods to extract, amplify, detect, type and sequence nucleic acids of plant pathogens from infected samples collected in the field. The analysis of the diversity and genetic structuration of populations is a direct source of information for disease control: for example, it allows us to know the global diversity of the pathogen in order to adapt the deployment of resistance and anticipate breakdowns, or to decipher the mechanisms of pathogenesis in order to choose sustainable resistance.

It also allows in some cases the identification of source populations, dissemination pathways or the estimation of biotic and abiotic parameters associated with a plant pathogen. More recently, the rise of high-throughput sequencing (HTS) methods combined with numerous methodological developments in the fields of molecular epidemiology have made it possible to reconstruct the history of the emergence and evolution of pathogens in time and space with increased precision (Croucher and Didelot 2015). However, the vast majority of studies to date have generally been based only on the analysis of contemporary samples collected over the past 40 years. Although potentially very informative, the analysis of such samples can in some situations show its limitations, especially in terms of temporal resolution.

13.4. Contributions of the herbarium samples

As mentioned above, the herbaria of natural history collections are very valuable sources of historical biological material that will complement the data from modern samples. More generally, museum collections of biological samples are becoming an increasingly important source for reconstructing the evolutionary history of life. Although not as old as "ancient" samples in the strict sense (from archeological or paleontological excavations, for example), such specimens have particular characteristics and limitations that must be taken into account or even corrected (representativeness, degradation, contamination, incomplete metadata), but they are also generally well classified and characterized (Wandeler *et al.* 2007). They cover the last 200 years, which include the last wave of major global agricultural changes.

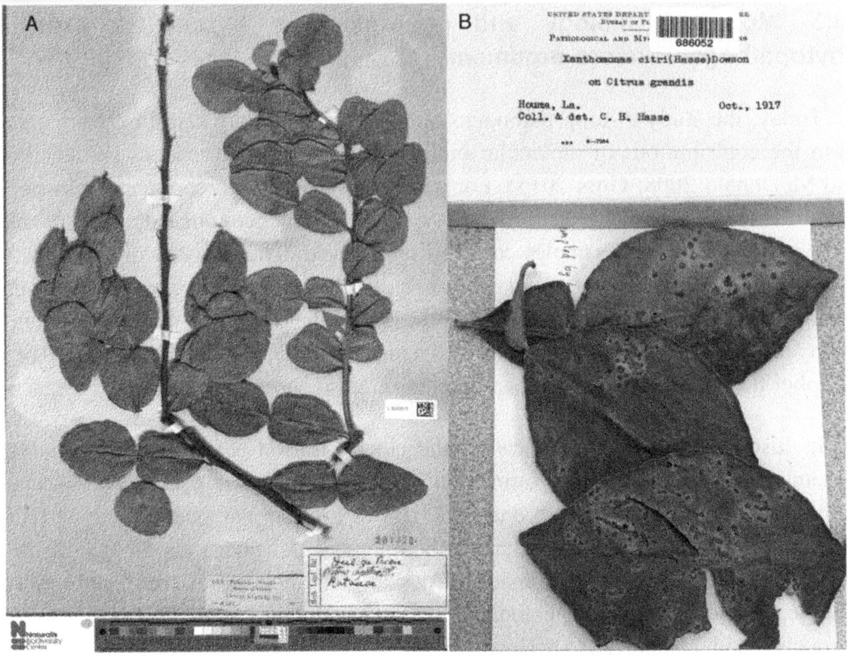

a) b)

Figure 13.2. *Herbarium plate with symptoms: a) collected by a botanist in 1860; b) collected by a plant pathologist in 1917 (photos: Naturalis Biodiversity Center and L. Gagnevin)*

COMMENT ON FIGURE 13.2.– *a) L.2019517 plate of symptomatic* Citrus *preserved at the Naturalis Biodiversity Center Herbarium (Leyden, The Netherlands), potential symptoms are indicated by arrowheads. b) Symptomatic* Citrus *sample preserved by the USDA Quarantine Service (Beltsville, MD, USA). Sample identified as symptomatic (brown spots with a chlorotic halo) upon harvest by Ms. Clara Hasse.*

Herbarium collections represent a source of information to improve our understanding of the mechanisms of emergence and evolution of crop pathogens (Yoshida *et al.* 2014) by means of the elements discussed below.

13.4.1. *Direct evidence*

First of all, a herbarium specimen displaying pathological symptoms characteristic of a disease represents "direct evidence" of the presence of this disease on a host plant, at a given date and location. Such information can be used to supplement and sometimes refine historical descriptions of plant pathogens. For example, Fawcett and Jenkins in 1933 used samples of *Citrus* preserved in British and American herbaria to date the first samples of the species infested with the bacterium *Xanthomonas citri* pv. *citri* (*Xci)* to the 1830s and to speculate on an Indian or Indonesian origin of the citrus canker disease. More recently, by studying samples of Silene species (family Caryophyllaceae) infected with flower anthrax (caused by the basidiomycete *Microbotryum violaceum*) in American herbarium samples, researchers were able to reconstruct the spatial and temporal distribution of this disease in the eastern United States as well as to identify possible host shifts between several plant species (Antonovics *et al.* 2003). More broadly, even recent herbarium samples can provide information (and genomes) of pathogens in areas or on hosts that are not easily accessible in the field (hard-to-reach areas, vanished hosts, etc.).

13.4.2. *Molecular analyses*

It is possible to extract, amplify and sequence nucleic acids from pathogens that infected the plants at the time of sampling and collection. Such genetic data enable first of all, via a "comparative genomics" approach, the differences in genetic architecture to be determined (e.g. variation in gene content, presence of insertions/deletions of nucleotide sequence fragments, point mutations, accessory genomes, ploidy) existing between modern and historical strains and potentially associated with the emergence of phytopathogenic agents. For example, in a study focusing on the potato (*Solanum tuberosum* L.) parasitic oomycete[4] *Phytophthora infestans*, which was responsible for the Great Irish Famine in 1845, analysis of herbarium samples revealed an increase in ploidy in modern strains of the pathogen (Yoshida *et al.* 2013). Such an evolutionary mechanism is advantageous for asexual species such as *P. infestans* since it increases genetic diversity and can induce the rearrangement of adaptive genes or alleles that favor the emergence and virulence of a pathogenic clone.

Historical genetic data of plant pathogens obtained from herbarium specimens also offer a great opportunity to make phylogenetic inferences. Indeed, to establish such inferences, it is necessary to have a "temporal signal", that is to be able to measure the progressive accumulation of mutations in the genomes of samples obtained at different dates. The larger the time window between ancient and modern individuals, the better the detection of the temporal signal and the more feasible the

---

4 Oomycetes are a class of eukaryotic micro-organisms resembling fungi (and long classified with them) but phylogenetically close to plants and brown algae.

phylogenetic inference of parameters. Typically, herbarium samples can be used as calibration points to jointly estimate substitution rate (the rate at which mutations accumulate in genomes) and divergence time within a phylogenetic tree (Rieux and Balloux 2016).

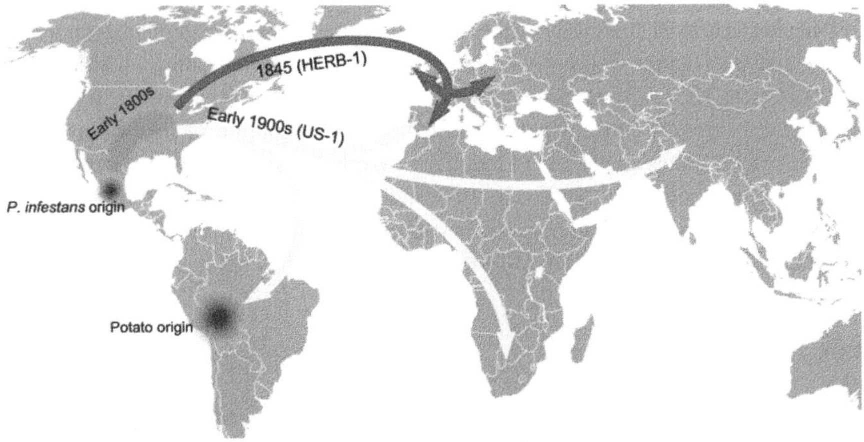

Figure 13.3. *Example of reconstructing Phytophthora* infestans *invasion routes from genomic data from herbarium samples (from Yoshida* et al. *(2013))*

Genetic divergence dates can then be compared with historical information in an attempt to elucidate the causes associated with epidemiological dynamics. For example, the common ancestor of *P. infestans* strains has been dated to the early 16th century, coinciding with the period of the Spanish conquest of Mexico. However, the pandemic that started in 1845 and the subsequent epidemics were either the result of successive introductions-replacements by *P. infestans* populations overcoming the resistant varieties deployed to control the disease, or the result of rapid diversification and adaptation to the new varieties from a limited gene pool. In contrast, modern populations are thought to have arisen from a later introduction (Figure 13.3) and also have significant evolutionary potential fostered by polyploidy, increased genetic plasticity and intense selection pressures (Martin *et al.* 2013).

Finally, genetic data obtained from herbarium specimens also allow access to sequences from the plant and thus enable us to specifically study host-pathogen interactions. Using the emblematic example of work on *P. infestans*, researchers were able to characterize the diversity of resistance genes present in historical potato specimens and show that most of them were introgressed[5] into cultivated varieties from wild species in the early 20th century (Yoshida *et al.* 2013).

13.5. How to explore a herbarium

As they appear at the beginning of the 21st century, the herbaria of natural history museums or botanical gardens are buildings where the botanical plates are most often arranged by botanical family, genus, geographical region and then species (Figure 13.4). The botanists did not choose exclusively healthy and "beautiful" specimens, but specimens representative of the reality in the field, including those infected or partially deteriorated by pests. Moreover, a large number of specimens are cultivated plants or related to cultivated plants, in relation to the initial objective of inventory and agricultural valorization. The family-genus-region-species ordering method is well adapted to our objectives since it allows for observing all specimens of a region (e.g. where the disease has been described), and then quickly extending to other regions and to other species of the genus, which is a source of information to study the host range of a pathogen and potentially to reconstruct its host jumps.

One can roughly distinguish between two ways of sampling: one is "by admission", for diseases whose symptoms are not visible when the plants are dried, and the other by orienting the exploration using criteria of location, date and plant species. The second is simply to look for characteristic symptoms of the disease, which can even be done ante-prospectively on digitized photographs increasingly available in museum databases. As an example, the New York Botanical Garden Steere Herbarium contains 7,600 rutaceae specimens, including 330 specimens of the genus *Citrus*. Symptoms of citrus bacterial canker were observed on 38 *Citrus* specimens and similar symptoms were observed on about 10 Rutaceae species not listed as hosts.

The exchange of scientific information and material in the 19th century was no less intense than today, and the various herbaria around the world constantly exchanged botanical plates (botanists usually created several plates for a single sample), making it possible not only to increase the robustness of analyses with redundant samples, but also, if necessary, to maintain a "backup copy" of a sample.

5 In biology, the term introgression refers to the transfer of a gene (e.g. for resistance) into a variety of high agronomic interest by successive crosses with another variety that is naturally resistant but of lesser agronomic interest.

Figure 13.4. *A room in the National Herbarium at the Muséum national d'Histoire naturelle de Paris (photo: R. Garrouste, MNHN)*

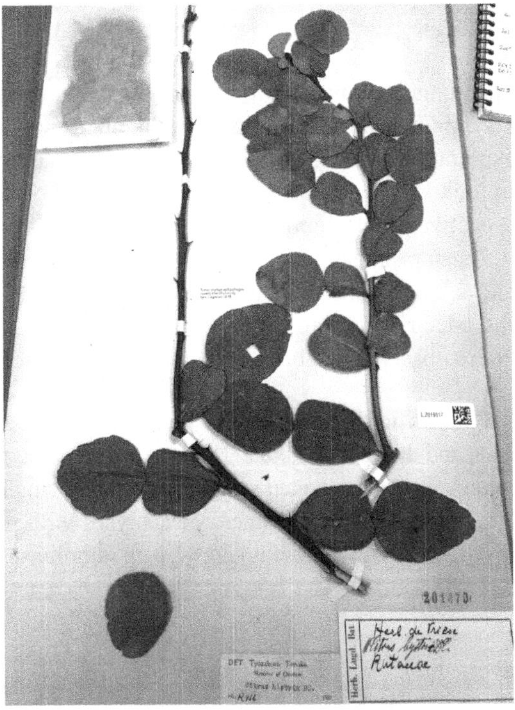

Figure 13.5. *Herbarium plate in Figure 13.2(a) after sampling (rectangular excision of about 1 cm² in the center of the plate) (photo: L. Gagnevin)*

The tricky part of processing herbarium specimens is, of course, the actual collection, since it is necessary to cut out enough potentially infected plant tissue to purify the nucleic acids for analysis (Figure 13.5). This is generally not a problem for cultivated specimens that are multiple and redundant, but may be a limiting factor for plants whose specimens must be preserved or for very old, non-redundant specimens. Tests are underway to evaluate nucleic acid collection techniques that preserve as much of the integrity of the botanical specimen as possible (by minimizing the amount of material collected or even by single smearing of the sample; e.g. Shepherd (2017)).

13.6. Characteristics of old nucleic acids and their treatment

The rise in the study of ancient DNA has been facilitated by massive parallel sequencing (HTS, see section 13.3). This technology is based on a first step of constructing nucleic acid libraries involving ligation, on both sides of the target DNA, of small specific artificial DNA segments (synthetic oligonucleotides), allowing an amplification of the target DNA (by PCR, Polymerase Chain Reaction) before a second step of large-scale sequencing. As early as the 2000s, using HTS, degraded DNA (such as ancient DNA) could thus be sequenced from whole-tissue extraction (Briggs *et al.* 2007; Burrell *et al.* 2015). A number of features make the processing and study of ancient DNAs (especially those preserved in herbarium plants) difficult: these will be described below for classical pathogens (DNA micro-organisms) extracted from the leaves of herbarium samples, and also for viruses, whose genetic carrier may not be DNA.

Analysis of numerous ancient sequences of diverse origins (Pääbo *et al.* 2004; Briggs *et al.* 2007) has revealed:

– DNA fragmentation: this is partially explained by the presence of a purine, whose depurination (hydrolysis of the purine bases of DNA, G[uanine] or A[denine]) facilitates the breakage of its DNA backbone (Lindahl and Andersson 1972). Thus, the two processes, depurination and fragmentation, are linked; however, some breaks occur independently of any depurination;

– an accumulation of nucleotide substitutions at the ends of the sequenced fragments: this accumulation is mainly due to the natural deamination process of the pyrimidine base C[ytosine], preferentially on single-stranded DNA. For reasons inherent to library preparation and sequencing, this deamination is most often characterized at the 5' ends by a substitution from C to T[hymin], and at the 3' ends of the same strand by a substitution from G to A.

Based on the analysis of 86 herbarium samples collected during the last three centuries (*Arabidopsis* and *Solanum* genera, infected or not by the oomycete *P. infestans*), an overview of the characteristics of ancient DNA has recently been established. The two main types of degradation observed, fragmentation and deamination, reflect a gradual and cumulative process (a significant decrease in fragment size with age is observed). Fragmentation affects an average of 1.7% of nucleotide bonds every 100 years, and deamination reaches nearly 5.0% of nucleotides at the end of fragments for the oldest samples (from the 1730s). A more detailed comparative analysis was performed for two categories of DNA (nuclear and chloroplast). It showed that the same degradations took place for both DNA categories but the rate is up to twice as high for nuclear DNA (Weiss *et al.* 2016).

Due to its fragmentation and low concentration, and also due to the presence of inhibitors in the leaves of host plants (phenolic compounds and polysaccharides, detrimental to the library construction step), ancient DNA of phytopathogens is difficult to purify using the most common commercial kits. Conventional extraction protocols have been optimized with a reduction in the number of steps, the addition of inhibitor-neutralizing molecules, and finally the use of purification columns adapted to the small sizes of DNA fragments (Kistler 2012; Healey *et al.* 2014; Gutaker *et al.* 2017). The quality of this extraction also depends on the taxonomy of the host plant, as well as the drying and storage conditions of the herbarium samples (Sarkinen *et al.* 2012; Choi *et al.* 2015). Similarly, pre-sequencing library construction must be adapted for degraded DNA samples (Sarkinen *et al.* 2012; Gansauge and Meyer 2013; Carøe *et al.* 2017). Finally, capture strategies (based on hybridization) can be developed to enrich samples with target nucleic acids (e.g. pathogen vs. plant), but they require a good knowledge of the micro-organism genome (Gasc *et al.* 2016).

13.6.1. *The particular case of viral nucleic acids*

Similar to the work cited above that explored the evolutionary histories of bacteria or oomycetes, recent advances in HTS technologies have also enabled the reconstruction of viral genomes dating back centuries (Smith *et al.* 2014; Yoshida *et al.* 2014). One of the specifics of applying HTS to characterize viruses, whether contemporary or ancient, is the very nature of the nucleic acids. Unlike the aforementioned studies, DNA is not the sole target of virology work. Moreover, there is no universal gene that can be used to reconstruct a phylogenetic tree of all viruses. In fact, a range of different viral metagenomics approaches without *a priori* knowledge has been described to date, each approach targeting a specific form of viral nucleic acid (RNA and/or DNA, single or double stranded) with its advantages and disadvantages (Massart *et al.* 2014; Roossinck *et al.* 2015).

A first approach is to extract total DNA or RNA from a sample and sequence the nucleic acids. The small proportion of genomic sequences of circular DNA viruses can also be enriched by rolling circle amplification (Wyant *et al.* 2012), an approach that was successfully used to reconstruct the total genome of a geminivirus responsible for an emerging disease in the United States from a grapevine sample at the University of Davis Herbarium in California (Al Rwahnih *et al.* 2015). Similarly, RNA preparations can be enriched in viral sequences by removing large amounts of ribosomal RNA from the plant (ribodepletion) (Adams *et al.* 2009).

A second approach is to extract viral nucleic acids from a preliminary semi-purification step of the viral particles (Virus Associated Nucleic Acids, VANA) (Candresse *et al.* 2014).

A third approach, which is limited to the analysis of RNA viruses, is to purify double-stranded RNA (dsRNA) molecules that accumulate in plant cells during virus replication (Roossinck *et al.* 2010). This approach has been used to identify a chrysovirus from Arizona maize cobs dating back approximately 1,000 years (Peyambari *et al.* 2019).

Finally, a fourth approach is the isolation and massive sequencing of small RNAs from the antiviral defense pathway known as RNA interference in plants (Pooggin 2018). These small interfering RNAs (siRNAs) are 21, 22 or 24 nucleotides in length and their combined sequences span most, if not all, of the viral genomes accumulated during infection (Ding and Voinnet 2007). This strategy has the advantage of being able to detect a broad spectrum of viruses regardless of their nature (RNA or DNA) and structure (linear, circular, single or double stranded).

The latter approach may be becoming central to the research and analysis of ancient viruses. Indeed, several recent studies have applied it to ancient plant material, such as archeological plant remains or herbarium specimens. RNA has long been considered a lost cause in archeovirology[6], due to its vulnerability to hydrolytic attack, but this work suggests that it can in fact be preserved under certain arid conditions for hundreds of years (Guy 2014). Furthermore, siRNAs are thought to be more stable than long RNAs or DNAs due in part to the secondary structures of RNA. For example, Smith *et al.* (2014) used this approach and successfully identified and reconstituted the ancient RNA genome of an isolate of Barley stripe mosaic virus (*Hordeivirus*, *Virgaviridae*) from 700-year-old barley seeds.

6 Discipline that studies viruses from prehistory to the contemporary era.

13.7. Xanthomonas *citri* pv. *citri* and its emergence in the Indian Ocean

Asiatic citrus canker is a bacterial disease that affects citrus in almost all tropical and subtropical areas. It is caused by the γ-proteobacterium *Xanthomonas citri* pv. *citri* (*Xci*); canker symptoms on leaves, fruits and stems have a major economic impact. A better understanding of the mechanisms underlying the emergence, evolution and spread of *Xci* appears to be an important prerequisite for a better management of this disease and to anticipate new emergences.

The *Xci* bacterium probably emerged in Asia, the host's area of origin (the genus *Citrus*), before expanding outside the continent (to America, Africa and Oceania) during the first half of the 20th century (Pruvost *et al.* 2014). In the Indian Ocean region, the disease was first described in 1937 in Mauritius and Rodrigues and then from the late 1960s in Reunion, the Comoros Archipelago and the Seychelles. For nearly 35 years, CIRAD researchers have collected strains of the *Xci* bacterium from different islands of the Indian Ocean as well as from other regions of the world. The genomes of a number of these strains have been sequenced in order to improve understanding of the emergence mechanisms of this bacterium in the Indian Ocean islands. For this purpose, a dataset consisting of 150 genomes sampled between 1980 and 2015 in the different islands of the region as well as in the rest of the world (including of course Asia, the area of origin of *Citrus*) has been constituted. In order to increase the temporal dimension of this dataset, the herbaria of the University of Reunion and Mauritius were explored. Two historical samples of *Citrus* sp. isolated in Mauritius, respectively in 1937 and 1974, displaying typical canker symptoms were selected for global DNA sequencing following the specific protocols presented in section 13.6. The historical sequences thus generated were analyzed in several ways.

First, we sought to determine the taxonomic origin of the sequence reads[7] using a classical taxonomic assignment approach, obtained by searching for homologous sequences of sequenced organisms deposited in public databases. For the two herbarium samples analyzed, the majority (approximately 60%) of the reads were homologous to citrus sequences, an expected result since the bulk of the tissue from which DNA was extracted is from the plant. Second, a fraction (about 15%) of the reads were homologous to the human genome, a result explained by contamination of the herbarium sample from the initial sampling by the explorer to the collection and all subsequent manipulations, including those within the herbarium. Finally, a minority (less than 5%) of the readings were attributed to *Xci*, the bacterium responsible for citrus canker, and to a mixture of various micro-organisms (bacteria, virus and fungi) present on or in the herbarium sample (also 5%).

7 The set of individual sequences generated by high-throughput sequencing.

In a second step, the reads assigned to *Xci* were analyzed in order to measure their degradation levels and in particular to compare them with those observed on contemporary samples. To do so, two statistics were estimated using the *mapDamage* tool (Jonsson *et al.* 2013): (1) the fragment size distribution and (2) the deamination intensity of Cytosines (C) according to their position in the read. Reads from historical samples revealed reduced fragment sizes (60 bp on average). In addition, deamination was observed on a significant fraction of the Cs of the first five bases present at the ends of the historical reads, in contrast to the modern reads which showed no trace of deamination. These results lead us to two conclusions: (1) DNA from relatively young herbarium samples (dating from 1937 and 1974 in this case) can be partially degraded (fragmentation and deamination) and (2) these traces of degradation attest to the authenticity of the historical sequences and prove that they are not the result of contemporary contamination during sample handling.

Finally, the relationships between the two historical strains and the 150 contemporary *Xci* genomes were reconstructed by a phylogenetic tree, the detailed analysis of which allowed us to hypothesize a first introduction of bacteria in Mauritius during the period following the abolition of slavery (1835). Thus, citrus canker could have been introduced by plants imported from Asia, during the recruitment of indentured workers (often of Indian origin) in the plantations of the Indian Ocean islands. An evolutionary rate for the *Xci* bacterium could also be inferred for the first time ever (Richard *et al.* 2020). Future analysis of additional historical herbarium samples from other geographic origins (including the Asian area of origin or other regions where the disease was introduced), other time periods (19th century) and other plant species (*Rutaceae* other than *Citrus* sp.) should allow for the reconstruction of the evolutionary history of the disease on a broader temporal and geographic scale.

13.8. Emergence and evolutionary history of plant pathogenic viruses: the geminivirus model

Geminiviruses are responsible for many emerging diseases worldwide with a major economic impact on vegetable and food crops, very important for food security, especially in sub-Saharan Africa, such as cassava (*Manihot esculenta*) and sweet potato (*Ipomoea batatas*). Their high evolutionary potential, with high mutation and recombination rates, makes these viruses an ideal model for understanding the epidemiological and evolutionary processes associated with viral emergence.

13.8.1. *Case of a species complex responsible for an emerging disease*

Cassava geminiviruses form a species[8] complex responsible for cassava mosaic disease, the most severe and damaging of this food crop in sub-Saharan Africa. To investigate the evolutionary processes behind the epidemiological success of these viruses, we undertook sequencing historical viral genomes from cassava samples from the Herbarium of the *Muséum national d'Histoire naturelle* (Paris) showing apparent symptoms of the disease (Figure 13.6).

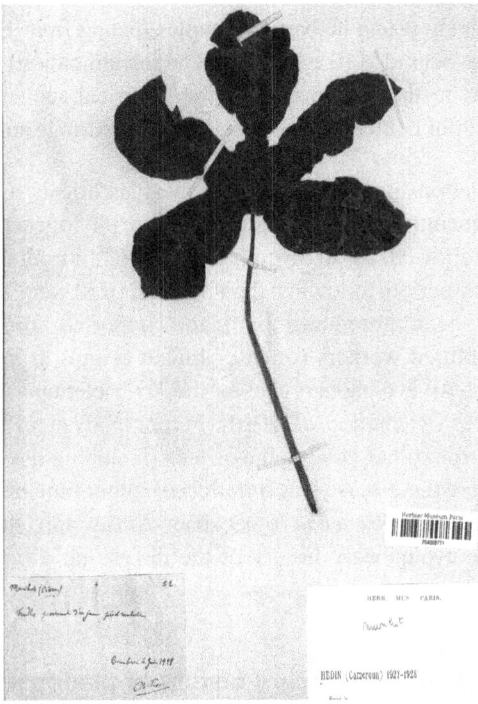

Figure 13.6. *Plate P04808771 from the MNHN Herbarium before sampling, including a cassava leaf (*Manihot glaziovii*) collected in Bambari in Central Africa in 1928 and showing a suspicion of cassava mosaic disease with symptoms of leaf deformation (photo: MNHN)*

As previously described, cassava geminiviruses trigger a plant immune-like defense response with the accumulation of 21, 22 and 24-nucleotide siRNAs

8 Several distinct but closely related species responsible (together or separately) for a disease and difficult to distinguish other than by genetics.

(Pooggin 2018). Using sequences of recovered small RNAs (harboring the features of such siRNAs), we were able to reconstruct the near-complete sequence of these geminiviruses from 50- and 90-year-old herbarium specimens. Complementing contemporary sequences, these ancient sequences can be used in phylogenetic, phylogeographic and comparative genomics studies to elucidate the emergence and evolution of the cassava mosaic agent.

13.8.2. *Case of a cryptic geminivirus*

Sweet potato, whose center of origin appears to be Central and South America (Clark *et al.* 2012), ranks among the 10 most important food crops, especially for sub-Saharan Africa. Because of its vegetative propagation by tubers or cuttings, sweet potato is prone to the accumulation and transport of pathogens, including various cryptic viruses[9] that are transmitted to progeny and will accumulate over generations to form a viral complex (an "association" of various viral elements of different origins) that can lead to severe forms of disease through synergistic phenomena with repercussions on plant growth, yield and quality (Paprotka *et al.* 2010). More than 30 viruses have been described in sweet potato. Half of these viruses are DNA viruses, described very recently using HTS approaches, most of which are associated with asymptomatic infections. This is notably the case for the species *Sweet potato symptomless virus 1* (SpSV-1, *Mastrevirus*, *Geminiviridae*), for which 14 complete genomes have been characterized in Asia, Africa and South America (Cao *et al.* 2017; Souza *et al.* 2018), following the first description of a partial sequence in a cultivar from Peru (Kreuze *et al.* 2009). This seminal work, using the siRNA approach in indexing[10] sweet potato germplasm, has indeed identified many cryptic viruses never before described but potentially participating in epidemic viral complexes. In order to study the evolutionary processes at the origin of this asymptomatic association between a virus and a host plant, we undertook to reconstruct the historical viral genomes of SpSV-1, from about 20 historical sweet potato herbarium samples dating back to 1820 for the oldest ones, with the siRNA approach. The first analyses confirmed our ability to reconstruct almost complete sequences of SpSV-1 dating back nearly 200 years, allowing us to have a picture of the viral community hosted by this plant and to hypothesize on the structure and epidemiological dynamics of the viral complexes.

9 The presence of the virus is not associated with symptoms of disease.

10 Evaluation of the viral species present in a plant.

13.9. Discussion

The approaches and results presented in this chapter illustrate not only the immense potential of historical herbarium collections in understanding the emergence and evolution of plant pathogenic micro-organisms, but also the recent development of new methodological approaches to robustly exploit these resources. New studies, on other diseases and other crops, are currently underway and will improve our knowledge on the history and evolution of phytopathogenic micro-organisms, an essential prerequisite for better management of these organisms.

Of course, it is important to bear in mind that the genetic information and data obtained from these samples have characteristics and limitations that must be considered throughout their processing.

On the one hand, the temporal and geographical distribution of the collections of natural history museums is biased by several factors: more or less intense periods of prospection according to the historical and economic context, aspects of scientific policy (often linked to the colonial or economic priorities of the country), botanical preferences of the field collectors, etc.

However, the fact that we work on plants of agronomic interest offers certain advantages: the samples are numerous and have a lower botanical and conservation value, they come from regions where plant diversity is valuable (e.g. area of origin), and the survey campaigns can be relatively systematic and rigorous because their objective is often to search for new biological resources for local and global agriculture.

On a strictly technical level, it is important to note that the genetic study of herbarium samples is still far from being equivalent to that of living strains: in particular, the fragmentation of ancient nucleic acids prevents the sequencing of long fragments and limits certain very informative comparative genomics approaches (genome architecture, study of repeated regions, identification of genes or regions that have disappeared from modern genomes, recombination events, etc.). Moreover, the vast majority of studies have so far focused on the fraction of nucleic acids assigned to the pathogen, leaving aside nucleic acids assigned to other micro-organisms or to the host plant, the analysis of which appears promising and necessary to understand epidemiological dynamics and co-evolution mechanisms in a more integrated manner.

13.10. Acknowledgments and funding

We would like to thank the herbarium curators Deby Arifiani (Herbarium Bogoriense, Indonesia), Jan Wieringa (Naturalis Biodiversity Center, Dutch Herbaria, The Netherlands), Timm Utteridge (Royal Botanical Garden, Kew, UK) for their hospitality and assistance, Mark Carine (Natural History Museum Herbarium, UK), Shannon Dominick (National Fungus Collection, USDA, USA), Matthew Pace (New York Botanical Garden, Steere Herbarium, USA), Meghann Toner (US National Herbarium, Smithsonian, USA) and Terry H. Trinder-Smith (Bolus Herbarium, University of Cape Town, South Africa).

Our work has been funded by the COST (European Cooperation in Science and Technology) action CA16107 EuroXanth, the SYNTHESYS program[11] funded by the Research Infrastructure action of the FP7 program "Capacities" of the European Community, CIRAD (AI-CRESI-3/2016), Agropolis Fondation (convention N° 1600-014), the *Agence nationale pour la recherche* (JCJC MUSEOBACT contract ANR-17-CE35-0009-01), ED 227, *Muséum national d'Histoire naturelle* and Sorbonne Université, the French Ministry of Higher Education, Research and Innovation, as well as the European Union (European Regional Development Fund, ERDF contract GURDT I2016-1731-0006632) and the Regional Council of Reunion Island.

13.11. References

Achtman, M. (2016). How old are bacterial pathogens? *Proceedings of the Royal Society B*, 283(1836).

Adams, I.P., Glover, R.H., Monger, W.A., Mumford, R., Jackeviciene, E., Navalinskiene, M., Samuitiene, M., Boonham, N. (2009). Next-generation sequencing and metagenomic analysis: A universal diagnostic tool in plant virology. *Molecular Plant Pathology*, 10(4), 537–545.

Al Rwahnih, M., Rowhani, A., Golino, D. (2015). First report of grapevine red blotch-associated virus in archival grapevine material from Sonoma County, California. *Plant Disease*, 99(6), 895.

Almeida, R.P.P., De La Fuente, L., Koebnik, R., Lopes, J.R.S., Parnell, S., Scherm, H. (2019). Addressing the new global threat of *Xylella fastidiosa*. *Phytopathology*, 109(2), 172–174.

Anderson, P.K., Cunningham, A.A., Patel, N.G., Morales, F.J., Epstein, P.R., Daszak, P. (2004). Emerging infectious diseases of plants: Pathogen pollution, climate change and agrotechnology drivers. *Trends in Ecology & Evolution*, 19(10), 535–544.

11 Available at: http://www.synthesys.info/.

Antonovics, J., Hood, M.E., Thrall, P.H., Abrams, J.Y., Duthie, G.M. (2003). Herbarium studies on the distribution of anther-smut fungus (*Microbotryum violaceum*) and Silene species (Caryophyllaceae) in the eastern United States. *American Journal of Botany*, 90(10), 1522–1531.

Briggs, A.W., Stenzel, U., Johnson, P.L., Green, R.E., Kelso, J., Prufer, K., Meyer, M., Krause, J., Ronan, M.T., Lachmann, M., Paabo, S. (2007). Patterns of damage in genomic DNA sequences from a Neandertal. *Proceedings of the National Academy of Sciences of the United States of America*, 104(37), 14616–14621.

Burrell, A.S., Disotell, T.R., Bergey, C.M. (2015). The use of museum specimens with high-throughput DNA sequencers. *Journal of Human Evolution*, 79, 35–44.

Candresse, T., Filloux, D., Muhire, B., Julian, C., Galzi, S., Fort, G., Bernardo, P., Daugrois, J.H., Fernandez, E., Martin, D.P., Varsani, A., Roumagnac, P. (2014). Appearances can be deceptive: Revealing a hidden viral infection with deep sequencing in a plant quarantine context. *PLoS One*, 9(7), e102945.

Cao, M., Lan, P., Li, F., Abad, J., Zhou, C., Li, R. (2017). Genome characterization of sweet potato symptomless virus 1: A mastrevirus with an unusual nonanucleotide sequence. *Archives of Virology*, 162(9), 2881–2884.

Carøe, C., Gopalakrishnan, S., Vinner, L., Mak, S.S.T., Sinding, M.H.S., Samaniego, J.A., Wales, N., Sicheritz-Pontén, T., Gilbert, M.T. (2017). Single-tube library preparation for degraded DNA. *Methods in Ecology and Evolution*, 9, 410–419.

Choi, J., Lee, H., Shipunov, A. (2015). All that is gold does not glitter? Age, taxonomy, and ancient plant DNA quality. *PeerJ*, 3, e1087.

Clark, C.A., Davis, J.A., Abad, J.A., Cuellar, W.J., Fuentes, S., Kreuze, J.F., Gibson, R.W., Mukasa, S.B., Tugume, A.K., Tairo, F.D., Valkonen, J.P.T. (2012). Sweet potato viruses: 15 years of progress on understanding and managing complex diseases. *Plant Disease*, 96(2), 168–185.

Croucher, N.J. and Didelot, X. (2015). The application of genomics to tracing bacterial pathogen transmission. *Current Opinion in Microbiology*, 23, 62–67.

Ding, S.W. and Voinnet, O. (2007). Antiviral immunity directed by small RNAs. *Cell*, 130(3), 413–426.

Fawcett, H.S. and Jenkins, A.E. (1933). Records of citrus canker from herbarium specimens of the genus Citrus in England and the United States. *Phytopathology*, 23(10), 820–824.

Gansauge, M.T. and Meyer, M. (2013). Single-stranded DNA library preparation for the sequencing of ancient or damaged DNA. *Nature Protocols*, 8(4), 737–748.

Gasc, C., Peyretaillade, E., Peyret, P. (2016). Sequence capture by hybridization to explore modern and ancient genomic diversity in model and nonmodel organisms. *Nucleic Acids Research*, 44(10), 4504–4518.

Goss, E.M. (2015). Genome-enabled analysis of plant-pathogen migration. *Annual Review of Phytopathology*, 53, 121–135.

Gutaker, R.M., Reiter, E., Furtwangler, A., Schuenemann, V.J., Burbano, H.A. (2017). Extraction of ultrashort DNA molecules from herbarium specimens. *Biotechniques*, 62(2), 76–79.

Guy, P.L. (2014). Prospects for analyzing ancient RNA in preserved materials. *Wiley Interdisciplinary Reviews: RNA*, 5(1), 87–94.

Hasse, C.H. (1915). *Pseudomonas citri*, the cause of citrus canker. *Journal of Agricultural Research*, 4(1), 97–99.

Healey, A., Furtado, A., Cooper, T., Henry, R.J. (2014). Protocol: A simple method for extracting next-generation sequencing quality genomic DNA from recalcitrant plant species. *Plant Methods*, 10, 21.

Jonsson, H., Ginolhac, A., Schubert, M., Johnson, P.L., Orlando, L. (2013). mapDamage 2.0: Fast approximate Bayesian estimates of ancient DNA damage parameters. *Bioinformatics*, 29(13), 1682–1684.

Kistler, L. (2012). Ancient DNA extraction from plants. In *Ancient DNA. Methods and Protocols*, Shapiro, B. and Hofreiter, M. (eds). Humana Press, Hatfield.

Kreuze, J.F., Perez, A., Untiveros, M., Quispe, D., Fuentes, S., Barker, I., Simon, R. (2009). Complete viral genome sequence and discovery of novel viruses by deep sequencing of small RNAs: A generic method for diagnosis, discovery and sequencing of viruses. *Virology*, 388(1), 1–7.

Larson, G., Piperno, D.R., Allaby, R.G., Purugganan, M.D., Andersson, L., Arroyo-Kalin, M., Barton, L., Vigueira, C.C., Denham, T., Dobney, K., Doust, A.N., Gepts, P., Gilbert, M.T.P., Gremillion, K.J., Lucas, L., Lukens, L., Marshall, F.B., Olsen, K.M., Pires, J.C., Richerson, P.J., Rubio de Casas, R., Sanjur, O.I., Thomas, M.G., Fuller, D.Q. (2014). Current perspectives and the future of domestication studies. *Proceedings of the National Academy of Sciences of the United States of America*, 111(17), 6139–6146.

Lindahl, T. and Andersson, A. (1972). Rate of chain breakage at apurinic sites in double-stranded deoxyribonucleic acid. *Biochemistry*, 11(19), 3618–3623.

Martin, M.D., Cappellini, E., Samaniego, J.A., Zepeda, M.L., Campos, P.F., Seguin-Orlando, A., Wales, N., Orlando, L., Ho, S.Y.W., Dietrich, F.S., Mieczkowski, P.A., Heitman, J., Willerslev, E., Krogh, A., Ristaino, J.B., Gilbert, M.T.P. (2013). Reconstructing genome evolution in historic samples of the Irish potato famine pathogen. *Nature Communications*, 4, 2172.

Massart, S., Olmos, A., Jijakli, H., Candresse, T. (2014). Current impact and future directions of high throughput sequencing in plant virus diagnostics. *Virus Research*, 188, 90–96.

McDonald, B.A. and Linde, C. (2002). Pathogen population genetics, evolutionary potential, and durable resistance. *Annual Review of Phytopathology*, 40, 349–379.

Mira, A., Pushker, R., Rodriguez-Valera, F. (2006). The Neolithic revolution of bacterial genomes. *Trends in Microbiology*, 14(5), 200–206.

Pääbo, S., Poinar, H., Serre, D., Jaenicke-Despres, V., Hebler, J., Rohland, N., Kuch, M., Krause, J., Vigilant, L., Hofreiter, M. (2004). Genetic analyses from ancient DNA. *Annual Review of Genetics*, 38, 645–679.

Paprotka, T., Boiteux, L.S., Fonseca, M.E., Resende, R.O., Jeske, H., Faria, J.C., Ribeiro, S.G. (2010). Genomic diversity of sweet potato geminiviruses in a Brazilian germplasm bank. *Virus Research*, 149(2), 224–233.

Peltier, G.L. and Frederich, W.J. (1924). Further studies on the relative susceptibility to citrus canker of different species and hybrids of the genus *Citrus*, including the wild relatives. *Journal of Agricultural Research*, 28(3), 227–239.

Peyambari, M., Warner, S., Stoler, N., Rainer, D., Roossinck, M.J. (2019). A 1,000-year-old RNA virus. *Journal of Virology*, 93, e01188-18.

Pooggin, M.M. (2018). Small RNA-omics for plant virus identification, virome reconstruction, and antiviral defense characterization. *Frontiers in Microbiology*, 9, 2779.

Pruvost, O., Magne, M., Boyer, K., Leduc, A., Tourterel, C., Drevet, C., Ravigné, V., Gagnevin, L., Guérin, F., Chiroleu, F., Koebnik, R., Verdier, V., Vernière, C. (2014). A MLVA genotyping scheme for global surveillance of the citrus pathogen *Xanthomonas citri* pv. *citri* suggests a worldwide geographical expansion of a single genetic lineage. *PLoS One*, 9(6), e98129.

Richard, D., Pruvost, O., Balloux, F., Boyer, C., Rieux, A., Lefeuvre, P. (2020). Time-calibrated genomic evolution of a monomorphic bacterium during its establishment as an endemic crop pathogen. *Molecular Ecology*, 00, 1–13.

Rieux, A. and Balloux, F. (2016). Inferences from tip-calibrated phylogenies: A review and a practical guide. *Molecular Ecology*, 25(9), 1911–1924.

Roossinck, M.J., Martin, D.P., Roumagnac, P. (2015). Plant virus metagenomics: Advances in virus discovery. *Phytopathology*, 105(6), 716–727.

Roossinck, M.J., Saha, P., Wiley, G.B., Quan, J., White, J.D., Lai, H., Chavarria, F., Shen, G., Roe, B.A. (2010). Ecogenomics: Using massively parallel pyrosequencing to understand virus ecology. *Molecular Ecology*, 19, 81–88.

Sarkinen, T., Staats, M., Richardson, J.E., Cowan, R.S., Bakker, F.T. (2012). How to open the treasure chest? Optimising DNA extraction from herbarium specimens. *PLoS One*, 7(8), e43808.

Savary, S., Willocquet, L., Pethybridge, S.J., Esker, P., McRoberts, N., Nelson, A. (2019). The global burden of pathogens and pests on major food crops. *Nature Ecology and Evolution*, 3(3), 430–439.

Shepherd, L.D. (2017). A non-destructive DNA sampling technique for herbarium specimens. *PLoS One*, 12(8), e0183555.

Smith, O., Clapham, A., Rose, P., Liu, Y., Wang, J., Allaby, R.G. (2014). A complete ancient RNA genome: Identification, reconstruction and evolutionary history of archaeological Barley Stripe Mosaic Virus. *Scientific Reports*, 4, 4003.

Souza, C.A., Rossato, M., Melo, F.L., Boiteux, L.S., Pereira-Carvalho, R.C. (2018). First report of sweet potato symptomless virus 1 infecting *Ipomoea batatas* in Brazil. *Plant Disease*, 102, 2052.

Stukenbrock, E.H. and Bataillon, T. (2012). A population genomics perspective on the emergence and adaptation of new plant pathogens in agro-ecosystems. *PLoS Pathogens*, 8(9), e1002893.

Stukenbrock, E.H. and McDonald, B.A. (2008). The origins of plant pathogens in agro-ecosystems. *Annual Review of Phytopathology*, 46, 75–100.

Wandeler, P., Hoeck, P.E., Keller, L.F. (2007). Back to the future: Museum specimens in population genetics. *Trends in Ecology & Evolution*, 22(12), 634–642.

Weiss, C.L., Schuenemann, V.J., Devos, J., Shirsekar, G., Reiter, E., Gould, B.A., Stinchcombe, J.R., Krause, J., Burbano, H.A. (2016). Temporal patterns of damage and decay kinetics of DNA retrieved from plant herbarium specimens. *Royal Society Open Science*, 3(6), 160239.

Wyant, P.S., Strohmeier, S., Schäfer, B., Krenz, B., Assunção, I.P., Silvestre de Andrade Lima, G., Jeske, H. (2012). Circular DNA genomics (circomics) exemplified for geminiviruses in bean crops and weeds of northeastern Brazil. *Virology*, 427(2), 151–157.

Yoshida, K., Schuenemann, V.J., Cano, L.M., Pais, M., Mishra, B., Yoshida, K., Schuenemann, V.J., Cano, L.M., Pais, M., Mishra, B., Sharma, R., Lanz, C., Martin, F.N., Kamoun, S., Krause, J., Thines, M., Weigel, D., Burbano, H.A. (2013). The rise and fall of the *Phytophthora infestans* lineage that triggered the Irish potato famine. *Elife*, 2, e00731.

Yoshida, K., Burbano, H.A., Krause, J., Thines, M., Weigel, D., Kamoun, S., Kamoun, S. (2014). Mining herbaria for plant pathogen genomes: Back to the future. *PLoS Pathogens*, 10(4), e1004028.

Zeder, M.A. (2017). Domestication as a model system for the extended evolutionary synthesis. *Interface Focus*, 7(5), 20160133.

14

The Yellow-Legged Asian Hornet: Prediction of the Risk of Invasion and the Study of its Color Variations

Claire VILLEMANT[1], Quentin ROME[2] and Adrien PERRARD[3]

[1] *ISYEB – CNRS, Muséum national d'Histoire naturelle, Sorbonne Université, École Pratique des Hautes Études, Université des Antilles, Paris, France*

[2] *PatriNat, IRD – OFB – CNRS, Muséum national d'Histoire naturelle, Paris, France*

[3] *iEES Paris, CNRS – INRAE – IRD, Université de Paris, Sorbonne Université, Paris, France*

The yellow-legged Asian hornet *Vespa velutina* was accidentally introduced into France before 2004. Present today in almost all of France, it has also reached neighboring countries. Like all hornets, its colonies are annual and die in winter, except for the future queens which take shelter to overwinter. Hidden, they can be accidentally transported over great distances via international trade. *Vespa velutina* has a wide natural range in Asia, from China to Indonesia, which includes many populations with sometimes very different coloration patterns. The invasive lineage introduced into France (like the one introduced into Korea) belongs to the *nigrithorax* form, originally distributed from Nepal to China. The studies detailed here show the essential role played by natural history collections, combined with

For a color version of all figures in this chapter, see www.iste.co.uk/pellens/natural.zip.

Natural History Collections in the Science of the 21st Century,
coordinated by Roseli PELLENS.

molecular analysis of fresh material, in answering questions concerning the diversity of a widely distributed species and its potential for expansion when it becomes invasive. The Chinese origin of the invasive lineages was confirmed. Coloration is not a reliable marker of the degree of relatedness between populations of *V. velutina*, although they have a strong genetic structure. Populations of *V. v. nigrithorax* appear to be genetically poorly structured, and they are distant cousins of the island populations of Indonesia. The invasive lineages are thus derived from a diverse origin population and are genetically mixed with other continental populations of the species. This genetic admixture supports the inclusion of these different populations in the climate niche models to correctly predict the outcome of the invasion.

14.1. Introduction

Vespa velutina, the yellow-legged Asian hornet, is the first exotic Vespidae species that has acclimated in Europe (Beggs *et al.* 2011). Individuals of the *nigrithorax* form were first observed in France in Lot-et-Garonne in 2004. In 16 years, the hornet spread throughout the country and, between 2010 and 2017, became established in neighboring countries: Spain, Portugal, Italy, Germany, Belgium, the United Kingdom and the Netherlands (Figure 14.1) (Rome and Villemant 2020). The hornet was also introduced into Korea in the 2000s (Choi *et al.* 2012) and into Tsushima Island in Japan in 2013 (Ueno 2014). Its presence in Europe has very quickly worried beekeepers and the general public due to the imposing size of its colonies (Rome *et al.* 2015) and the damages it causes on apiaries (Arca *et al.* 2014). *Vespa velutina* is indeed a significant predator of honey bees although it also preys on a wide variety of other prey such as insects or spiders (Villemant *et al.* 2014). Although varroa mites (parasitic mites of bees), viral diseases, scarcity of floral resources and pesticides are greater threats to beekeeping, this predator represents an additional factor of decline for European honey bees (Requier *et al.* 2019). Moreover, it was classified in France as a "2nd category health risk" under the rural code in 2012 and as an "invasive alien species" (*espèce exotique envahissante*, EEE) under the environmental code in 2013; it has also been part of the list of EEEs of concern for the European Union since 2016.

Since 2010, the rapid spread of *V. velutina* in France has raised many questions, particularly concerning the origin and expansion capabilities of the lineage that has become invasive. However, *V. velutina* has a very large natural range which includes many populations, some of which show obvious morphological differences, particularly in their coloration patterns (Figure 14.2). It was therefore important to verify whether all these populations belong to the same species and whether the observed differences reflect reproductive isolations. Indeed, the predictions concerning the invasive potential of the Asian hornet in Europe are not the same if the population of origin is isolated and dependent on a particular climate, or if it is diversified thanks to frequent migrations between different geographical areas. The study of specimens from natural history collections and molecular analyses made from recently collected specimens has made it possible to answer several questions: (1) What is the origin of the invasive lineages of *V. velutina* in France and Korea? (2) How far can the potential invasion area of *V. velutina nigrithorax* extend? (3) What is the structure of the diversity of *V. velutina* in its area of origin, and where do the morphological differences observed between its populations come from?

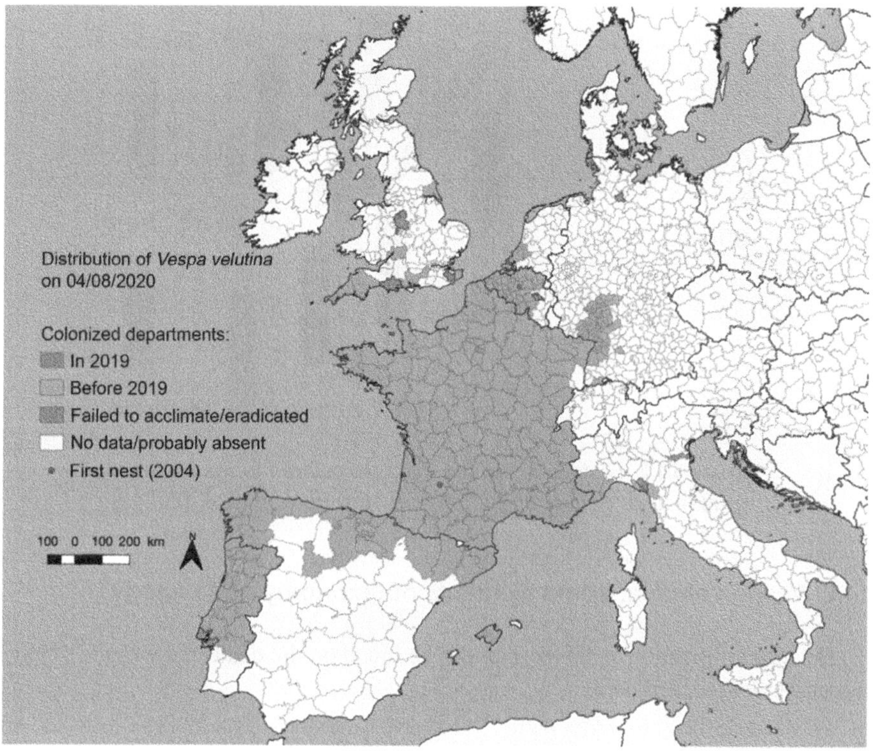

Figure 14.1. *Distribution of Vespa velutina in Europe (August 2020). Source: Q. Rome – MNHN-INPN;EuroGeographics for administrative boundaries*

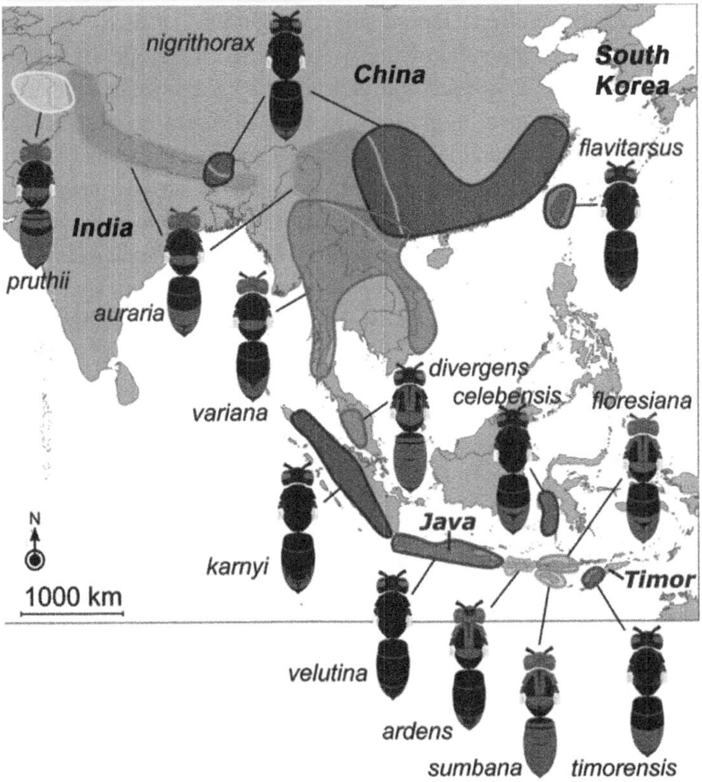

Figure 14.2. *Distribution of the different colored forms of Vespa* velutina *in its native range. Source: from Perrard* et al. *(2014a)*

14.2. *Vespa velutina*: some elements of taxonomy and biology

Detailed information on this species can be found on the dedicated website[1]. Only a few points are recalled here.

14.2.1. *A species: 13 colored forms*

The genus *Vespa* includes 22 species, all of which live in the Asian region (Central and Southeast Asia). Only two species have a range that extends from Asia to

1 Available at: http://frelonasiatique.mnhn.fr/.

Europe: the Oriental Hornet *Vespa orientalis* Linnaeus, 1771 and the European Hornet *Vespa crabro* Linnaeus, 1758. The latter was the only species present in France before the arrival of the yellow-legged hornet.

Vespa velutina was described by Lepeletier in 1836 from the island of Java (Indonesia). Its range extends from the eastern mountains of Afghanistan through China to the Indonesian island of Timor (north of Australia). Thirteen color forms can be distinguished throughout this range (Figure 14.2), each with a delimited distribution and a color pattern that differs to a greater or lesser extent from that of other populations (Perrard *et al.* 2011, 2014a). The species was long divided into two, with the western populations considered another species: *Vespa auraria* Smith, 1852, but in 2006, a study showed that these two populations could hybridize in some localities (Nguyen *et al.* 2006). *Vespa auraria* is therefore considered synonymous with *V. velutina* and as the 13th colored form of this species. As for the lineage that has become invasive, *V. v. nigrithorax*, it was described in Darjeeling (India) in 1905 by du Buysson and is found from northern India (Darjeeling, Sikkim) to eastern Nepal, Bhutan and China (Carpenter and Kojima 1997; Perrard *et al.* 2014a).

14.2.2. *One nest per year*

As with other hornets, each colony of *V. velutina* is founded in the spring by a queen emerging from the winter period. Her descendants are workers who feed the larvae, enlarge, maintain and defend the nest while the queen devotes herself to laying eggs until the fall. Two-thirds of colonies relocate their nests during the summer: the primary nest founded by the queen is abandoned for a larger secondary nest built by workers atop a tree. A large colony can produce up to 13,000 individuals in a season (April to November), including more than 500 future queens and as many males (Rome *et al.* 2015). After the reproductive swarm in the fall, the entire colony dies except for the future queens. These reproductive females spend the winter sheltered in the ground or in a cavity. The colony is annual and never survives the winter; the nest is never reused. As the future founders are able to fly over long distances, the invasion front in Europe progresses by about 80 km per year. However, as with other social wasps, the species can also move over much greater distances through the transport of overwintering queens (Beggs *et al.* 2011; Robinet *et al.* 2017).

14.2.3. *Insectivore, but not exclusively*

The species is mainly insectivorous. The workers capture prey in flight and then cling to a material to cut it up and only bring its thorax back to the nest to feed the larvae; the adults only consume sweet liquids (nectar, sap, honey and honeydew) and the protein liquid regurgitated by the larvae. In fall, the predation exerted by *V. velutina* on bees can locally threaten apiaries, but above all it is the presence of hornets hovering in front of the hive that stresses the bees and causes them to stop foraging. The resulting decrease in honey reserves can lead to the death of the bee colony during the winter (Requier *et al.* 2019).

14.3. Sampling of specimens

In order to answer the questions mentioned above, it is necessary to collect a large number of specimens and amount of information on the occurrence of each population. Collecting specimens across a range as large as that of *V. velutina* is very labor-intensive, and quickly becomes very expensive. It took several funding requests, three expeditions (each costing several thousand euros) and the contribution of dozens of entomologists who had also collected in Asia to gather specimens from the entire range of the hornet over 2 years. Other collections allowed specimens in alcohol and data regarding the presence of the invasive lineage in France to be obtained, and also, thanks to the collaboration of a Korean colleague, data regarding the invasive lineage in Korea. However, these samples and data proved insufficient to better understand the distribution of this species and to evaluate its potential for expansion in Europe or to study its morphological variability.

In this context, the use of collection specimens offered an economical and ecological complement to the sampling: in total, in addition to the 125 specimens collected in the field, nearly 360 specimens from six public collections and three private collections (Table 14.1) were used for the morphological study and/or to provide data on the presence of the species in its native range. Finally, data on the presence of the *nigrithorax* form in France were obtained thanks to a monitoring protocol based on citizen science and on networks of local actors. Records of nests were reported via a form available on the Internet[2] which has made it possible to monitor the progress of the invasion in France since 2004. Each year, however, important verification work (based on photos requested from observers) must be undertaken beforehand, as about one-third of the reports are erroneous (confusion with European hornet nests or other wasp species) or correspond to duplicates (the same nest reported several times) (Rome and Villemant 2020).

2 Available at: http://frelonasiatique.mnhn.fr/.

National collections	Modeling	Coloring
Institut royal des Sciences naturelles de Belgique (Brussels, Belgium)	0	12
Muséum national d'Histoire naturelle (Paris, France)	86	178
Museum Zoologicum Bogoriense (Bogor, Indonesia)	0	53
Nationaal Natuurhistorisch Museum Naturalis (Leyden, The Netherlands)	0	93
Natural History Museum (London, Great Britain)	7	64
Zhegjiang University (Hangzhou, China)	22	40
Private collections	0	8

Table 14.1. *Number of specimens from natural history collections used for invasion risk modeling of the* Vespa velutina nigrithorax *form and study of* Vespa velutina *species coloration patterns*

14.4. The origin of invasive lineages of *V. velutina* in France and Korea

Molecular analyses and invasion scenarios carried out within the framework of a community program to support European beekeeping by a team of researchers from the laboratory *Évolution, Génomes, Comportement, Écologie* (CNRS/IRD/Université Paris Sud) of Gif-sur-Yvette, the *Institut de systématique, évolution, biodiversité* (CNRS/MNHN/UPMC/EPHE/UA) and the *Institut Sophia Agrobiotech* (CNRS/INRA/UNICE) have determined the origin of the invasive lineages of the hornet and identified biological factors that may explain its successful acclimatization in France.

14.4.1. *The history of the invasion explained by genetics*

The 2010 and 2011 collections provided specimens that were preserved in alcohol for molecular analyses. A total of 170 *Vespa velutina* individuals – 83 from France, 8 from Korea and 79 from other Asian countries (Vietnam, Indonesia and China, Yunnan and Zhejiang/Jiangsu regions) – were used. The genetic diversity of these six populations was measured from microsatellite DNA sequences (22 loci) and their maternal origin, identified using mitochondrial DNA (CO1). Various invasion scenarios were then tested by combining these genetic data with demographic parameters available for this species (Arca *et al.* 2015).

14.4.2. *A single queen at the origin of the invasive lineage in France*

The geographical origin of the invasive lineages from France and Korea was located in an area between the Chinese provinces of Zhejiang and Jiangsu, where the *nigrithorax* form is the only one present. This area includes the metropolis of Shanghai and the city of Yixing, internationally renowned for its pottery production. These results support the hypothesis that the Asian hornet may have arrived in France through international trade, in containers of bonsai pottery from China. They also show that the genetic diversity of the invasive population in France is drastically lower than that of the original population. Genetic analysis of nymphs from various nests collected in France showed that most colonies included several paternal lineages. A female Asian hornet can therefore mate with several males, a behavior rarely observed in species of the genus *Vespa*. According to these data, the most likely scenario is that the invasive lineage in France is the result of a single queen fertilized by several males (Arca *et al.* 2015). This lineage thus manifests surprising multiplication and dispersal capabilities despite its very low genetic diversity. The small sample size from Korea did not allow us to learn more about the invasive population in this country, apart from its origin from the same region of China.

14.5. Expansion risks in Europe and worldwide

The prediction of the potential acclimation areas of *V. v. nigrithorax* in Europe and worldwide was an essential preliminary step to implement specific surveillance and control measures. This modeling work was carried out thanks to the active collaboration of researchers from three MNHN teams: ISYEB, CESCO and UMS PatriNat. It also benefited from the contribution of all the individuals and organizations that sent nest reports[3].

14.5.1. *Data and methods for inferring range and predicting invasion risk*

By combining the localities of presence of a species and the climatic variables of these localities, the niche models (species distribution models, SDM) make it possible to define the climatic niche of a species and to predict its potential acclimatization range from a geographical projection of this niche. Eight niche models each using the same eight climate variables and based on the occurrence data of *V. v. nigrithorax* in its areas of origin and invasion (France and Korea) were used to construct a consensus map of potential invasion risks. The climatic variables used to

3 Available at: http://frelonasiatique.mnhn.fr/.

model its distribution were: mean annual temperature, temperatures of the warmest and coldest months, seasonal temperature, annual precipitation, precipitation of the wettest and driest months and precipitation seasonality (BioClim variables[4]).

In 2010, the invasion concerned 39 French departments, and about 1,700 georeferenced pieces of data of *V. v. nigrithorax* presence between 2004 and 2010 were available, as well as one piece of data from Korea (Busan) and 20 pieces of data from China, from collections made in the Zhejiang/Jiangsu area. Sampling in Asia was supplemented by 15 pieces of data from the literature and by examination of specimens of *V. v. nigrithorax* in the collections of the MNHN, the British Museum and Zhejiang University (Hangzhou, China), adding 34 additional localities to the inventory. Because the data from the invasion area were much more numerous than those from the area of origin, only two localities drawn at random from each invaded French department were retained so that the samples considered in the models were of the same size (n = 69) in both areas (Villemant *et al.* 2011).

This same dataset and niche models were subsequently used to predict the evolution of the potential acclimation range of the hornet in 2100 under 13 climate change scenarios (Barbet-Massin *et al.* 2013). Finally, by comparing *V. v. nigrithorax* presence data in France between 2011 and 2015 with those from 2004 to 2010, we tested the reliability of niche models to predict the risks of expansion of this species (Barbet-Massin *et al.* 2018).

14.5.2. *Strong expansion in Europe and the Northern Hemisphere*

Consensus maps obtained in 2011 from niche models show that acclimation of *V. v. nigrithorax* in southwestern France could be predicted from native range data, while combined native range and invaded range data provide a better prediction of its potential for acclimation worldwide. Figure 14.3 shows that most European countries have a significant risk of acclimation by the yellow-legged hornet, with more limited risks in the drier regions of southern Europe. If accidentally introduced, this invasive hornet could also acclimate to many other parts of the world, including areas already invaded by the common wasp *Vespula germanica*, which has a similar climatic niche (Villemant *et al.* 2011). Climate change is also expected to greatly increase the potential for acclimation of *V. v. nigrithorax* in 2100 in the Northern Hemisphere (Figure 14.4), particularly in central and eastern Europe and the southern United States, but also in the Southern Hemisphere (Barbet-Massin *et al.* 2013).

4 Available at: http://www.worldclim.org/.

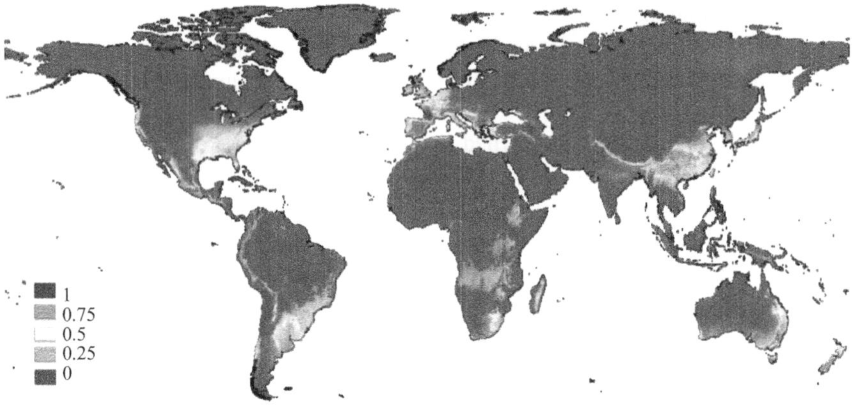

Figure 14.3. *Consensus map of acclimation probabilities of* Vespa velutina *worldwide in 2011. Colors show the probability of acclimation ranging from 0 = nil to 1 = highly likely*

Finally, although the invasive lineage in France has not yet reached climatic equilibrium, the niche models provided with the 2004–2010 data create a map of acclimation potentialities that correctly predicts the invasion area observed in 2015. The fact that the SDM models provide correct predictions shows that the invasion progress of *V. v. nigrithorax* is at least partially dependent on climatic conditions (Barbet-Massin *et al.* 2018). However, there remains a need to acquire more biological and ecological data to improve these predictions.

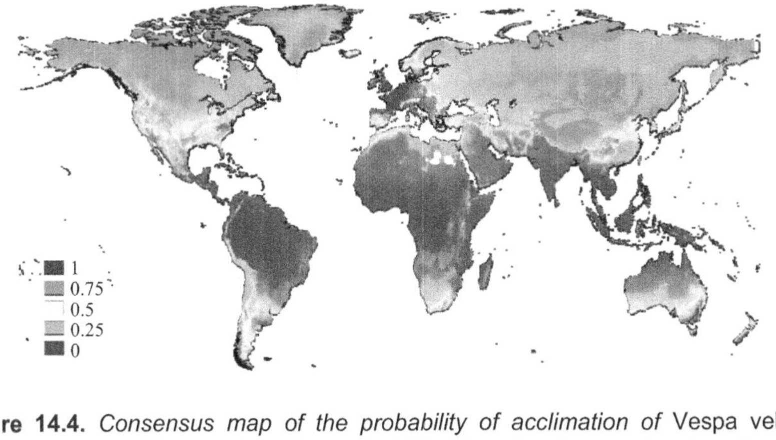

Figure 14.4. *Consensus map of the probability of acclimation of* Vespa velutina *worldwide in 2100. Colors show the probability of acclimation ranging from 0 = nil to 1 = highly likely. Comparing with Figure 14.3, note the strong trend of expansion of favorable areas especially in the Northern Hemisphere*

14.6. Origin of color and shape variations

The most glaring distinction between *Vespa velutina* populations comes from their color patterns (Perrard *et al.* 2011). These hornets are yellow in some areas, coppery in other localities, and nearly black in several areas (Figure 14.2). Taking into account populations with intermediate patterns and those that bring together several types of coloration, it is sometimes difficult to envision that these different insects all belong to the same species. Determining what links these populations, what distinguishes them and how these differences arose not only helps to delineate the species, but also to understand what mechanisms lead to variation between populations and potentially to the appearance of new species.

Two main factors may be involved in the coloration of an insect: the environmental conditions under which the insect grows and genetic differences. The variation in color between specimens was therefore quantified and compared to three pieces of data: geographical variation, climatic variation between collection areas and genetic variation. This study was part of Adrien Perrard's thesis at the MNHN (Perrard *et al.* 2014a).

14.6.1. *The importance of collection specimens*

The missions in Asia had made it possible to collect 125 specimens from 69 localities, but this result remained poor compared to the ambition of the project: out of 18 identified populations, only 10 were sampled and only half of them were represented by at least 5 individuals. The data provided by the collections proved to be crucial: thanks to 100 entomologists who collected this species between 1876 and the present day, we were able to triple the number of localities sampled (Figure 14.5) and almost quadruple the number of specimens analyzed.

However, there are several limitations to the collection data: first, molecular analysis of old specimens does not always yield results due to high DNA degradation. For these reasons, specimens more than 10 years old were not used. However, as molecular techniques are continuously improving (see Chapter 13 by Gagnevin *et al.*), it is likely that such specimens will also be able to be considered in the near future. Second, knowing that the color of insects can fade with time and with the collection and preservation methods used, the quantification of colors, especially by image analysis, may be problematic. Finally, despite this increased sampling, a bias remains in the areas sampled: some populations are oversampled while others are represented by only a small number of specimens.

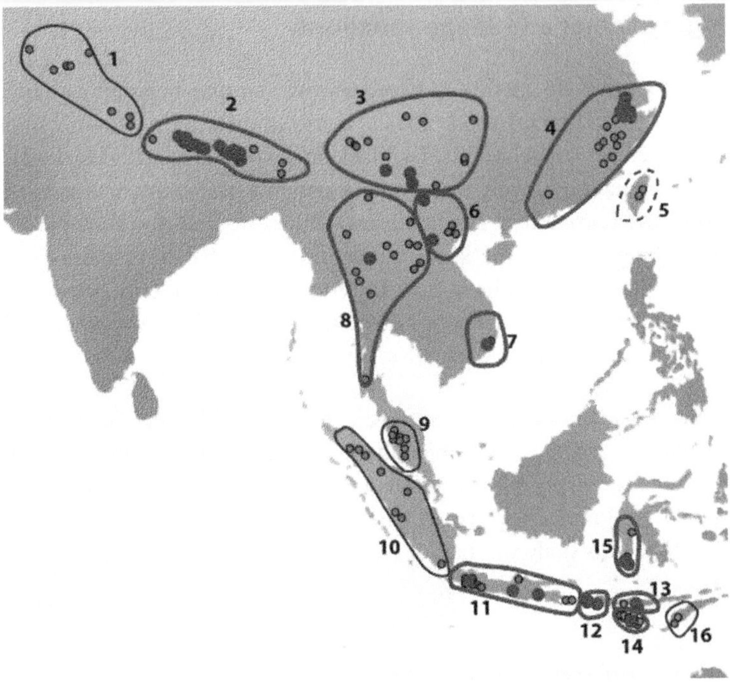

Figure 14.5. *Sampling of* Vespa velutina *populations*

COMMENT ON FIGURE 14.5.– *The green and red dots represent the localities of the specimens used for the morphological study; only those in the red localities were used for the molecular analyses. The numbers correspond to the 16 populations surveyed. Population 5 was not included in the analysis due to lack of sufficient sample size.*

Taking into account the phenotypic diversity of all known populations of the species, each color variation was coded according to its position on the insect's body and its degree of melanization. Thus, 23 variations were detected among the 448 specimens studied, resulting in 73 different color patterns across the *V. velutina* species. A correspondence analysis was used to quantify the similarity between these different patterns (Figure 14.6). These data were then compared with geographical data, by estimating the geodesic distance between collection sites and climatic similarity, from eight climatic variables estimated for these same sites[5] and, for the subsample of recent specimens, with genetic data from CO1 sequences and 11 microsatellite markers.

5 Available at: http://www.worldclim.org.

14.6.2. *Discordance between genetic lineages and colored forms*

While color analysis showed a fairly continuous variation in coloration patterns, genetic data strongly suggest that the species can be divided into two major groups: on the one hand, the genetically weakly differentiated continental populations, including *V. v. nigrithorax* and *V. v. auraria* (confirmed to be part of this species), and on the other hand, a group consisting of the Indonesian island populations, which are well differentiated both from each other and their continental cousins.

Color similarities do not correspond to genetic similarities: some distant populations show similar colorations despite strong genetic differences. In addition, individuals from populations with very different color patterns do not show significant genetic differences. In the same way, the geographical distance or the climatic conditions are not sufficient to explain the variations of colors, which are sometimes very marked between very close geographical populations.

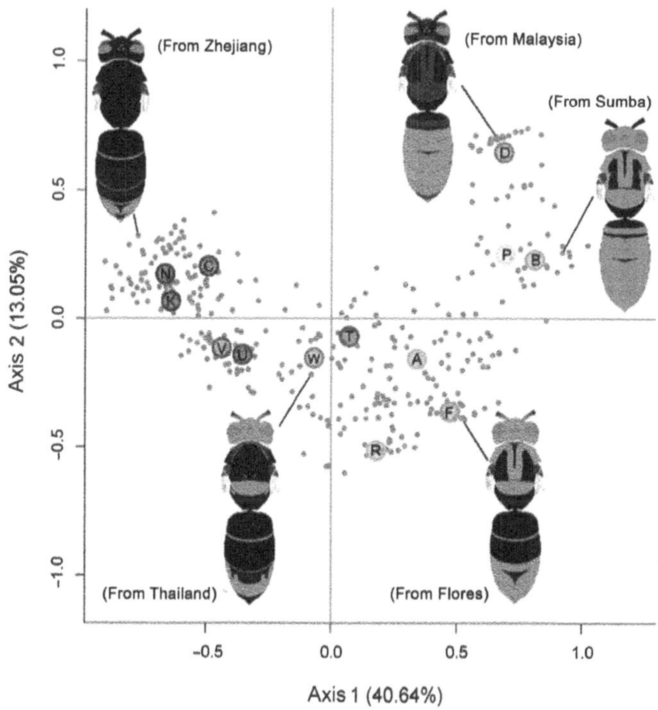

Figure 14.6. *Distribution of different coloration patterns of* Vespa velutina *on the first plane of the correspondence analysis. N =* nigrithorax*; K =* karnyi*; C =* celebensis*; V =* velutina*; U =* flavitarsus*; W =* variana*; T =* timorensis*; R =* ardens*; A =* auraria*; F =* floresiana*; D =* divergens*; P =* pruthii*; S =* sumbana

These analyses show that a distinction must be made between color forms and genetic links: color is not a good estimator of genetic differences between populations because, like other phenotypes, it can be selected and influenced by different factors. This does not exclude, however, color variation from having a genetic basis; indeed, the constancy of a given color pattern within a population, even for individuals such as those of the invasive lineage of *V. velutina* living in regions that are very different from their native area, suggests the contrary. One hypothesis that may explain the high color variation observed in this species is a mixture of founder effects and natural selection. The founder effect is related to the colonization of new territories, during which rare variations in the original population may become fixed in the colonial population. Moreover, the coloration of hornets is not neutral: it sends a visual signal to potential predators that the insect can defend itself with its sting. It therefore depends on natural selection, which itself varies from one territory to another according to the predators present, and also according to the dominant color patterns of the prey that the predators have learned to reject because of their toxicity. These effects could explain how very different patterns could be fixed in relatively close but isolated populations.

14.7. Conclusion

The studies detailed here show the essential role that natural history collections have played in answering questions about the diversity of a widely distributed species and its potential for expansion once it has become invasive. Combined with data from molecular analyses, they have shown that coloration is not a reliable marker of the degree of relatedness between populations, which nevertheless have a strong genetic structure in their area of origin. This population structure was also found through the study of another morphological trait, wing shape, using the same material (Perrard *et al.* 2014b). These data also confirmed the synonymy between *V. velutina* and the former species *V. auraria*, originally described on the basis of color criteria (Perrard *et al.* 2014a).

The populations of *V. v. nigrithorax* thus appear to be distant cousins of the island populations of Indonesia and the invasive lineages are thus derived from a population of fairly diversified origin and genetically mixed with other populations from the north of the Asian continent. This admixture supports the choice to have taken these populations into account in the climate niche models used (Villemant *et al.* 2011), which correctly predicted the outcome of the invasion over several years (Barbet-Massin *et al.* 2018).

14.8. References

Arca, M., Papachristoforou, A., Mougel, F., Rortais, A., Monceau, K., Bonnard, O., Tardy, P., Thiéry, D., Silvain, J.-F., Arnold, G. (2014). Defensive behaviour of *Apis mellifera* against *Vespa velutina* in France: Testing whether European honeybees can develop an effective collective defence against a new predator. *Behavioural Processes*, 106, 122–129.

Arca, M., Mougel, F., Guillemaud, T., Dupas, S., Rome, Q., Perrard, A., Muller, F., Fossoud, A., Capdevielle-Dulac, C., Torres-Leguizamon, M., Chen, X.X., Tan, J.L., Jung, C., Villemant, C., Arnold, G., Silvain, J.-F. (2015). Reconstructing the invasion and the demographic history of the yellow-legged hornet, *Vespa velutina*, in Europe. *Biological Invasions*, 17, 2357–2371.

Barbet-Massin, M., Rome, Q., Muller, F., Perrard, A., Villemant, C., Jiguet, F. (2013). Climate change increases the risk of invasion by the yellow-legged hornet. *Biological Conservation*, 157, 4–10.

Barbet-Massin, M., Rome, Q., Villemant, C., Courchamp, F. (2018). Can species distribution models really predict the expansion of invasive species? *PLoS ONE*, 13, e0193085.

Beggs, J.R., Brockerhoff, E.G., Corley, J.C., Kenis, M., Masciocchi, M., Muller, F., Rome, Q., Villemant, C. (2011). Ecological effects and management of invasive alien Vespidae. *BioControl*, 56, 505–526.

Carpenter, J.M. and Kojima, J. (1997). Checklist of the species in the subfamily Vespinae (Insecta: Hymenoptera: Vespidae). *Natural History Bulletin of Ibaraki University*, 1, 51–92.

Choi, M.B., Martin, S.J., Lee, J.-W. (2012). Distribution, spread, and impact of the invasive hornet *Vespa velutina* in South Korea. *Journal of Asia-Pacific Entomology*, 15(3), 473–477.

Nguyen, L.T., Saito, F., Kojima, J., Carpenter, J.M. (2006). Vespidae of Vietnam (Insecta: Hymenoptera) 2. Taxonomic notes on Vespinae. *Zoological Science*, 23(1), 95–104.

Perrard, A., Muller, F., Rome, Q., Villemant, C. (2011). Observations sur le frelon asiatique à pattes jaunes, *Vespa velutina* Lepeletier, 1836 (Hymenoptera, Vespidae). *Bulletin de la Société entomologique de France*, 116(2), 159–164 [Online]. Available at: https://lasef.org/wp-content/uploads/BSEF/116-2/5-1535_Perrard_et_al.pdf.

Perrard, A., Arca, M., Rome, Q., Muller, F., Tan, J., Bista, S., Nugroho, H., Baudoin, R., Baylac, M., Silvain, J.-F., Carpenter, J.M., Villemant, C. (2014a). Geographic variation of melanisation patterns in a hornet species: Genetic differences, climatic pressures or aposematic constraints? *PLoS ONE*, 9(4), e94162.

Perrard, A., Baylac, M., Carpenter, J.M., Villemant, C. (2014b). Evolution of wing shape in hornets: Why is the wing venation efficient for species identification? *Journal of Evolutionary Biology*, 27(12), 2665–2675.

Requier, F., Rome, Q., Chiron, G., Decante, D., Marion, S., Menard, M., Muller, F., Villemant, C., Henry, M. (2019). Predation of the invasive Asian hornet affects foraging activity and survival of honey bees in Western Europe. *Journal of Pest Science*, 92, 567–578.

Robinet, C., Suppo, C., Darrouzet, E. (2017). Rapid spread of the invasive yellow-legged hornet in France: The role of human-mediated dispersal and the effects of control measures. *Journal of Applied Ecology*, 54(1), 205–215.

Rome, Q. and Villemant, C. (2016). Invasive species: Bee-hawking hornet already in line of fire. *Nature*, 534(7605), 37.

Rome, Q. and Villemant, C. (2020). Le frelon asiatique *Vespa velutina* – Inventaire national du Patrimoine naturel. *Muséum national d'Histoire naturelle* [Online]. Available at: http://frelonasiatique.mnhn.fr [Accessed 4 August 2020].

Rome, Q., Muller, J.F., Touret-Alby, A., Darrouzet, E., Perrard, A., Villemant, C. (2015). Caste differentiation and seasonal changes in *Vespa velutina* (Hym: Vespidae) colonies in its introduced range. *Journal of Applied Entomology*, 139(10), 771–782.

Ueno, T. (2014). Establishment of the invasive hornet *Vespa velutina* (Hymenoptera: Vespidae) in Japan. *International Journal of Chemical, Environmental & Biological Sciences*, 2(4), 220–222.

Villemant, C., Barbet-Massin, M., Perrard, A., Muller, F., Gargominy, O., Jiguet, F., Rome, Q. (2011). Predicting the invasion risk by the alien bee-hawking yellow-legged hornet *Vespa velutina nigrithorax* across Europe and other continents with niche models. *Biological Conservation*, 144(9), 2142–2150.

Villemant, C., Muller, F., Rome, Q., Perrard, A., Barbet-Massin, M., Jiguet, F. (2014). Estimating the potential range expansion and environmental impact of the invasive bee-hawking hornet, *Vespa velutina nigrithorax*. In *In Silico Bees*, Devillers, J. (ed.). CRC Press, Boca Raton, FL.

15

Exploring Temporal Changes in the Composition of Macroalgal Communities by Using Collections

Marine ROBUCHON[1], Éric FEUNTEUN[2], Romain JULLIARD[3], Florence ROUSSEAU[1] and Line Le GALL[1]

[1] *ISYEB – CNRS, Muséum national d'Histoire naturelle, Sorbonne Université, École Pratique des Hautes Études, Université des Antilles, Paris, France*

[2] *BOREA – IRD, CNRS, Muséum national d'Histoire naturelle, Sorbonne Université, Université des Antilles, Université de Caen-Normandie, Dinard, France*

[3] *CESCO – CNRS, Muséum national d'Histoire naturelle, Sorbonne Université, Paris, France*

The objective of this study was to assess the feasibility of using natural history collections to trace temporal changes in species distribution and community composition using the example of macroalgae that are preserved as herbaria. The collections constitute a time series since it is possible to establish where and when the specimens were collected. Moreover, compared to a species list, the main advantage of collections is that specimens constitute scientific evidence that can be re-evaluated. However, because harvesting practices and sampling density vary greatly over time, care must be taken before analyzing trends in either species distribution or community composition. The range of possible analyses is compared with the available data and recommendations are made so that current and future

For a color version of all figures in this chapter, see www.iste.co.uk/pellens/natural.zip.

Natural History Collections in the Science of the 21st Century, coordinated by Roseli PELLENS.

sampling strategies can contribute to the establishment of long-term series necessary for the analysis of biodiversity dynamics.

15.1. On the constitution of macroalgal collections

15.1.1. *Large seaweeds*

Seaweeds are usually preserved in a herbarium like most flowering plants. However, not all seaweeds are plants. The dichotomy between plant and animal, inherited from the vision of Aristotle, is now outdated. Classifications of eukaryotes that have developed on the basis of phylogenetic analyses, count for nearly two dozen deep lineages into which, what was once considered as algae, are distributed (Adl *et al.* 2019). With respect to macroalgae, brown algae are classified as Stramenopiles (also called Heterokontophyta) while red and green algae form a lineage with land plants. Despite their phylogenetic differences, the three major groups of seaweeds perform similar functions in coastal ecosystems (primary production) and provide habitats for other species that find refuge there (they are called engineer species). Seaweeds play a crucial role in rocky coastal ecosystems, where their photosynthetic activity ensures one of the highest productivities recorded: a coastal zone dominated by algae produces more than a terrestrial forest or a cultivated field (Mann 1973).

15.1.2. *Algal herbaria*

The preservation of plants in herbaria began during the Renaissance. This technique required paper and became widespread in the 18th century thanks to technical advances in paper production. It was in the 19th century that the realization of herbaria was the most prolific. This scientific activity was carried out not only by trained botanists and professionals, but also by many enlightened amateurs, generally from the bourgeoisie among whom this activity was particularly fashionable. Algae, which were more difficult to access, were only studied by a small number of botanists and amateurs who often maintained close relations through correspondence or during the organization of study excursions. However, herbarium conservation is particularly well adapted to algae. Indeed, because of their mucilaginous walls, they adhere well to paper, allowing a relatively easy handling of the herbaria and especially a very good conservation of their anatomy over time. Because of their relative rarity compared to plant herbaria, algal herbaria have generally been well preserved and are now kept in institutions.

15.1.3. *Data associated with the herbaria*

With the first herbaria, the information concerning the collections was often laconic, but from the beginning of the 19th century, it became good practice to note the place of collection by means of toponyms as well as the date of collection (Figure 15.1). On the basis of this information, it is possible to establish where and when the harvests took place. The coordinates (latitude and longitude) are generally not mentioned on the label, and are inferred from the names of the places, in particular by means of toponymic databases such as *GeoNames*[1].

The authors of the herbaria were also often authors of works that allow us today to better understand the conceptual context in which the collections were made. The work of the Crouan brothers, pioneers in the study of the algae of Brittany, is a particularly emblematic example. The specimens they collected, as early as the 1830s, served as the foundation for all their subsequent work, which led to the publication of the *Florule du Finistère* (Crouan and Crouan 1867), a work that is still a reference today.

15.1.4. *Specimens and scientific evidence*

Algae are organisms whose tissue structure is quite simple; thus, the characters to identify them are few and taxonomic ambiguities are frequent. Plant classifications have been developed on the basis of the diversity of forms of reproductive organs. However, in algae, the reproductive organs are hidden, which is why Linnaeus called them cryptogams (Linnaeus 1753). The reliable identification of algae is therefore an often complex activity requiring sections and microscopic observations and therefore requires a high level of expertise. Since the last decade of the 20th century, the advent of molecular systematics has dramatically changed the classifications of algae and other phyletic groups, and has revealed the presence of many cryptic species (Leliaert *et al.* 2014). New species of macroalgae are being described, including along the coasts of the English Channel (Robuchon *et al.* 2014, 2015), even though these coasts have been studied the most intensively and for the longest period of time.

1 Available at: http://www.geonames.org/.

Figure 15.1. *Voucher from the Dinard Herbarium representing a specimen of* Halarachnion ligulatum *from Eugène Malard's collection (1859–1916) when he was creating the Phycological Herbarium of the English Channel from the* Muséum's *maritime laboratory located at the time in Tatihou. The label data specify that it was collected on the island of Tatihou on September 10, 1895*

Indeed, the discipline of phycology came about at the same time in France and England, thanks to the establishment of marine stations. The regime of proof of identification then took on its full meaning. In contrast to species lists or even drawings (Daston and Galison 2012), herbarium specimens allow for the re-examination of taxonomic identification in light of anatomo-morphological characters and characterization by new methods, as is the case for species identified from molecular characters. They are therefore a source of analysis that allows us to make verifiable and refutable hypotheses that should be mobilized, particularly for diachronic studies of the distribution of species and the composition of communities.

15.1.5. *The herbarium of the Dinard maritime laboratory*

The algal herbarium of the Dinard maritime laboratory has been recently transferred to the *Herbier national du Muséum national d'Histoire naturelle de Paris* (National Herbarium of the National Museum of Natural History in Paris) in the cryptogamy section. This collection of 2,959 specimens collected from 1826 to 1978 by at least 25 phycologists of different status and practice (Figure 15.2) has been fully computerized and digitized. Among these specimens, the label data make it possible to locate 2,281 in Brittany and Normandy (Figure 15.3). However, some of the data and taxonomic identification were not always precise enough to be integrated into diachronic studies of species distribution and community composition, so 84% of the specimens (1928) were included in the analysis.

15.2. Exploring temporal changes in species distribution

The study conducted by Robuchon *et al.* (2016) on the 1,928 specimens of the Dinard Herbarium, including taxonomic identity, date and place of collection, mapped the presence of 345 species of macroalgae at the three most sampled time periods: T1 (1843–1866), T2 (1910–1931) and T3 (1949–1967). This diachronic information on species occurrence can be used to trace temporal changes in species distribution in two ways. The first analysis performed was to compare the species composition of locations that were sufficiently sampled at different time periods to highlight potentially locally extinct species, that is species collected at a location in one period and not collected at that same location in subsequent periods. The study of the Dinard Herbarium has highlighted that, in the Saint-Malo region, 27 species collected at T2 were not collected at T3.

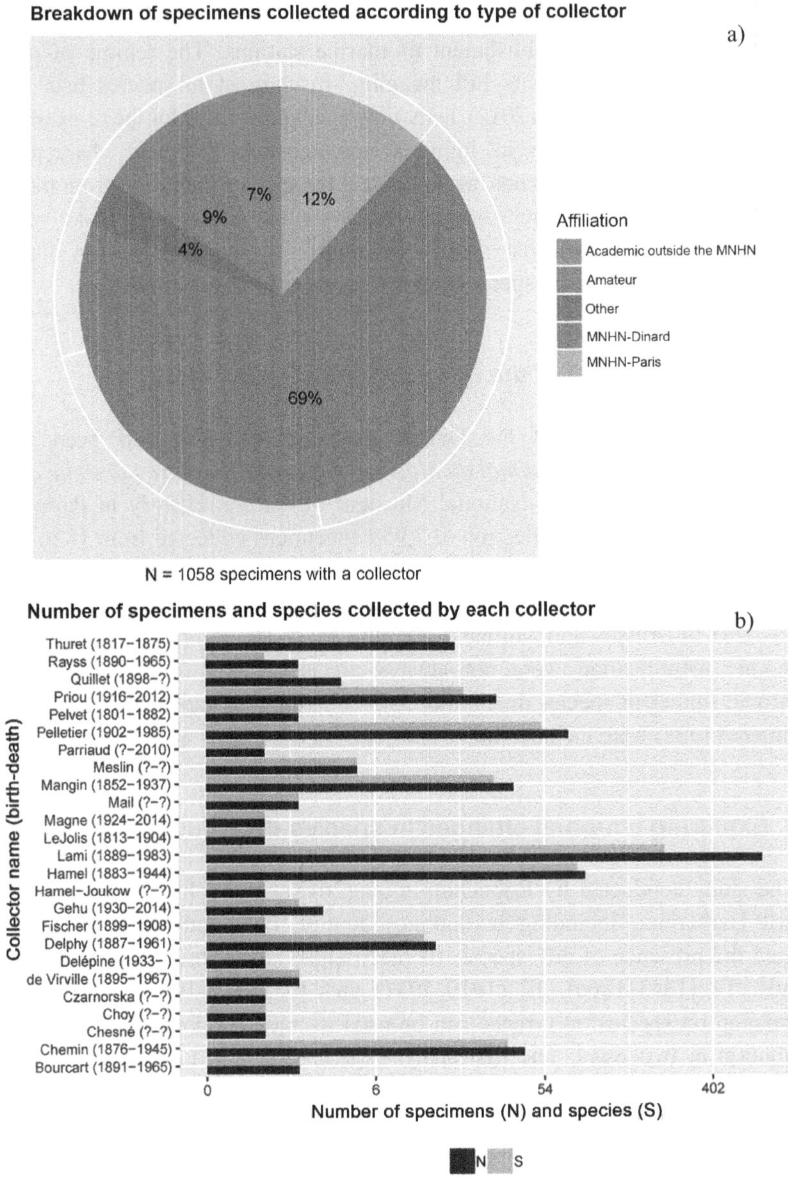

Figure 15.2. *Contribution from collectors to the Dinard Herbarium's algal collection. a) The distribution of the 1,058 specimens with the collector information by the collectors' institution of origin. b) The contribution of the different collectors in terms of number of specimens and species (represented in logarithmic scale for a better readability) brought to this collection*

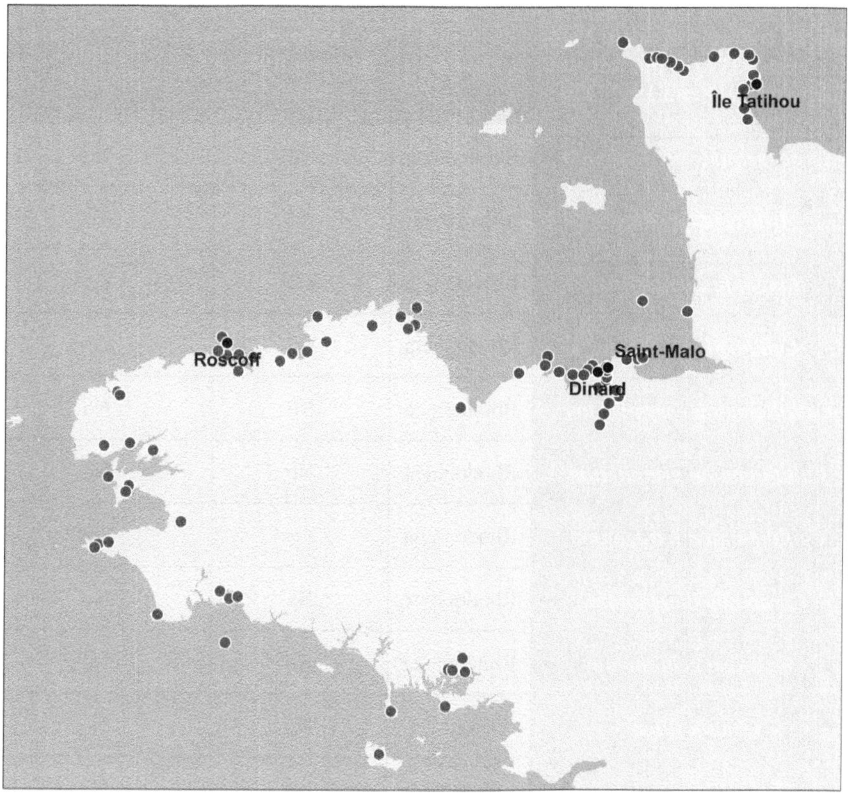

Figure 15.3. *Collection locations of the Dinard Herbarium specimens in Brittany and Normandy*

Because the intensity of sampling at this site is not comparable across time periods, this study alone is not sufficient to draw conclusions about the local extinction of these 27 species. However, it has allowed the initiation of new studies. Thus, a bibliographic study coupled with the study of algal specimens collected between 2007 and 2015 in this region showed that of these 27 species, 8 had been spotted while the remaining 19 had not been spotted since T2 (personal communication from Robuchon and Le Gall) (Table 15.1). These initial results motivated the implementation of new sampling to verify if these 19 species were really locally extinct. The first sampling in this series took place in April 2016. Result: only one of the 19 species was found, the brown algae *Scytosiphon lomentaria* (Table 15.1).

Species collected between 1910 and 1931 (T2) but not between 1949 and 1967 (T3)	Phylum	Spotted between 1931 and 2015	Sampled in 2016
Acrochaetium subtilissimum	Rhodophyta	No	No
Arthrocladia villosa	Ocrophyta	No	No
Bangia fuscopurpurea	Rhodophyta	Yes	Yes
Calliblepharis jubata	Rhodophyta	Yes	Yes
Callithamnion corymbosum	Rhodophyta	No	No
Callithamnion granulatum	Rhodophyta	No	No
Callithamnion tetragonum	Rhodophyta	Yes	Yes
Ceramium diaphanum	Rhodophyta	No	No
Ceramium echionotum	Rhodophyta	No	No
Chorda filum	Ocrophyta	No	No
Cladophora flexuosa	Chlorophyta	No	No
Cutleria multifida	Ocrophyta	Yes	Yes
Cystoseira foeniculacea	Ocrophyta	No	No
Dasya ocellata	Rhodophyta	No	No
Dasya punicea	Rhodophyta	No	No
Desmarestia aculeata	Ocrophyta	No	No
Dictyopteris polypodioides	Ocrophyta	Yes	Yes
Ectocarpus siliculosus	Ocrophyta	Yes	Yes
Eudesme virescens	Ocrophyta	No	No

Halopteris filicina	Ocrophyta	Yes	Yes
Halopteris scoparia	Ocrophyta	No	No
Halurus equisetifolius	Rhodophyta	Yes	Yes
Mesogloia vermiculata	Ocrophyta	No	No
Polysiphonia stricta	Rhodophyta	No	No
Scytosiphon lomentaria	Ocrophyta	No	Yes
Taonia atomaria	Ocrophyta	No	No
Tilopteris mertensii	Ocrophyta	No	No

Table 15.1. *The 27 species of the Dinard Herbarium collected in the Saint-Malo area between 1910 and 1931 (T2) but not between 1949 and 1967 (T3), indicating their phylum of affiliation, whether they were spotted between 1931 and 2015, and whether they were sampled in 2016*

This initial verification sampling thus suggests that 18 species may have become locally extinct in the Saint-Malo area since 1931, including large brown algae such as *Chorda filum*, *Cystoseira foeniculacea* and *Tilopteris mertensii* (Figure 15.4). While the validity of this finding still remains to be confirmed by additional sampling, it is nevertheless consistent with changes in distribution area of red algae species in Brittany (Gallon *et al.* 2014) and the causes of this disappearance remain to be explored.

Another possible analysis would be to use geo-referenced and dated taxonomic occurrences to construct species distribution models (Newbold 2010) and project them to different time periods in order to model temporal evolution in the distribution of these species. Although this approach has been considered with specimens from the Dinard Herbarium, it relies on the availability of historical environmental data (e.g. seawater temperature in the late 19th/early 20th century) at sufficiently fine spatial resolution, which are currently unavailable.

Figure 15.4. *Dinard Herbarium voucher representing three species of large brown algae that have not been documented since 1931: a)* Chorda filum, *b)* Cystoseira foeniculacea *and c)* Tilopteris mertensii

15.2.1. *Perspectives for exploring temporal changes in species distribution*

It is possible to explore temporal changes in species distribution from collections under certain conditions, either by using only the observations or by using those observations to model species distribution. In both cases, only specimens associated with a date and collection location can be used (Figure 15.5).

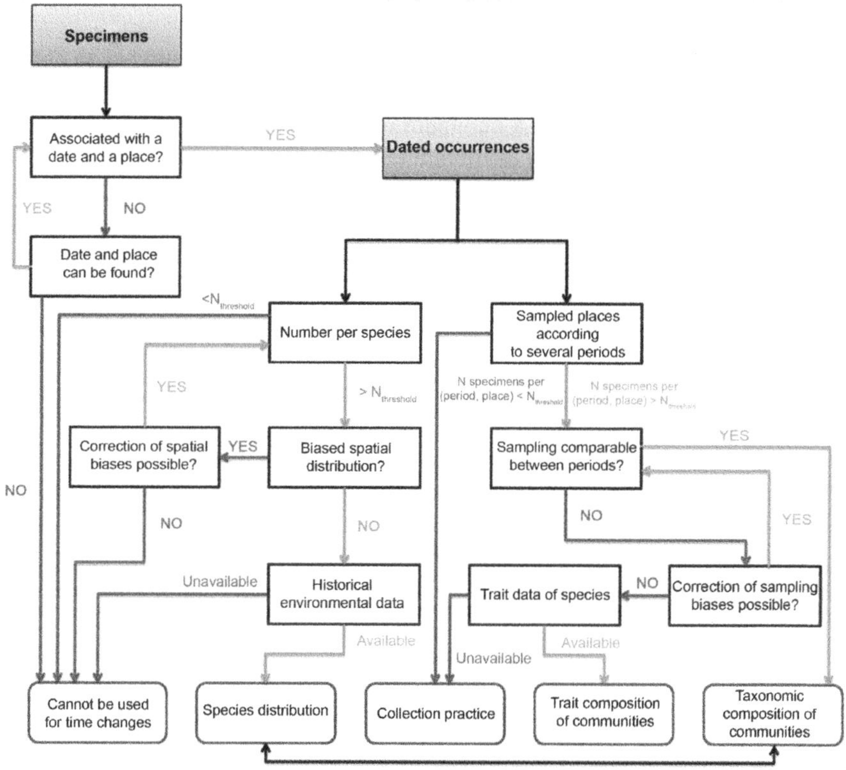

Figure 15.5. *Decision tree showing how collections can be used to trace temporal changes in species distribution and community composition; $N_{thresholds}$ to be defined for each study*

Exploring temporal changes in species distribution using only observations is only possible if the collections can accurately trace temporal changes in community composition (Figure 15.5). In this case, the simultaneous study of temporal changes in community composition in multiple locations enables documentation of local species extinctions or occurrences. Historical collections such as the Dinard Herbarium were established to maximize the number of species globally, not to constitute a reference of algal diversity at given locations. Thus, most of these collections often consist of specimens collected in a very heterogeneous manner and generally do not allow for accurate tracing of temporal changes in community composition unless a very large number of specimens are collected and a selection of specimens is applied to limit sampling bias (Wernberg *et al.* 2011). Ensuring that collections accurately track temporal changes in community composition is therefore a key criterion to consider when defining future sampling strategies.

Exploring temporal changes in species distribution using occurrence data from collections to model species distribution requires several conditions (Figure 15.5). Classically, it is considered that a minimum of 30 occurrences is needed to be able to build and evaluate the species distribution model(s) (Wisz *et al.* 2008). However, analyses can now be performed for smaller samples. Next, the collected data should be checked for spatial sampling bias and a data selection procedure should be applied to correct for this bias if it exists. For example, collection data often have a spatial bias toward locations most accessible to collectors (Newbold 2010). The risk of using such spatially biased data without further correction is to incorrectly estimate the environmental conditions in which species occur and therefore to incorrectly model their niches and distributions (Guillera-Arroita *et al.* 2015). Finally, environmental data corresponding to the sampling periods of the specimens must be available. If these three conditions are met, it will then be possible to build species distribution models from geo-referenced and dated taxonomic occurrences (Newbold 2010) and to project them to different time periods in order to model temporal evolution in the distribution of these species. Jointly projecting the distribution of different species can also trace temporal changes in the taxonomic composition of communities (Figure 15.5).

15.3. Exploring temporal changes in community composition

15.3.1. *Example of the study of the Dinard Herbarium*

In the Dinard Herbarium study, locations sampled at multiple time periods were searched in an effort to trace temporal changes in community composition (Robuchon *et al.* 2016). This search identified one location sampled at both T1 and T2, 13 locations sampled at both T2 and T3, and no common locations sampled during T1 and T3. However, of the 13 locations sampled at both T1 and T2, only Saint-Malo was sampled with more than 30 specimens per period. Since the study of a single location over two periods was not sufficiently informative to trace temporal changes in community composition, Robuchon *et al.* (2016) concluded that the

Dinard Herbarium alone could not trace temporal changes in the composition of algal communities in Brittany.

However, the study of the Dinard Herbarium has revealed the practices of algal collection and the temporal changes in these practices between the late 19th and mid-20th centuries (Robuchon *et al.* 2016). Thus, the different locations of intense collecting identified in the Dinard Herbarium reflect the activity of the itinerant marine laboratory now located in Dinard (Lamy *et al.* 2016). During the first period of intense collecting T1 (1843–1866), the maritime laboratory did not yet exist, but the phycologists Gustave Thuret and Edouard Bornet built up one of the largest collections of Channel coast algae. They collected only one or two specimens per taxon, which reveals that the method of the time was to establish an exhaustive inventory of the species found in the Channel and not to faithfully represent the composition of the algal communities in a given place. The second period of intense collection T2 (1910–1931) followed the construction of the maritime laboratory in Tatihou (1888) and included the move of the maritime laboratory from Tatihou to Saint-Servan, at the entrance to the Rance estuary (1923). During this second period, the specimens were mainly collected by the phycologist Robert Lami near the maritime laboratory of Saint-Servan. The large proportion of common taxa (compared to rarer taxa) in his collections as well as the larger number of specimens collected per taxon (compared to the previous period) indicate that the collecting practices of the time were more concerned with describing the composition of algal communities rather than the search for rare taxa. Robert Lami was also the main collector of specimens from the Dinard Herbarium during the third period of intense collecting (1949–1967) which followed the move of the marine laboratory between Saint-Servan and Dinard, a few kilometers on the other bank of the Rance Maritime (a canal) (1935). This third period of intense collecting was characterized by a low number of specimens per taxon and an interest in rare taxa, an approach similar to that of the first period. This was consistent with the reception of taxonomic specialists at the Dinard maritime laboratory, who preferentially sampled rare species.

15.3.2. *Perspectives for exploring temporal changes in community composition*

As with exploring temporal changes in species distribution, tracing temporal changes in communities from collections requires a number of conditions (Figure 15.4). The first is to have dated taxonomic occurrences. Next, it is necessary to ensure that the number of specimens sampled per time period and location is sufficiently large. Then, it is important to check that the sampling is comparable between periods (e.g. methods used, communities targeted and seasonality). If this is the case, comparing the taxonomic composition of the communities between periods for each location allows one to directly trace the temporal changes in the taxonomic composition of the communities. If this is not the case (as is most often the case with collection data), then either a sampling bias correction procedure must be applied in order to be able to compare the taxonomic composition of communities between time periods, or communities must be compared between time periods via indicators (other than taxonomic composition) that are less sensitive to sampling bias. This is the case for certain indicators based on species traits that provide information on, for example, their distribution, habitat, shape or life history. For instance, the indicator of average community temperature (Devictor *et al.* 2008), which is based on the thermal niche of species, is one of the indicators that can be used. Indeed, there is no reason for collectors to consciously choose to collect warm water species or cold water species. Thus, tracing temporal changes in the mean temperature of communities is a way to characterize the rate at which communities have "warmed up", which can then be compared to the rate of global warming to study how communities are keeping pace with the issue (Devictor *et al.* 2008, 2012). However, this approach requires trait information (e.g. thermal niche for the average community temperature indicator) for all species in the communities.

In cases where collections do not meet any of the conditions for studying changes in community composition, studying the spatiotemporal dynamics of collection can provide information on the history of collection practices (Robuchon *et al.* 2016).

15.4. Conclusion: sampling and analysis strategies for the future

Algal community composition and species distribution are being impacted by global change (Wernberg *et al.* 2011; Gallon *et al.* 2014) and in particular by increasing seawater temperatures and heat waves. Using algal collections to track these changes when these collections have not been established to address this question is proving to be particularly challenging.

In order to monitor these changes, one solution would be to set up long-term series with associated specimens to enrich the collections. Maintaining tissue collections to preserve DNA would allow us to consider using the tools of molecular systematics and population genomics to characterize the dynamics of these organisms in the face of climate change. Failing that, floristic surveys could document changes in species distribution and species composition of communities, while being aware of the limitations of reliable identification, which is particularly difficult for algae and could introduce bias. To limit this bias, sampling should be concentrated on the most easily identifiable species, especially for those whose distribution is changing due to global change. Environmental DNA could also be a new collection item, but the spatial scale it covers and the remanence times still need to be further characterized.

An alternative way to track changes in algal community composition would be to use analytical methods that allow the study of algal community responses from a large number of collections of diverse origin. These methods involve either correcting for sampling bias (Isaac *et al.* 2014) or using indicators that are not very sensitive to sampling bias (Devictor *et al.* 2008, 2012). They have already been used to identify signals of change in community composition from participatory science data, and their use in analyzing heterogeneous collections remains to be tested.

15.5. References

Adl, S.M., Bass, D., Lane, C.E., Lukeš, J., Schoch, C.L., Smirnov, A., Agatha, S., Berney, C., Brown, M.W., Burki, F., Cárdenas, P., Čepička, I., Chistyakova, L., del Campo, J., Dunthorn, M., Edvardsen, B., Eglit, Y., Guillou, L., Hampl, V., Heiss, A.A., Hoppenrath, M., James, T.Y., Karnkowska, A., Karpov, S., Kim, E., Kolisko, M., Kudryavtsev, A., Lahr, D.J.G., Lara, E., Le Gall, L., Lynn, D.H., Mann, D.G., Massana, R., Mitchell, E.A.D., Morrow, C., Park, J.S., Pawlowski, J.W., Powell, M.J., Richter, D.J., Rueckert, S., Shadwick, L., Shimano, S., Spiegel, F.W., Torruella, G., Youssef, N., Zlatogursky, V., Zhang Q. (2019). Revisions to the classification, nomenclature, and diversity of eukaryotes. *Journal of Eukaryotic Microbiology*, 66, 4–119.

Crouan, P.L. and Crouan, H.M. (1867). *Florule du Finistère*. F. Klincksieck, Paris.

Daston, L. and Galison, P. (2012). *Objectivité*. Les presses du réel, Dijon.

Devictor, V., Julliard, R., Couvet, D., Jiguet, F. (2008). Birds are tracking climate warming, but not fast enough. *Proceedings of the Royal Society B: Biological Sciences*, 275, 2743–2748.

Devictor, V., van Swaay, C., Brereton, T., Brotons, L., Chamberlain, D., Heliölö, J., Herrando, S., Julliard, R., Kuussaari, M., Lindström, Å., Reif, J., Roy, D.B., Schweiger, O., Settele, J., Stefanescu, C., van Strien, A., van Turnhout, C., Vermouzek, Z., WallisDeVries, M., Wynhoff I., Jiguet F. (2012). Differences in the climatic debts of birds and butterflies at a continental scale. *Nature Climate Change*, 2, 121–124.

Gallon, R.K., Robuchon, M., Leroy, B., Le Gall, L., Valero, M., Feunteun, E. (2014). Twenty years of observed and predicted changes in subtidal red seaweed assemblages along a biogeographical transition zone: Inferring potential causes from environmental data. *Journal of Biogeography*, 41, 2293–2306.

Guillera-Arroita, G., Lahoz-Monfort, J.J., Elith, J., Gordon, A., Kujala, H., Lentini, P.E., McCarthy, M.A., Tingley, R., Wintle B.A. (2015). Is my species distribution model fit for purpose? Matching data and models to applications. *Global Ecology and Biogeography*, 24, 276–292.

Isaac, N.J.B., van Strien, A.J., August, T.A., de Zeeuw, M.P., Roy, D.B. (2014). Statistics for citizen science: Extracting signals of change from noisy ecological data. *Methods in Ecology and Evolution*, 5, 1052–1060.

Lamy, D., Robuchon, M., Kervran, L., Dennetiere, B., Feunteun, E., Le Gall, L. (2016). Dinard Herbarium: History of a marine station. *Cryptogamie, Algologie*, 37, 29–45.

Leliaert, F., Verbruggen, H., Vanormelingen, P., Steen, F., López-Bautista, J.M., Zuccarello, G.C., De Clerck O. (2014). DNA-based species delimitation in algae. *European Journal of Phycology*, 49, 179–196.

Mann, K.H. (1973). Seaweeds: Their productivity and strategy for growth. *Science*, 182, 975–981.

Newbold, T. (2010). Applications and limitations of museum data for conservation and ecology, with particular attention to species distribution models. *Progress in Physical Geography*, 34, 3–22.

Robuchon, M., Le Gall, L., Gey, D., Valero, M., Vergés, A. (2014). Kallymenia crouaniorum (Kallymeniaceae, Rhodophyta), a new red algal species from the *Laminaria hyperborea* understorey community. *European Journal of Phycology*, 49, 493–507.

Robuchon, M., Valero, M., Gey, D., Le Gall, L. (2015). How does molecular-assisted identification affect our estimation of α, β and γ biodiversity? An example from understory red seaweeds (Rhodophyta) of *Laminaria* kelp forests in Brittany, France. *Genetica*, 143, 207–223.

Robuchon, M., Lamy, D., Kervan, L., Dennetiere, B., Julliard, R., Le Gall, L. (2016). Dinard Herbarium: A source of information to infer temporal changes in seaweed communities. *Cryptogamie, Algologie*, 37, 47–60.

Wernberg, T., Russell, B.D., Thomsen, M.S., Gurgel, C.F.D., Bradshaw, C.J.A., Poloczanska, E.S., Connell, S.D. (2011). Seaweed communities in retreat from ocean warming. *Current Biology*, 21, 1828–1832.

Wisz, M.S., Hijmans, R.J., Li, J., Peterson, A.T., Graham, C.H., Guisan, A., NCEAS Predicting Species Distributions Working Group (2008). Effects of sample size on the performance of species distribution models. *Diversity and Distributions*, 14, 763–773.

16

Herbaria, Witnesses of the Stakes of Biodiversity Conservation and the Impacts of Global Changes

Serge MULLER, Vanessa R. INVERNÓN and Germinal ROUHAN
ISYEB – CNRS, Muséum national d'Histoire naturelle, Sorbonne Université, École Pratique des Hautes Études, Université des Antilles, Paris, France

In addition to their fundamental role in systematics and taxonomy, herbaria (annotated collections of pressed and dried plant specimens) have great potential for environmental studies. They can contribute to the evaluation of the floristic knowledge of territories and to the identification of "hotspots" of plant biodiversity; to the characterization of floristic regressions of certain species; and to the study of the history of introductions and expansion of invasive exotic species or of symbionts (viruses, bacteria, fungi and insects) collected unintentionally with the plant samples.

Herbarium specimens can also be used to characterize environmental changes in territories such as air, water and soil quality (linked to local pollution or more global changes), as well as the impacts of climate change on the phenology and possibly the morphology of plant species.

For a color version of all figures in this chapter, see www.iste.co.uk/pellens/natural.zip.

Natural History Collections in the Science of the 21st Century, coordinated by Roseli PELLENS.

A condition for the continued interest of herbarium collections in the 21st century is that they continue to be "alive", that is regularly enriched by new plant collections.

16.1. Introduction

NOTE.– *What is a herbarium?*

A herbarium is a collection of plants pressed, dried and fixed on a paper medium. The infrastructure that preserves this collection is also called a "Herbarium". Specifically, a herbarium sample or specimen is a scientific object if it contains, in addition to the plant itself (including as many organs as possible: roots, stems, leaves, flowers, fruits, seeds, etc.), associated data that are usually recorded on labels. These data associated with the gathering, inform about the place and date of gathering and the name of the collector; these basic data can be completed with the altitude, the habitat, the type of soil, the phytosociology and any other type of information that could be potentially useful with regard to the gathering. Once in the collection, a herbarium specimen is determined by a botanist and thus receives a scientific name; this name can eventually be questioned, on the one hand, following later studies on this sample, and on the other hand following the progression of knowledge on the classification of plants. The history of observations and identifications is preserved in the form of labels glued successively on the herbarium sheet. The scientific life of a herbarium specimen never ends.

Herbaria are of great interest because they are custodians of the richness and variations of the flora of a given locality. Moreover, the herbaria collected over the centuries testify with their labels to the evolution of the diversity of plants in space (by exploiting the localities of collection) and in time (by exploiting the collection dates). Thus, herbaria are the only verifiable data that document the presence of a species at a given place and time. This is why herbarium data are fundamental for biodiversity research.

The first herbaria were created in Italy in the 16th century. These collections of pressed and dried plants then underwent a significant development during the following centuries. According to the *Index Herbariorum*, there are approximately 3,000 active public herbaria currently inventoried in 176 countries, comprising over 380 million herbarium specimens (Nualart *et al.* 2017; Thiers 2018).

Their use undoubtedly suffered at the end of the 20th century from the rise of new and promising molecular tools for studying flora. However, numerous works carried out in recent decades have restored their usefulness, both for taxonomy and for the conservation of plant material allowing genetic analyses (Besnard *et al.* 2018).

The classical use of herbaria remains, in fact, their use for the description and the study of the variability of taxa, and their classification. In addition to the morpho-anatomical characters of plants, herbarium specimens, even dry and old ones, can also provide DNA and be used in molecular phylogeny studies (Savolainen *et al.* 1995; Wandeler *et al.* 2007). They can also be used to study the evolution of genetic diversity, as has been well illustrated in Italy with the species *Anacamptis palustris* (Jacq.) R.M. Bateman, Pridgeon and M.W. Chase (Cozzolino *et al.* 2007). It is thus increasingly recommended that the collection of herbarium samples be accompanied by the systematic collection and long-term conservation (in freezer rooms) of plant material allowing these genetic studies (Gaudeul and Rouhan 2013).

In addition to this fundamental use in taxonomy and systematics, herbaria have been the subject of numerous studies over the last few decades on their use in understanding changes in biodiversity and environmental conditions.

It is mainly English-speaking countries that have developed this valorization of herbaria until now. Among the first important publications on the subject, we can mention Shetler's synthesis "The herbarium: past, present and future" (Shetler 1969).

The study by Teece *et al.* (2002) on the use of the Lewis and Clark herbarium, collected between 1804 and 1806 (two centuries ago) between St. Louis (Missouri) and the Pacific Ocean, is a real plea for the use of herbariums for environmental applications. The enthusiasm for this valorization of herbaria led Funk to write an article entitled "100 uses of an herbarium (well at least 72)" (Funk 2003). A remarkable synthesis of the use of herbaria for biogeographic and environmental studies was more recently published by Lavoie (2013).

The digitization of herbaria, undertaken on a large scale since that of the Paris Herbarium in the early 2010s, makes images of millions of plants available to all (Le Bras *et al.* 2017).

This chapter repeats, completes and updates a first note on the use of herbaria in highlighting environmental change presented at the international conference *Botanists of the Twenty-first Century – Roles, Challenges and Opportunities*, organized by UNESCO in September 2014 and published in the proceedings of this conference (Muller 2016).

16.2. Evaluation of the floristic richness and conservation issues of territories

Herbarium samples are associated with more or less precise initial data on the collection (in particular, date and place), classically recorded on a label. These chorological data allow us to evaluate the floristic richness of territories (e.g. Peat *et al.* (2007) for Antarctic flora), and to define the potential distribution areas of the species (Hernandez and Navarro 2007). These data are very useful in ecology and biogeography, notably in orienting searches for rare or unknown taxa in under-surveyed areas, as well as in defining conservation priorities and delimiting the relevant perimeters of areas to be protected (e.g. McDougal *et al.* (1998) in New Brunswick; Funk *et al.* (1999) in Guyana; Gimaret-Carpentier *et al.* (2003) in India; and Ungricht *et al.* (2005) in New Caledonia).

The data from the herbaria can also be very useful to orientate the research of species considered to be extinct, like *Nesogenes orerensis* (Cordem.) Marais [Orobanchaceae] (Figure 16.1), a taxon endemic to Reunion Island, which was found in October 2005 by botanists of the *Conservatoire botanique national de Mascarin* (CBN), within the framework of an emergency plan, using information from the herbarium plates of the only two collections made by L.H. Boivin in 1851 and E.J. de Cordemoy some years later.

In French Guiana, data from the Cayenne Herbarium (CAY) have been used extensively in the inventory of ZNIEFFs (*zones naturelles d'intérêt écologique, faunistique et floristique*, natural areas of ecological, faunistic and floristic interest) in this territory and were used to produce zonings for the SDOM (*Schéma départemental d'orientation minière*, departmental mining guidance plan) in 2008. Herbaria data are also mobilized for the monitoring of Natura 2000 sites by niche modeling established from herbarium samples (Amici *et al.* 2014).

From the old data provided by the herbaria, it is also possible to evaluate, by comparison with current data, the losses of biodiversity and the extinction phenomena of species, as illustrated by Lienert *et al.* (2002) for *Swertia perennis* L. in Switzerland or by Case *et al.* (2007) for *Panax quinquefolius* L. in the United States.

Figure 16.1. *Herbarium plate of* Nesogenes orerensis *collected by L.H. Boivin in 1851 (P00227968, MNHN Herbarium)*

For example, the collection of *Botrychium multifidum* (S.G. Gmel.) Rupr. made on August 17, 1913 by L. de Vergnes (Figure 16.2) in the Tanet massif attests to the presence of this species a century ago in the Hautes-Vosges (France), where it has not been mentioned since this collection and is considered extinct (Muller 2006).

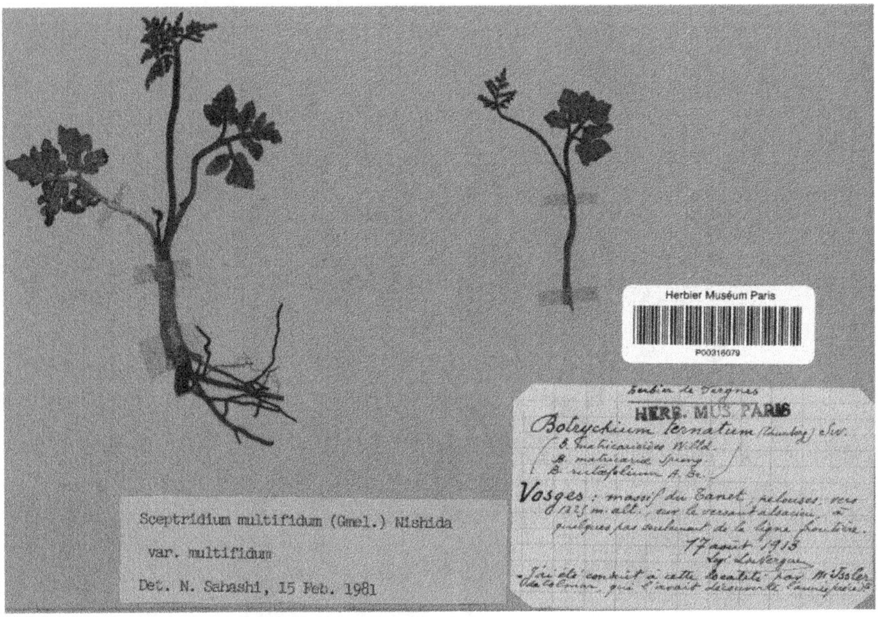

Figure 16.2. *Herbarium sheet of* Botrychium multifidum *(S.G. Gmel.) Rupr. collected in the Tanet massif (Vosges, France) by L. de Vergnes on August 17, 1913 (P00316079, MNHN Herbarium)*

In some cases, the herbarium samples provide information on other components of the biocenoses that may have been collected, usually unintentionally, with the botanist's target plant samples. For example, studies have been carried out in the Netherlands on diatom communities attached to samples of aquatic macrophyte beds, such as *Potamogeton pectinatus* L., *Najas marina* L., *Chara hispida* L. (van Dam and Mertens 1993), or in the Czech Republic on diatoms attached to sphagnum mosses (Poulickova *et al.* 2013), to characterize habitat quality. More recently, herbarium samples have been used to characterize plant mycorrhizal communities (Heberling and Burke 2019).

In France, Bachelard and Chabrol (2010) have highlighted the presence of eggs deposited by the butterfly *Phengaris alcon alcon* D. & S. on herbarium samples of *Gentiana pneumonanthe* L., attesting to the ancient presence (at the time of the plant's collection) of the butterfly and thus, in some cases, its persistence over an interval period of more than 130 years. Hembry (2013), on the other hand, used herbarium samples of *Phyllanthus wilderi* (J. Florence) W.L. Wagner and Lorence to confirm the disappearance of butterflies with leaf miner caterpillars of the family *Gracillariidae* as a result of deforestation on the island of Mangareva in the Gambier Archipelago (French Polynesia).

Beauvais *et al.* (2017) highlighted the value of herbaria for assessing the impact of deer grazing on plants based on the calculation of the leaf area of 692 herbarium specimens of white trillium (*Trillium grandiflorum* (Michx.) Salisb.) collected in Quebec, Canada, and then comparing them with modern specimens sampled in 2013 and 2014 at sites with white-tailed deer (*Odocoileus virginianus* Zimmermann) densities greater than eight individuals per square kilometer or without deer.

16.3. Studies of introduction pathways and colonization of invasive exotic plants and pathogens

Data on the location and date of collection of samples of introduced and invasive exotic species can be used to trace the history of their introduction and to investigate the underlying mechanisms of their establishment and colonization. This information can be used for several purposes: (1) the reconstruction of the history of the discovery of new exotic species in territories, (2) the study of the dynamics of colonization by exotic plants, (3) the characterization of the habitats of exotic species in the areas of introduction and possible modifications of these habitats, (4) the analysis and comparison of genetic variability of populations of introduced exotic species in territories of origin and naturalization, and (5) the evaluation of the importance of exotic species in relation to native species in floristic corteges and their temporal modifications (Muller 2015).

This topic has been the subject of numerous studies in recent decades, which have made it possible, thanks to the information provided by herbarium samples, to trace the history of colonization by species that have become invasive, such as *Heracleum mantegazzianum* Somm. and Lev. in Central Europe (Pysek 1991), *Ambrosia artemisiifolia* L. in France (Chauvel *et al.* 2006) (Figure 16.3) and in Europe (Gaudeul *et al.* 2011), *Fallopia japonica* (Houtt.) Ronse Decr in America (Barney 2006) and three exotic species of *Solidago* L. in Europe (Weber 1998).

Figure 16.3. *Collection of* Ambrosia artemisiifolia *L. by F. Gagnepain on September 21, 1898 in a field in Cercy (France) (P02435743, MNHN Herbarium)*

Herbarium data have also established and discussed the notion of a lag phase between the introduction and expansion of a species (Delisle *et al.* 2003; Aikio *et al.* 2010). A recent synthesis (Antunes and Schamp 2017) emphasizes the value of herbarium data in being able to predict invasions.

Herbaria have also been used to study the history and characterize pathogens and pests of exotic origin in certain plants. For example, as early as 1933, Jenkins and Fawcett (1933) used herbarium samples of *Citrus* L. preserved in English and American herbaria to date the first samples of the species infested by the bacterium *Pseudomonas citri* Hasse to the 1830s (Fawcett and Jenkins 1933; Jenkins and Fawcett 1933). But perhaps the most famous case of using herbarium samples to study pathogens is that of potato (*Solanum tuberosum* L.), whose parasitic oomycete (*Phytophthora infestans (*Mont.) de Bary) that caused the Irish famine could be clearly identified genetically through analyses of 28 herbarium samples collected between 1845 and 1847 in Ireland (Ristaino *et al.* 2001). In Australia, the examination of more than 500 herbarium samples of two species of *Eucalyptus* L'Hér. has allowed for the detection and study of the histology of the spread of "minor" butterflies of the genus *Perthida* Common (Abbott *et al.* 1999). More recently in Europe, Lees *et al.* (2011) were able to locate in the Balkans the territory of origin of the horse-chestnut leaf miner (*Cameraria ohridell* Deschka and Dimic), a species that strongly impacts horse-chestnut trees, thanks to caterpillars of the butterfly preserved in herbarium samples collected in 1879 and that could be identified by DNA amplification and "barcoding". Finally, a study of 42,000 herbarium specimens for flower anther-smut disease (caused by the basidiomycete *Microbotryum violaceum*) led to the understanding that the disease was correlated with certain species life history traits (lifespan) rather than with an evolutionarily consistent group of plants (Hood *et al.* 2010). Other such studies could still be cited, all based on the search for herbarium samples infected with pathogens or pests.

16.4. Analysis of the impact of pollution and changes in air quality

The use of herbarium specimens to assess the environmental conditions of the stations from which the herbaria were collected has been studied extensively since the 1970s. In some cases, chemical analyses have been carried out to reveal natural metal contents of the substrates from which the plants were collected (Brooks *et al.* 1977; Faucon *et al.* 2010).

Numerous studies have involved the determination of heavy metals in herbarium samples to characterize local pollution. In particular, bryophytes and lichens are used as bio-accumulators of these heavy metals. Thus, Herpin *et al.* (1997) compared the heavy metal content (arsenic, cadmium, chromium, copper, lead, nickel, vanadium and zinc) of herbarium samples collected between 1845 and 1901 with samples collected in 1991 in order to analyze the increasing air pollution in the

19th century, then its recent decrease due to efforts to reduce emissions of these metals. Similar analyses have been carried out in Spain (Penuelas and Filella 2002) and the United Kingdom (Shotbolt *et al.* 2007). In Switzerland, Weiss *et al.* (1999) used herbarium samples of sphagnum mosses to study changes in lead isotopes over the last 130 years.

Studies have also been conducted on changes in nitrogen levels of bryophytes in relation to atmospheric deposition, for example (Baddeley *et al.* 1994) on *Racomitrium lanuginosum* (Hedw.) Brid. in Great Britain from herbarium samples collected from the mid-19th century onward. Wilson *et al.* (2009) compared nitrogen levels in bryophyte tissues collected in the vicinity of the Cape Town metropolis in South Africa from 1875 onward with recent samples to assess increases in atmospheric nitrogen deposition over the 20th century.

A decrease in stomatal density, following an increase in CO_2 content, was demonstrated in the United Kingdom on leaf samples of woody species (Woodward 1987; Beerling and Chaloner 1993). Similar results were then obtained in North America (Kouwenberg *et al.* 2003) and in tropical forests in Australia (Greenwood *et al.* 2003). More recently, a study was conducted in French Guiana on the functional response of two tropical forest woody species based on C and O isotope analysis of leaf herbarium samples collected over a 200-years period (Bonal *et al.* 2011). Herbarium samples thus appear to be valuable witnesses of the anatomical and physiological responses of species to changes in CO_2 levels.

Finally, an increase in flavonoid concentrations in *Bryum argenteum* Hedw. moss was established in response to ozone depletion and increased UVB in herbarium samples collected in Antarctic regions (Ryan *et al.* 2009).

16.5. Study of phenological changes in flora as a result of climate change

The collection dates of herbarium samples provide data for the study of variations in the phenology of flora over the last two centuries, which may reveal an effect of climate change on the flora (Figure 16.4). This research theme has developed significantly in recent years, following evidence in the 1990s of the reality of such climate change-induced phenological changes in flora (Bradley *et al.* 1999; Fitter and Fitter 2002).

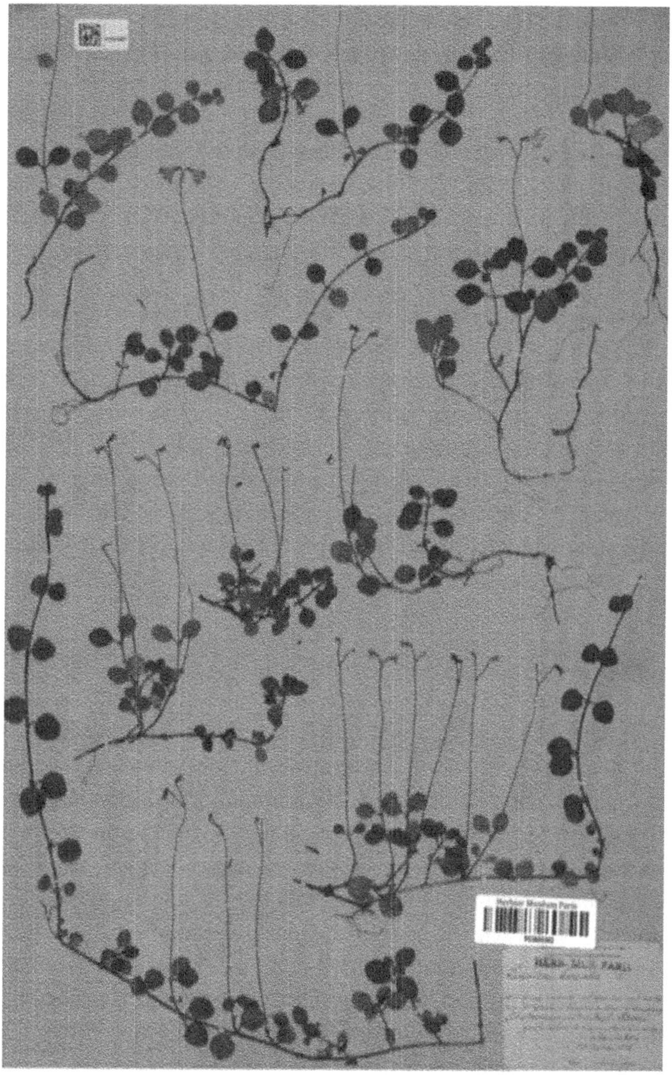

Figure 16.4. *Herbarium plate of a collection of* Linnaea borealis *L. made on August 10, 1908 in Champagny-en-Vanoise (France) in a well-identified station, still existing and thus allowing a comparison a century later of the phenological stage at the same date (P03686362, MNHN Herbarium)*

The first studies using herbarium samples date back about 20 years. First, Primack *et al.* (2004) compared the flowering times during 2003 of 229 plants from the Arnold Arboretum in Boston (Massachusetts), the oldest in the United States, with flowering data from herbarium samples collected in the same arboretum since 1885. They showed an average of eight days earlier flowering over the last century.

A few years later, Houle (2007) compared the flowering dates of vernal plants in deciduous forests in Quebec and found that they flowered two to six days earlier at the end of the 20th century than at the beginning. More recently, a study in the Australian mountains used herbarium samples to identify alpine (Australian) species whose altered phenology may reflect climate change (Gallagher *et al.* 2009).

Much work has been published in recent years (Davis *et al.* 2015; Park and Schwartz 2015; Willis *et al.* 2017a; Park *et al.* 2018) that has confirmed the value of herbaria in characterizing phenological changes. A synthesis of methods and applications of these has been published recently. Jones and Daehler (2018) highlight the increasing importance of herbaria following their digitization. Spellman and Mulder (2016) used phenological observations collected in a participatory science program to evaluate phenological models developed on the basis of herbarium samples. Some other participatory science programs focus on herbarium phenology to contribute to climate change studies (Willis *et al.* 2017b).

16.6. Conclusion

Herbaria appear to be very good witnesses of the flora present at the time of the collection of plants, as well as of the modifications of these floristic communities over time, but also as indicators of the ecological and environmental changes in these territories. Thus, herbaria constitute a gigantic library, with their metadata increasingly accessible in digital form and the possibility of sequencing the DNA of preserved samples, which make them an invaluable resource for understanding the ecological and evolutionary responses of plants to global environmental changes. Herbaria provide information on long-term effects of at least four drivers of these global changes: pollution, habitat modification, climate change and invasive species (Lang *et al.* 2018). These findings argue for heightened interest in these showcases of flora and human activities of past centuries.

Yet, it has been established that collections of herbarium samples have significantly decreased in North America in recent decades (McDougall *et al.* 1998; Prather *et al.* 2004; Beauvais *et al.* 2017). The same is likely to be true in Europe and France.

It would thus certainly be very useful to relaunch campaigns for the collection of herbarium samples, perhaps more targeted according to identified objectives. These operations could be carried out, for the French territory, in partnership with the national and regional botanical associations, as well as with the national botanical conservatories, with the objective of maintaining "living" herbaria, that is regularly fed by plant collections.

16.7. References

Abbott, I., Wills, A., Burbidge, T. (1999). Historical incidence of *Perthida* leafminer species (Lepidoptera) in southwest Western Australia based on herbarium specimens. *Australian Journal of Ecology*, 24, 144–150.

Aikio, S., Duncan, R.P., Hulme, P.E. (2010). Lag-phases in alien plant invasions: Separating the facts from artefacts. *Oikos*, 119, 370–378.

Amici, V., Geri, F., Bonini, I., Rocchini, D. (2014). Ecological niche modelling with herbarium data: A framework to improve Natura 2000 habitat monitoring. *Applied Ecology and Environmental Research*, 12(3), 645–659.

Antunes, P.M. and Schamp, B. (2017). Constructing standard invasion curves from herbarium data – Toward increased predictability of plant invasions. *Invasive Plant Science and Management*, 10, 293–303.

Bachelard, P. and Chabrol, L. (2010). Témoignages de la présence de Phengaris alcon alcon (Lep. Lycaenidae) dans les herbiers anciens. *Oreina*, 9, 4–7.

Baddeley, J.A., Thompson, D.B.A., Lee, J.A. (1994). Regional and historical variation in the nitrogen-content of Racomitrium lanuginosum in Britain in relation to atmospheric nitrogen deposition. *Environmental Pollution*, 84, 189–196.

Barney, J.N. (2006). North American history of two invasive plant species: Phytogeographic distribution, dispersal vectors, and multiple introductions. *Biological Invasions*, 8, 703–717.

Beauvais, M.-P., Pellerin, S., Dubé, J., Lavoie, C. (2017). Herbarium specimens as tools to assess the impact of large herbivores on plant species. *Botanique*, 95(2), 153–162.

Beerling, D.J. and Chaloner, W.G. (1993). The impact of atmospheric CO2 and temperature change on stomatal density: Observations from Quercus robur lammas leaves. *Annals of Botany*, 71, 231–235.

Besnard, G., Gaudeul, M., Lavergne, S., Muller, S., Rouhan, G., Sukhorukov, A.P., Vanderpoorten, A., Jabbour, F. (2018). Editorial: Herbarium-based science in the 21st century. *Bot. Letters*, 165(3–4), 323–327.

Bonal, D., Ponton, S., Le Thiec, D., Richard, B., Ningre, N., Herault, B., Ogee, J., Gonzalec, S., Pignal, M., Sabatier, D., Guehl, J.M. (2011). Leaf functional response to increasing atmospheric CO2 concentrations over the last century in two northern Amazonian tree species: a historical $\partial^{13}C$ and $\partial^{18}O$ approach using herbarium samples. *Plant, Cell & Environment*, 34, 1332–1344.

Bradley, N.L., Leopold, A.C., Ross, J., Huffaker, W. (1999). Phenological changes reflect climate change in Wisconsin (climate warming/phenology/flowering/bird migration/ photoperiod). *Proceedings of the National Academy of Sciences of the USA*, 96, 9701–9704.

Brooks, R.R., Lee, J., Reeves, R.D., Jaffré, T. (1977). Detection of nickeliferous rocks by analysis of herbarium specimens of indicator plants. *Journal of Geochemical Exploration*, 7, 49–57.

Case, M.A., Flinn, K.M., Jancaitis, J., Alley, A., Paxton, A. (2007). Declining abundance of American ginseng (Panax quinquefolius L.) documented by herbarium specimens. *Biological Conservation*, 134, 22–30.

Chauvel, B., Dessaint, F., Cardinal-Legrand, C., Bretagnolle, F. (2006). The historical spread of Ambrosia artemisiifolia in France from herbarium records. *Journal of Biogeography*, 33, 665–673.

Cozzolino, S., Cafasso, D., Pellegrino, G., Musacchio, A., Widmer, A. (2007). Genetic variation in time and space: The use of herbarium specimens to reconstruct patterns of genetic variation in the endangered orchid *Anacamptis palustris*. *Conservation Genetics*, 8, 629–639.

Crawford, P.H.C. and Hoagland, B.W. (2009). Can herbarium records be used to map alien species invasion and native species expansion over the past 100 years? *Journal of Biogeography*, 36, 651–661.

van Dam, H. and Mertens, A. (1993). Diatoms on herbarium macrophytes as indicators for water quality. *Hydrobiologia*, 269–270, 437–445.

Daru, B.H., Park, D.S., Primack, R.B., Willis, C.G., Barrington David, S., Whitfeld, T.J.S., Seidler, T.G., Sweeney, P.W., Foster, D.R., Ellison, A.M., Davis, C.C. (2017). Widespread sampling biases in herbaria revealed from large-scale digitization. *New Phytologist*, 217, 939–955.

Davis, C.C., Willis, C.G., Connolly, B., Kelly, C., Ellison, A.M. (2015). Herbarium records are reliable sources of phenological change driven by climate and provide novel insights into species' phenological cueing mechanisms. *American Journal of Botany*, 102(10), 1–11.

Delisle, F., Lavoie, C., Jean, M., Lachance, D. (2003). Reconstructing the spread of invasive plants: Taking into account biases associated with herbarium specimens. *Journal of Biogeography*, 30, 1033–1042.

Diskin, E., Proctor, H., Jebb, M., Sparks, T., Donnelly, A. (2012). The phenology of Rubus fruticosus in Ireland: Herbarium specimens provide evidence for the response of phenophases to temperature, with implications for climate warming. *International Journal of Biometeorology*, 56(6), 1103–1111.

Faucon, M.-P., Meersseman, A., Ngoy Shutcha, M., Mahy, G., Ngomgo Luhembwe, M., Malaisse, F., Meerts, P. (2010). Copper endemism in the Congolese flora: A database of copper affinity and conservational value of cuprophytes. *Plant Ecology and Evolution*, 143, 5–18.

Fawcett, H.S. and Jenkins, A.E. (1933). Records of citrus canker from herbarium specimens of the genus Citrus in England and the United States. *Phytopathology*, 23, 820–824.

Fitter, A.H. and Fitter, R.S.R. (2002). Rapid changes in flowering time in British plants. *Science*, 296, 1689–1691.

Funk, V.A. (2003). 100 uses for an herbarium (well at least 72). *American Society of Plant Taxonomists Newsletter*, 17, 17–19.

Funk, V.A., Zermoglio, M.F., Nasir, N. (1999). Testing the use of specimen collection data and GIS in biodiversity exploration and conservation decision making in Guyana. *Biodiversity and Conservation*, 8, 727–751.

Gallagher, R.V., Hughes, L., Leishman, M.R. (2009). Phenological trends among Australian alpine species: Using herbarium records to identify climate-change indicators. *Australian Journal of Botany*, 57, 1–9.

Gaudeul, M. and Rouhan, G. (2013). A plea for modern botanical collections to include DNA-friendly material. *Trends in Plant Science*, 18, 184–185.

Gaudeul, M., Giraud, T., Kiss, L., Shykoff, J.A. (2011). Nuclear and chloroplast microsatellites show multiple introductions in the worldwide invasion history of Common ragweed *Ambrosia artemisiifolia*. *PLoS ONE*, 6(3), e17658.

Gimaret-Carpentier, C., Dray, S., Pascal, J.-P. (2003). Broad-scale biodiversity pattern of the endemic tree flora of the Western Ghats (India) using canonical correlation analysis of herbarium records. *Ecography*, 26, 429–444.

Greenwood, D.R., Scarr, M.J., Christophel, D.C. (2003). Leaf stomatal frequency in the Australian tropical rainforest tree Neolitsea dealbata (Lauraceae) as a proxy measure of atmospheric pCO_2. *Palaeogeography, Palaeoclimatology, Palaeoecology*, 196, 375–393.

Hood, M.E., Mena-Alí, J.I., Gibson, A.K., Oxelman, B., Giraud, T., Yockteng, R., Arroyo, M.T.K., Conti, F., Pedersen, A.B., Gladieux, P., Antonovics, J. (2010). Distribution of the anther-smut pathogen Microbotryum on species of the Caryophyllaceae. *New Phytologist*, 187, 217–229.

Jones, C.A. and Daehler, C.C. (2018). Herbarium specimens can reveal impacts of climate change on plant phenology; a review of methods and applications. *PeerJ*, 6, e4576.

Heberling, J.M. and Burke, D.J. (2019). Utilizing herbarium specimens to quantify historical mycorrhizal communities. *Applications in Plant Sciences*, 7(4), e1223.

Hembry, D.H. (2013). Herbarium specimens reveal putative insect extinction on the deforested island of Mangareva (Gambier Archipelago, French Polynesia). *Pacific Science*, 67, 553–560.

Hernandez, H. and Navarro, M. (2007). A new method to estimate areas of occupancy using herbarium data. *Biodiversity and Conservation*, 16, 2457–2470.

Herpin, U., Markert, B., Weckert, V., Berlekamp, J., Friese, K., Siewers, U., Lieth, H. (1997). Retrospective analysis of heavy metal concentrations at selected locations in the Federal Republic of Germany using moss material from a herbarium. *Science of the Total Environment*, 205, 1–12.

Houle, G. (2007). Spring-flowering herbaceous plant species of the deciduous forests of eastern Canada and 20th century climate warming. *Canadian Journal of Forest Research*, 37, 505–512.

Jenkins, A.E. and Fawcett, H.S. (1933). Records of citrus scab mainly from herbarium specimens of the genus Citrus in England and the United States. *Phytopathology*, 23, 475–482.

Johnson, K.G., Brooks, S.J., Fenberg, P.B., Glover, A.G., James, K.E., Lister, A.M., Michel, E., Spencer, M., Todd, J.A., Valsami-Jones, E., Young, J., Stewart, J.R. (2011). Climate change and biosphere response: Unlocking the collections vault. *Bio Science*, 61(2), 147–153.

Kouwenberg, L.L.R., McElwain, J.C., Kürschner, W.M., Wagner, F., Beerling, D.J., Mayle, F.E., Visscher, H. (2003). Stomatal frequency adjustment of four conifer species to historical changes in atmospheric CO_2. *American Journal of Botany*, 90, 610–619.

Lang, P.L.M., Willems, F.M., Scheepens, J.F., Burbano, H.A., Bossdorf, O. (2018). Using herbaria to study global environmental. *New Phytologist*, 221(1), 110–122.

Lavoie, C. (2013). Biological collections in an ever changing world: Herbaria as tools for biogeographical and environmentals studies. *Perspectives in Plant Ecology, Evolution and Systematics*, 15, 68–76.

Le Bras, G., Pignal, M., Jeanson, M.L., Muller, S., Aupic, C., Carré, B., Flament, G., Gaudeul, M., Gonçalves, C., Jabbour, F., Lerat, E., Lowry, P., Offroy, B., Pérez, E., Poncy, O., Rodrigues Invernon, V., Rouhan, G., Haevermans, T. (2017). The French Muséum national d'Histoire naturelle vascular plant herbarium collection dataset. *Scientific Data*, 4, 1–16.

Lees, D.C., Lack, H.W., Rougerie, R., Hernandez-Lopez, A., Raus, T., Avtzis, N.D., Augustin, S., Lopez-Vaamonde, C. (2011). Tracking origins of invasive herbivores through herbaria and archival DNA: The case of the horse-chestnut leaf miner. *Frontiers in Ecology and the Environment*, 9, 322–328.

Lienert, J., Fischer, M., Diemer, M. (2002). Local extinctions of the wetland specialist Swertia perennis L. (Gentianaceae) in Switzerland: A revisitation study based on herbarium records. *Biological Conservation*, 103, 65–76.

Loiselle, B.A., Jorgensen, P.M., Consiglio, T., Jimenez, I., Blake, J.G., Lohmann, L.G., Montiel, O.M. (2007). Predicting species distributions from herbarium collections: Does climate bias in collection sampling influence model outcomes? *Journal of Biogeography*, 35(1), 105–116.

McDougall, A.S., Loo, J.A., Clayden, S.R., Goltz, J.G., Hinds, H.R. (1998). Defining conservation priorities for plant taxa in southeastern New Brunswick, Canada using herbarium records. *Biological Conservation*, 86, 325–338.

Meineke, E.K., Davis, C.C., Davies, T.J. (2018). The unrealized potential of herbaria for global change biology. *Ecological Monographs*, 88(4), 505–525.

Miller-Rushing, A.J., Primack, R.B., Primack, D., Mukunda, S. (2006). Photographs and herbarium specimens as tools to document phenological changes in response to global warming. *American Journal of Botany*, 93(11), 1667–1674.

Muller, S. (2006). *Les plantes protégées de Lorraine, distribution, écologie, conservation.* Biotope, Mèze.

Muller, S. (2015). Intérêt des herbiers pour la connaissance des dynamiques spatiotemporelles des invasions biologiques. *Revue d'écologie (Terre et Vie)*, 70(12), 229–235.

Muller, S. (2016). L'intérêt des herbiers pour la mise en évidence des changements macroécologiques et environnementaux. In *UNESCO International Conference – Botanists of the Twenty-first Century: Roles, Challenges and Opportunities,* UNESCO (ed.). UNESCO, Paris.

Nualart, N., Ibanez, N., Soriano, I., Lopez-Pujol, J. (2017). Assessing the relevance of herbarium collections as tools for conservation biology. *Botanical Review*, 83(3), 303–325.

Park, I.W. and Schwartz, M.D. (2015). Long-term herbarium records reveal temperature-dependent changes in flowering phenology in the southeastern USA. *Int. J. Biometeorol.*, 59, 347–355.

Park, D.S., Breckheimer, I., Williams, A.C., Law, E., Ellison, A.M., Davis, C.C. (2018). Herbarium specimens reveal substantial and unexpected variation in phenological sensitivity across the eastern United States. *Phil. Trans. R. Soc. B*, 374, 20170394.

Peat, H.J., Clarke, A., Convey, P. (2007). Diversity and biogeography of the Antarctic flora. *Journal of Biogeography,* 34, 132–146.

Penuelas, J. and Filella, I. (2002). Metal pollution in Spanish terrestrial ecosystems during the twentieth century. *Chemosphere*, 46, 501–505.

Poulickova, A., Hájková, P., Kintrova, K., Bat'kova, R., Czudkova, M., Hájek, M. (2013). Tracing decadal environmental change in ombrotrophic bogs using diatoms from herbarium collections and transfer functions. *Environmental Pollution*, 179, 201–209.

Prather, L.A., Alvarez-Fuentes, O., Mayfield, M.H., Ferguson, C.J. (2004). The decline of plant collecting in the United States: A threat to the infrastructure of biodiversity studies. *Systematic Botany*, 29, 15–28.

Primack, D., Imbres, C., Primack, R.B., Miller-Rushing, A.J., Del Tredici, P. (2004). Herbarium specimens demonstrate earlier flowering times in response to warming in Boston. *American Journal of Botany*, 91, 1260–1264.

Pysek, P. (1991). *Heracleum mantegazzianum* in the Czech Republic: Dynamics of spreading from the historical perspective. *Folia Geobotanica et Phytotaxonomica*, 26, 439–454.

Ristaino, J.B. (1998). The importance of archival and herbarium materials in understanding the role of oospores in late blight epidemics of the past. *Phytopathology*, 88, 1120–1130.

Ristaino, J.B., Groves, C.T., Parra, G.R. (2001). PCR amplification of the Irish potato famine pathogen from historic specimens. *Nature*, 411, 695–697.

Ryan, K.G., Burne, A., Seppelt, R.D. (2009). Historical ozone concentrations and flavonoid levels in herbarium specimens of the Antarctic moss *Bryum argenteum*. *Global Change Biology*, 15, 1694–1702.

Savolainen, V., Cuénoud, P., Spichiger, R., Martinez, M.D.P., Crèvecoeur, M., Manen, J.F. (1995). The use of herbarium specimens in DNA phylogenetics – Evaluation and improvement. *Plant Systematics and Evolution*, 197, 87–98.

Shetler, S.G. (1969). The herbarium: Past, present and future. *Proceedings of the Biological Society of Washington*, 82, 687–758.

Shotbolt, L., Buker, P., Ashmore, M.R. (2007). Reconstructing temporal trends in heavy metal deposition: Assessing the value of herbarium moss samples. *Environmental Pollution*, 147, 120–130.

Spellman, K.V. and Mulder, C.P.H. (2016). Validating herbarium-based phenology models using citizen-science data. *Bioscience*, 66, 897–906.

Teece, M.A., Fogel, M.L., Tuross, N., McCourt, R.M., Spamer, E.E. (2002). The Lewis and Clark Herbarium of the Academy of Natural Sciences, Part 3. Modern environmental applications of a historic nineteenth century botanical collection. *Notulae Naturae*, 477, 1–20.

Thiers, B. (2018). The World's Herbaria 2017: A summary report based on data from Index Herbariorum. *Index Herbariorum*, 2.1.

Ungricht, S., Rasplus, J.Y., Kjellberg, F. (2005). Extinction threat evaluation of endemic fig trees of New Caledonia: Priority assessment for taxonomy and conservation with herbarium collections. *Biodiversity and Conservation*, 14, 205–232.

Wandeler, P., Hoeck, P.E.A., Keller, L.F. (2007). Back to the future: Museum specimens in population genetics. *Trends in Ecology and Evolution*, 22, 634–642.

Weber, E. (1998). The dynamics of plant invasions: A case study of three exotic goldenrod species (Solidago L.) in Europe. *Journal of Biogeography*, 25, 147–154.

Weiss, D., Shotyk, W., Kramers, J.D., Gloor, M. (1999). Sphagnum mosses as archives of recent and past atmospheric lead deposition in Switzerland. *Atmospheric Environment*, 33, 3751–3763.

Willis, C.G., Ellwood, E.R., Primack, R.B., Davis, C.C., Pearson, K.D., Gallinat, A.S., Yost, J.M., Nelson, G., Mazer, S.J., Rossington, N.L., Sparks, T.H., Soltis, P.S. (2017a). Old plants, new tricks: Phenological research using herbarium specimens. *Trends in Ecology & Evolution*, 32(7), 531–546.

Willis, C.G., Law, E., Williams, A.C., Franzone, B.F., Bernardos, R., Bruno, L., Hopkins, C., Schorn, C., Weber, E., Park, D.S., Davis, C.C. (2017b). CrowdCurio: An online crowdsourcing platform to facilitate climate change studies using herbarium specimens. *New Phytologist*, 215, 479–488.

Wilson, D., Stock, W.D., Hedderson, T. (2009). Historical nitrogen content of bryophyte tissue as an indicator of increased nitrogen deposition in the Cape Metropolitan Area, South Africa. *Environmental Pollution*, 157, 938–945.

Woodward, F.I. (1987). Stomatal numbers are sensitive to increases in CO_2 from pre-industrial levels. *Nature*, 327, 617–618.

Yoshida, K., Burbano, H.A., Krause, J., Thines, M., Weigel, D., Kamoun, S. (2014). Mining herbaria for plant pathogen genomes: Back to the future. *PLoS Pathog.*, 10(4), e1004028.

17

Digital Photography *In Natura* in Zoology: More Biology in Natural History Collections?

Romain GARROUSTE
ISYEB – CNRS, Muséum national d'Histoire naturelle, Sorbonne Université, École Pratique des Hautes Études, Université des Antilles, Paris, France

How can we improve our knowledge of living organisms and the dissemination of knowledge regarding biodiversity through the use of images *in natura* and digital media collections associated with natural history collections? And how can we give them even more weight in the framework of comparative biology, by reaffirming the importance of specimens, by innovating the structure and access to databases linking specimens and digital media, and more generally by insisting on the importance and accessibility of data (occurrences, distributions) associated with organisms in order to improve our knowledge of ecosystems? And especially in the context of participatory science, which is still being developed?

For a color version of all figures in this chapter, see www.iste.co.uk/pellens/natural.zip.

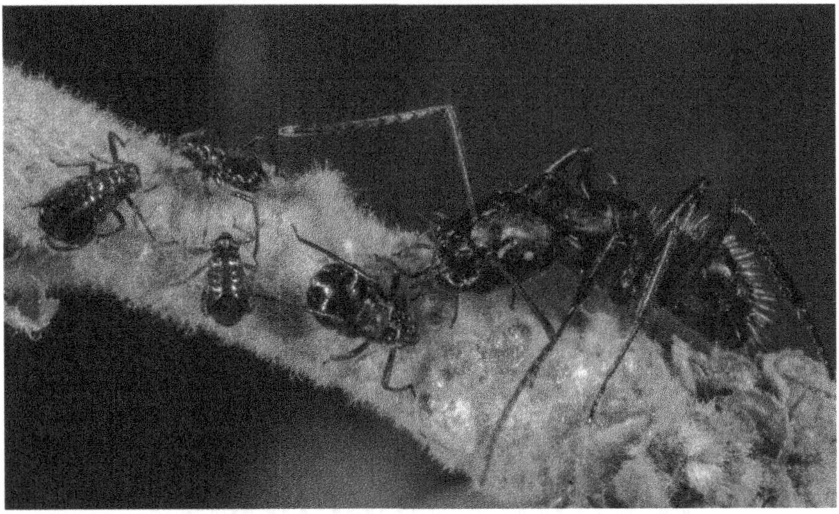

Figure 17.1. *Quadri trophic interactions (photo: Romain Garrouste)*

COMMENT ON FIGURE 17.1.– *These four taxa (the plant, a pubescent oak twig, the ant* Camponotus sp. *and two species of Hemiptera (a* Lachnus *aphid and a* Phylloxeridae *aphid)), show an example of complex relationships within biocenoses. Initially, the observer might want to illustrate the relationship between the* Lachnus *aphid and the ant. But the ant interacts with the* Phylloxeridae *aphid in this picture (pressed against the stem). The photo therefore provides evidence of interactions more complex than those highlighted by simple observation.*

17.1. Images and collections... for comparative biology

Comparative biology is needed to anchor biodiversity (and the study of the entirety of life on Earth) in the academic landscape and to move away from a biology that is often circumscribed to model organisms, which are becoming increasingly well known, but which limits a comprehensive view of living organisms and ecosystems (Garrouste and Roy 2017; Grandcolas 2017; Wheeler 2018). One consequence of this emerging comparative and integrative biology is a certain revival of taxonomy and systematics – resurrection of the "naturalistic" aspects (Pearse 2003) – necessary to delineate taxa, understand their organization and hierarchy, and complexity, in order to truly ground this both ancient and emerging discipline in a vital evolutionary perspective. This is one of the foundations of a new science of biodiversity in full revival. One of the consequences also lies in a certain interest in paleontology, where fossils are used for an evocative molecular biology, to integrate these comparative and temporal aspects into a historical dimension (Figure 17.2).

Figure 17.2. *The neotropical lizard* Uracentron azureus *(Mana area, French Guiana) is a good example to demonstrate the color change after specimen preservation and the link to taxonomy (photo: Romain Garrouste)*

COMMENT ON FIGURE 17.2.– *Collected specimens turn blue in the alcohol in which they are preserved. The name of this species,* azureus, *indicates that it was described from a discolored collection specimen and not from a freshly collected or observed individual with a marked green color.*

This comparative and therefore necessarily evolutionary biology has been called for by scientists for a long time (Nelson 1970; Wheeler 1995), but is limited by the resources allocated to these disciplines and paradoxically less and less endowed as the evidence of an immense biodiversity remains to be described in order to understand the functioning of life on our planet, to appreciate its usefulness through ecosystem services and to predict its future. There is a methodological debate, inevitably influenced by these economic considerations, on the modalities of the current inventory of living organisms, and some want to free themselves from the current rules for describing taxa, or propose interpretations of these modalities. One of these debates concerns the possibility of freeing oneself from the standard specimens (vouchers) for the descriptions of new taxa in order to allow, for example, the use of photographs to describe a new species (photography-based taxonomy [PBT]) (Pape 2016). After having briefly recalled (the debate is available on various media such as *Researchgate*[1] or a special issue of the journal *Bionomina*[2]) the main dangers of these new temptations in taxonomy, we propose a new way of using photography *in natura* ("wild shot") for the constitution of digital collections of images, associated with collection specimens, and other media (sounds, videos) to improve the description of taxa, and especially to better describe the biology of the species and to understand their ecology. One could also imagine that photography associated with specimens would partially supplement the lack of linkage that exists in primary biodiversity data (from 70% to 18% in GBIF data in almost 50 years; Troudet *et al.* 2017). Curiously, even the Society for Integrative and Comparative Biology, seeking to establish a standard for the use of video images (Brainerd *et al.* 2017), does not indicate the need for a voucher (Troudet *et al.* 2017). Photography and modern science are inextricably linked (Wilder 2009), and it is possible to summarize this intimate association in the natural sciences today (Figures 17.1 and 17.3).

17.2. Accelerating the process of the incomplete inventory of life

There is currently a real discussion about the need to adequately inventory the biodiversity on our planet in order to improve our knowledge of how ecosystems function and what will happen to them in the face of current and future changes. This long-term project is not questioned, or only marginally, and generally outside the scientific community. But there is a real methodological debate on the modalities of these inventories, and some want to break the current rules of taxon descriptions, or propose adjustments that do not meet the scientific criteria of systematics (Suarez and Tsutsui 2004; Rocha *et al.* 2014).

We will not repeat this debate, which, via taxonomy journals, sees more or less clear-cut opinions. Recently, a correspondence to *Zootaxa* (Ceríaco *et al.* 2016) with more than 450 signatories, including taxonomists, curators and other taxonomy users

1 Available at: https://www.researchgate.net/publication/317135329_The_wild_shot_Photography_for_more_biology_in_natural_history_collections_not_for_replacing_vouchers.

2 Available at: https://www.mapress.com/j/bn/issue/view/bionomina.12.1.

from all continents, has received wide attention and stimulated a broad controversy about possible interpretations of the Zoological Code of Nomenclature (International Commission on Zoological Nomenclature 1999) regarding the use of photography in taxonomy.

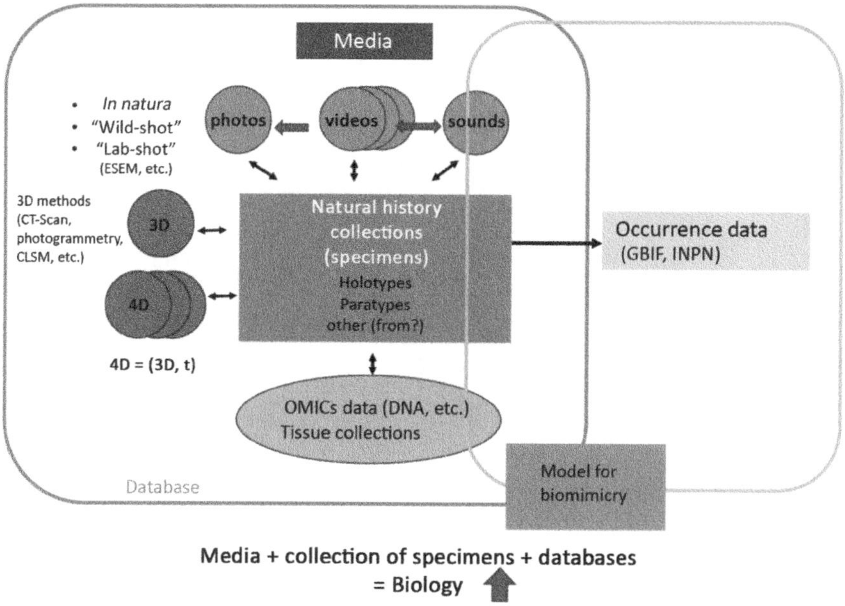

Figure 17.3. *Optimal relationships between specimens and the associated digital media set*

COMMENT ON FIGURE 17.3.– *The specimen provides the link between the digital databases and the occurrence data. Scientific information is enhanced by this relationship, which allows new inferences to be made from original observations. This architecture becomes an ideal structure to consider the comparative biology of the 21st century and, for example, to allow an optimized biomimetic exploration approach (after publications, not only for types of specimens). The heritage value of the specimens and their associated information could thus be increased, as it is the case of the figured specimens in paleontology for example, but above all by the increase in their informative value.*

These authors recalled the need to keep specimens as name-bearers and as refutable scientific evidence in response to the intentions of a group of taxonomists who would have liked to base the description of taxa on photography-based taxonomy, as defended by Pape *et al.* (2016) or Garraffoni and Freitas (2017), that is, the possibility of describing species by using good quality photographs, without the need for specimen deposit.

Figure 17.4. Panstrongylus geniculatus *(Hemiptera, Reduviidae: Triatominae) (French Guiana) feeding on an* Eacles *(Lepidoptera, Saturniidae) (photo: Romain Garrouste)*

COMMENT ON FIGURE 17.4.– *This behavior was first observed by Garrouste (2009) and this image served as "evidence" of the observation of this entomophagous behavior while Triatominae are considered fairly strict Haematophagous vertebrates.*

One of the reasons for this fierce defense of this unusual method lies in the urgency of describing life, the increasingly strict regulations for collecting samples (recently reinforced by the Nagoya protocol), the scarcity of certain species – including those not yet described – and the small number of rare and endangered organisms (mammals, birds and other vertebrates). Dubois and Nemésio (2007) had already concluded on the importance of specimens in taxonomy. But if the subject of these controversies is photography as the only way to document a new species, none of them really discussed what photography could represent for natural science knowledge based on specimen collections, including type series, and in combination with other media (video and sound), and other associated data. The debate has never extended to photography itself and its new contributions enabled by new techniques, whether in the wild (*in natura* or "wild-shot") or in the laboratory ("lab shot"), including new imaging techniques such as environmental scanning electron microscopy (ESEM) or confocal laser scanning microscopy (CLSM) (Weirauch *et al.* 2015), CT-scan (computed tomography scanning), and other non-invasive 3D technologies such as photogrammetry ("high tech photos") (Amorim *et al.* 2016).

Figure 17.5. *The predatory bug* Zelus renardii *(here specimen MNHN EH-24519, Var, France) set out to conquer the world from the American continent (photo: Romain Garrouste)*

COMMENT ON FIGURE 17.5.– *We do not know its impact in the biocenoses where it settles. The association between* in natura *photos and collection specimens allows traceability and the assurance of a perennial and secure connection between identification and observation, which is very useful for the detection of exotic species and more broadly for citizen science (extract from Garrouste 2019).*

17.3. Why more biology in natural history collections?

It is obvious that the new photonic technologies and digital media reinforce and expand the possibilities now available.

Illustration, especially photography, with new techniques that are more and more accessible and efficient, is essential for teaching and research and is developing steadily. There is also an increasing demand for new methods that allow for numerous high-quality scientific illustrations.

I wish to advocate a position on the use of photography and associated digital data management that would allow for more specimen-related illustrations in collection databases, thus allowing for "more biology" (in the broadest sense) in these databases that are often limited to a purely taxonomic interest. If taxonomy is to be integrative, more biological information is needed. These needs are obvious to complete the databases of natural history collections of museums, scientific departments and universities that host collections. This could also become a challenge for small institutions to be a repository of validated and useful information closer to local ecosystems (protected areas, etc.), especially in relation to academic societies, and constitute a new development of citizen science at the local scale, and integrable at a larger scale (GBIF, etc.).

The need to increase the quantity and quality of taxonomic illustrations has already been emphasized and advocated for by Amorim *et al.* (2016), as sources of information (high-tech photos) for taxonomy, and illustrated by Ang *et al.* (2013), or Garrouste and Roy (2017) with an example of behavior associated with a specimen with a collection number (Figure 17.6).

There is a critical lack of biological information on a very large number of species, many of which are represented by a limited number of specimens in natural history collections, or even by a single specimen (holotype) without any biological information. This is especially the case for invertebrates, such as arthropods. When massive collections are made, including in international programs, in most cases we are not able to document the biology of the collected species and this is regrettable. There are commendable exceptions and the MNHN's *Planète (Re)visitée* program[3] attempts to remedy this lack of information intrinsic to the method by using "intermediate" photography (after collection, in the field laboratory but before naturalization) (this information comes from my personal communication from Laure Corbari).

The use of photography in nature to improve understanding of the biology of organisms is obvious, but it must be organized and practiced rigorously and linked to collected specimens. This would, for example, overcome the problems related to the conservation of specimens whose color, shape and general appearance may change after death and/or naturalization (preparation for conservation). This may be the case in the presence of organic pigments, organic substances such as waxes, scales, feathers and other dander modified after death, or in preserving fluids for insects or other soft-bodied animals (Figure 17.1). There is a large number of such examples in all major animal taxa. Eggs, larval stages, chrysalids, cocoons, pupae, galls, biomineralizations, bioconstructions (nests, etc.) or constructions (burrow molds, etc.) could be incorporated into this process and linked to a specimen (Figures 17.2, 17.6,

3 Available at: http://www.laplaneterevisitee.org/fr.

and 17.5). One must also consider the importance of polymorphisms, for example, social insect castes and the many local variations in populations, changes due to the environment (change in phenotypes), including anthropogenic ones (Lees and Creed 1975). There are also cases of seasonal polymorphism and with other determinisms under the dependence of various genes in the sense of a single species (Gautier *et al.* 2018). The *in natura* photograph of the organism, with all these associated elements, interacting organisms (parasites, phoretic organisms, prey, etc.), is associated with the specimen that becomes the only objective link to the population concerned (Figures 17.1, 17.7, and 17.8).

One of the direct consequences would be to increase the taxonomic quality of emerging citizen sciences and "reference" websites, especially those concerning natural heritage and conservation (such as the INPN in France[4]).

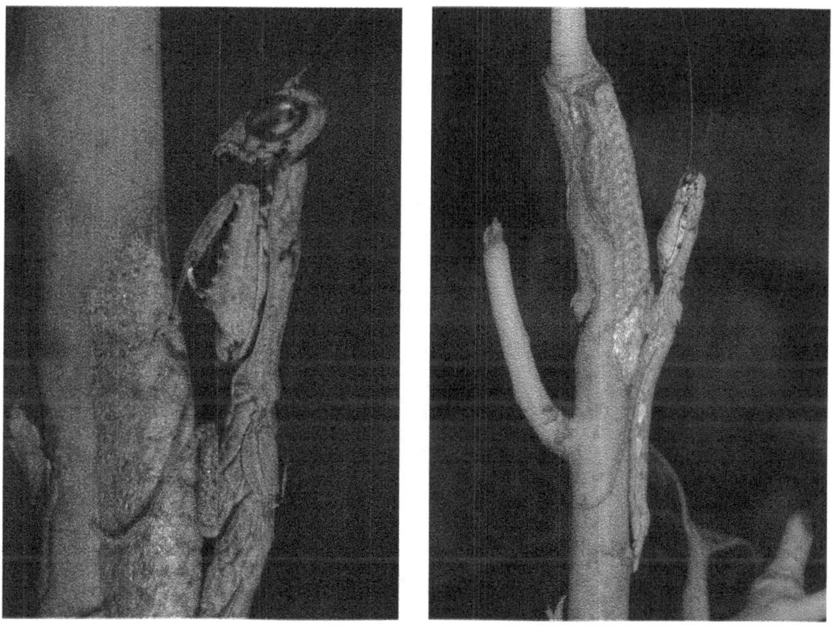

Figure 17.6. Paralygdamia madecassa *guarding its eggs on a twig and imitating a twig (photos: Romain Garrouste)*

COMMENT ON FIGURE 17.6.– *This species from Madagascar was known but its maternal behavior had not yet been observed. This type of observation makes it possible to attribute this behavior to 100% of the known species of this family (Tarachodidae)*

4 Available at: https://inpn.mnhn.fr/accueil/index?lg=en.

and to make it a valid criterion in systematics. This species is naturalized with its eggs, which is associated with the preserved specimen (specimen MNHN-EP-EP3557), in Garrouste and Roy (2017).

17.4. Images in the natural sciences: a collection like any other?

The idea would be to set up tools to manage a new data flow related to specimens in natural history collections, which would be derived from images taken in natural conditions without artifacts, with behaviors as close as possible to natural or undisturbed situations ("wild shot") or before naturalization of specimens (field laboratory images in "intermediate" situations, before conservation and preparation), as practiced by many taxonomists working on reptiles, amphibians or fishes (e.g. Ineich *et al.* 2016, on lizards). This approach should be developed for all taxa so as not to lose information regarding the specimen's shape, color (Figure 17.2) and general appearance, as close to the living appearance as possible, with the aim of combining as much information as possible, including through the use of video and/or sound.

Figure 17.7. *Predation of a* Chatogekko amazonicus *lizard (Amazonian dwarf gekko) by a wolf spider (Lycosidae) (French Guiana). Predation of vertebrates by arthropods is not common and is always interesting to document. These are functional interactions, here between predators (intraguild predation) (photo: Romain Garrouste)*

Note that from a good video from semi-professional to pro-professional equipment, such as modern 24 × 36 SLR cameras with 4K resolution (3,840 × 2,160 pixels, 16:9 format), it is possible to obtain usable photo files and sound files.

But for photography, to allow cropping (cropping of areas of interest while keeping a good resolution), it is worthwhile using SLR cameras with more than 24 megapixels (5,656 × 4,240 pixels) or images in the context of the dissemination of knowledge for any media.

Natural history collections must be ready for these new media to complement specimen databases (Figure 17.3). Many institutions appear ready for this workflow or are preparing for it. The MNHN has a commission, the *Commission d'orientation sur le numérique* (COSNum), which includes this theme of new collections necessary for research and the valorization of collections resulting from research and complementary to "traditional" collections of natural history specimens. Already, a scientific media library completes[5] a sound library and a Wiki allows the intervention on the contents by authorized users[6].

The specimen of our optimal digital environment is then the link to all media and carries an additional scientific value compared to the specimen alone, whether it is the holotype, a specimen of the typical series to the simple identified specimen.

At this point, it is also relevant to ask whether, in the manner of figured specimens in paleontology (figured specimens in publications, e.g. Reynard and Brilha 2018), even if they do not constitute an element of a typical series but a simple drawing or photo of an identified specimen, these specimens could possess an additional heritage value. This can also be seen as a desire to improve the digitization and thus the valorization of collections.

A possible ideal in this type of collection would be a holotype or specimen of the type series of a taxon that would have images *in natura* and a whole series of associated images of behaviors, sounds and videos and that would of course also be linked to a database of molecular information (DNA and "*-omics*" barcodes, tissues) relative to this specimen or to others (allotype of the opposite sex, ecological interactions with other taxa, bioconstructions, etc.) (see Chapter 18 by Monnet *et al.*). This is perhaps the new challenge for natural history collections in the 21st century, including for small structures in search of valorization and conservation objectives (regional museums, for example, university collections, academic societies), sometimes located near conservation areas or centers of endemism, particular ecosystems or with rare expertise in systematics of uncommon organisms.

In conclusion, it is necessary to defend and develop the use of classical photography, especially in the field, in combination with high-tech imaging

5 Available at: https://sonotheque.mnhn.fr/.

6 Available at: http://collections.mnhn.fr/wiki/Wiki.jsp?page=Accueil.

methods, as well as their development, to improve the quality of taxonomic descriptions and to add new original biological information on taxa in natural history collections and associated databases. In any case, images cannot replace the scientific value of a specimen, which will be available for all the new non-invasive analyses we are inventing and those that are not yet invented.

17.5. The Hemiptera of France: an exemplary iconography

Natural history guides are important elements for providing knowledge of the biodiversity of a region. For insects, this makes sense for species of agronomic, medical, etc., interest. However, identification, especially down to the species level, is often difficult, especially when using photographs, whether *in natura* or in the laboratory, where often collection specimens, which although carefully treated, cannot describe all the diagnostic characters of a taxon.

For the iconography of this type of work, with the aim of precise determination, but especially of initiation to the taxonomy of a group, it is thus necessary to set up a process of taking photographs associated with a collection of samples, integrating an associated numbering. The result is a traceability of the images, useful for taxonomic and nomenclatural updating (Figure 17.8).

This is the approach used in Garrouste's book (2015), which combines specimens collected as part of the book, use of specimens already in the collection, links with the *Muséum*'s sound library for sound-producing species (Flashcode), with the INPN (Flashcode), and instructions on taking images *in natura* (Figure 17.7).

17.6. Trait databases, query automation and bio-inspiration

Advances in automated image analysis, including in open-source software such as Image J (Schneider *et al.* 2012), make it possible to consider automated search procedures, and to consider using databases of specimen or specimen-associated images from mass digitization procedures of collections.

Thus, the massive digitization of collections is finding new uses through the constitution of morphological and functional trait databases.

Using calibrated images (e.g. Pantone-type color range), it is possible to set up algorithms to retrieve morphological traits of a large batch of species. This is underway for plants, especially herbarium specimens (Weigelt *et al.* 2019).

Figure 17.8. *Immature predatory bug* Picromerus bidens *feeding on a Lepidoptera* Charaxes jasius *caterpillar (Var, France). Interactions between organisms are documented by* in natura *images and provide information on the functioning of ecosystems (from (Garrouste 2015)) (photo: Romain Garrouste)*

There is no equivalence yet in zoology, even if trait databases are starting to be set up. For standardized digitizations of insect, arthropod or other animal specimens, it would probably be possible to achieve equivalent results.

Thus, the wild shot photos (motion, reproduction, and other behaviors) could help verify some of the functions that are often inferred from these images (Figure 17.1).

For example, the *Traits Database of European Bees* (SCALE) lists the important traits of all European bee species (sociality, nest type, tongue length, etc.)[7].

In all cases, the conjunction of functional traits and media databases are an important prerequisite for the implementation of a series of applications, from conservation to bio-inspiration (or biomimicry). The search for traits and

7 Available at: http://scales.ckff.si/scaletool/?menu=11.

functions that can be used in a biomimetic framework remains an application to be implemented and a real craze exists for this approach, which remains to be organized. Indeed, if the biomimetic approach often starts from engineering that seeks technological solutions, one could imagine an inverse approach where the functional traits of the organisms would be selected and then proposed to the developers within the framework of the transfer of bio-inspired technologies from the morpho-functional characteristics of the organisms.

17.7. Conclusion: a new challenge for natural history

The systematization of the use of digital images and the organization of these new media collections is essential to consider their use in taxonomy and systematics, population biology by functional traits and ecology of interactions, eco-ethology, biodiversity conservation, and to make them more informative, and thus increase tenfold the primary purpose (specimen bank and type specimens – "vouchers") of natural history work. Most of the applied uses are obvious (agronomy, medical and health biology, species conservation), but some emerging uses could benefit from it, such as bio-inspiration (or biomimicry), which needs to know the morpho-anatomical and biological characteristics of organisms to find relevant models, in particular through automated procedures (high-throughput phenotyping), and more broadly comparative biology, which is now more than ever necessary to understand the complexity of living organisms and their mechanisms.

The use of these media in the dissemination of knowledge and teaching at all levels remains obvious and will allow new uses and developments (e-teaching, web documentaries, participative sciences, etc.) and cultural applications for events (exhibitions, conferences) or publishing. Let us mention in this context the existence of an e-learning Sorbonne Université-MNHN INSECT (online course), which gives a large part to the image in its pedagogy[8].

While this is a new challenge (new digital storage media, consultation applications, legal problems to do with copyright and exploitation, etc.) for institutions that manage natural history collections (museums, universities), it also represents a tremendous renewal and the possibility of attracting new audiences to the natural sciences, biodiversity and the environment, through participatory sciences or the multiplication of media, including social networks.

It is necessary to defend the use of photography, especially in the field *in natura* ("wild-shot"), in combination with high-tech (non-conventional) imaging methods,

8 Available at: http://edu.mnhn.fr/course/view.php?id=327.

as well as their development, to improve the quality of taxonomic descriptions and to provide original biological information on taxa in natural history collections and associated databases. In any case, images cannot replace the scientific value of a specimen, which will be available for all the new non-invasive analyses we are inventing and those that have not yet been invented. This is a very dynamic sector, which is also expanding rapidly because of methods related to synchrotron imaging, methods using X-rays and developments in photonics in general.

17.8. References

Amorim, D.S., Santos, C.M.D., Krell, F.-T., Dubois, A., Nihei, S.S., Oliveira, O.M.P., Pont, A., Song, H., Verdade, V.K., Fachin, D.A. *et al.* (2016). Timeless standards for species delimitation. *Zootaxa*, 4137(1), 121–128 [Online]. Available at: https://doi.org/10.11646/zootaxa.4137.1.9.

Ang, Y., Wong, L.J., Meier, R. (2013). Using seemingly unnecessary illustrations to improve the diagnostic usefulness of descriptions in taxonomy – A case study on *Perochaeta orientalis* (Diptera, Sepsidae). *ZooKeys*, 355, 9–27 [Online]. Available at: https://doi.org/10.3897/zookeys.355.6013.

Brainerd, E.L, Hedrick, T.H., Creamer, A.T., Müller, U.K. (2017). Data management rubric for video data. *Organismal Biology Integrative and Comparative Biology*, 57(1), 33–47 [Online]. Available at: https://doi.org/10.1093/icb/icx060.

Ceríaco, L.M.P., Gutiérrez, E.E., Dubois, A. (2016). Photography-based taxonomy is inadequate, unnecessary, and potentially harmful for biological sciences, *Zootaxa*, 4196(3), 435–445 [Online]. Available at: https://doi.org/10.11646/zootaxa.4196.3.9.

Dubois, A. and Nemésio, A. (2007). Does nomenclatural availability of nomina of new species or subspecies require the deposition of vouchers in collections? *Zootaxa*, 1409, 1–22.

Garraffoni, A.R.S. and Freitas, A.V.L. (2017). Photos belong in the taxonomic code. *Science*, 355(6327), 805.

Garrouste, R. (2009). La première observation *in natura* de l'entomophagie de *Panstrongylus geniculatus* (Latreille 1811) hématophage vecteur de la maladie de Chagas (Hemiptera: Reduviidae). *Annales de la Société entomologique de France (N.S.)*, 45(3), 302–304.

Garrouste, R. (2015). *Hémiptères de France, de Belgique, du Luxembourg et de Suisse.* Delachaux et Niestlé, Muséum national d'Histoire naturelle, Paris.

Garrouste, R. (2019). *Zelus renardii* (Kolenati, 1856) : une Réduve nouvelle pour la France (Hemiptera, Reduviidae, Harpactorinae). *Bulletin de la Société entomologique de France*, 124(3), 335–336.

Garrouste, R. and Roy, R. (2017). Note sur le comportement maternel de la Mante *Paralygdamia madecassa* Saussure & Zehntner, 1895 (Dictyoptera, Tarachodidae), endémique de Madacascar. *Bulletin de la Société entomologique de France*, 122(1), 105–109.

Gautier, M., Yamaguchi, J., Foucaud, J., Loiseau, A., Ausset, A., Facon, B., Gschloess, B., Lagnel, J., Loire, E., Parrinello, H. *et al.* (2018). The genomic basis of color pattern polymorphism in the harlequin ladybird. *Current Biology*, 28, 1–7.

Grandcolas, P. (2017). Loosing the connection between the observation and the specimen: A by-product of the digital era or a trend inherited from general biology? *Bionomina*, 12, 57–62.

Ineich, I., Glaw, F., Vences, M. (2016). A new species of *Blaesodactylus* (Squamata: Gekkonidae) from Tsingy limestone outcrops in Namoroka National Park, north-western Madagascar. *Zootaxa*, 4109, 523–541 [Online]. Available at: https://doi.org/10.11646/zootaxa.4109.5.2.

International Commission on Zoological Nomenclature (1999). *International Code of Zoological Nomenclature*, 4th edition. International Trust for Zoological Nomenclature, London.

Lees, D.R. and Creed, E.R. (1975). Industrial melanism in *Biston betularia*: The role of selective predation. *J. Anim. Ecol.*, 44, 67–83.

Nelson, G.J. (1970). Outline of a theory of comparative biology. *Systematic Zoology*, 19, 373–384.

Pape, T. (2016). Taxonomy: Species can be named from photos. *Nature*, 537, 307–307.

Pearse, J. (2003). The promise of integrative biology: Resurrection of the naturalist. *Integrative and Comparative Biology*, 43, 276–277.

Reynard, E. and Brilha, J. (eds) (2018). *Geoheritage: Assessment, Protection and Management*. Elsevier, Amsterdam.

Rocha, L.A., Aleixo, A., Allen, G., Almeda, F., Baldwin, C.C., Barclay, M.V.L., Bates, J.M., Bauer, A.M., Benzoni, F., Berns, C.M. *et al.* (2014). Specimen collection: An essential tool. *Science*, 344(6186), 814–815.

Santos, C.M.D., Amorim, D.S., Klassa, B., Fachin, D.A., Nihei, S.S., De Carvalho, C.J.B., Falaschi, R., Mello-Patiu, C., Couri, M.S., Oliveira, S. *et al.* (2016). On typeless species and the perils of fast taxonomy. *Systematic Entomology*, 41(3), 511–515.

Schneider, C.A., Rasband, W.S., Eliceiri, K.W. (2012). NIH Image to ImageJ: 25 years of image analysis. *Nature Methods*, 9, 671–675.

Suarez, A.V. and Tsutsui, N.D. (2004). The value of museum collections for research and society. *Bioscience*, 54, 66–74.

Troudet, J., Grandcolas, P., Blin, A., Vignes-Lebbe, R., Legendre, F. (2017). Taxonomic bias in biodiversity data and societal preferences? *Scientific Reports*, 7, 9132 [Online]. Available at: https://doi.org/10.1038/s41598-017-09084-6.

Weigelt, P., König, C., Kreft, H. (2019). GIFT – A global inventory of floras and traits for macroecology and biogeography. *Journal of Biogeography*. DOI: 10.1111/ jbi.13623.

Weirauch, C. and Frankenberg, S. (2015). From "insect soup" to biodiversity discovery: Taxonomic revision of *Peloridinannus* Wygodzinsky, 1951 (Hemiptera: Schizopteridae), with description of six new species. *Arthropod Systematics and Phylogeny*, 73(3), 457–475.

Wheeler, Q.D. (1995). The "Old Systematics": Classification and phylogeny. In *Biology, Phylogeny, and Classification of Coleoptera. Papers Celebrating the 80th Birthday of Roy A. Crowson (Volume 1)*, Pakaluk, J. and Ślipiński, S.A. (eds). Museum and Institute of Zoology PAS, Warsaw.

Wheeler, Q.D. (2018). Blank canvas: The case for descriptive taxonomy. *Integrative and Comparative Biology*, 58(6), 1118–1121 [Online]. Available at: https://doi.org/10.1093/icb/ icy067.

Wilder, K. (2009). *Photography and Science*. Reaktion Books, London.

18

The Use of Large Natural History Datasets to Respond to Current Scientific and Societal Issues

Anne-Christine MONNET[1], Thomas HAEVERMANS[1], Anne-Sophie ARCHAMBEAU[2], Philippe GRANDCOLAS[1] and Roseli PELLENS[1]

[1] *ISYEB – CNRS, Muséum national d'Histoire naturelle, Sorbonne Université, École Pratique des Hautes Études, Université des Antilles, Paris, France*

[2] *PatriNat, IRD – OFB – CNRS, Muséum national d'Histoire naturelle, Paris, France*

18.1. Introduction

Natural history collections represent extraordinarily rich and diverse sets of objects. These collections also correspond to colossal datasets that allow us to respond to extremely varied scientific and societal issues. The computerization of natural history collections over the last few decades, and especially their aggregation in global portals, has provided direct and efficient access to the richness and diversity of more than 200 years of collection in more than 5,000 institutions (natural history museums, herbaria, universities, etc.) around the globe. The changes in the quantity and nature of the data available to researchers have transformed scientific practice in many scientific fields, notably biogeography, macroecology and conservation biology. Many questions that were addressed in a partial or localized manner can be revisited

For a color version of all figures in this chapter, see www.iste.co.uk/pellens/natural.zip.

Natural History Collections in the Science of the 21st Century, coordinated by Roseli PELLENS.

and re-analyzed, and thus, many contemporary societal issues can be addressed at finer spatial, temporal or taxonomic scales.

The challenge now is to appropriate these data to answer new questions. Collection data are unique in that they were generated before the questions that their users are asking today. These data were obtained for inventory purposes, by opportunity of collection or to answer a great diversity of questions. They are therefore potentially very heterogeneous in terms of sampling protocols and nature. In this context, how can we make the most of these data in the context of new and current uses, especially when they cover larger extents? How can we ensure a mutually enriching dialog among research programs, scientific questions, collection management and database management? In this chapter, we present several facets of data reuse from natural history collections. We make some suggestions to optimize the uses of past data and ensure that data from contemporary studies can be reused in the future.

18.2. Making data available: a revolution

We have entered an era of digitization and computerization of natural history collections, a precursor to the current dynamics of open science. The resulting data are thus directly available to a growing number of users. The ultimate goal is to make the information associated with all existing specimens in natural history collections available in open access, whether it is about their collection or subsequent uses. In practice, the mobilization of data is done according to the availability of specialists and/or the interest of each institution in a territory, natural resource, or group of organisms and, in more particular cases, to respond to specific societal demands.

Data are often made available through data access portals such as the Global Biodiversity Information Facility (GBIF) (Figure 18.1), consortia of specialists in different taxonomic groups (e.g. VertNet[1] for vertebrates) or the publication of data papers, articles that enable the description and publication of data from a study or a natural history collection (Monnet *et al.* 2021). The user can then export the data directly from the consultation and downloading platforms made available by the data providers. Depending on the case, the nature of the data will be significantly different, from the raw data to the data partially reinterpreted by the database designers. Users must therefore be warned about the nature of the data made available, and the collection method and possible processing must be indicated.

1 Available at: www.vertnet.org.

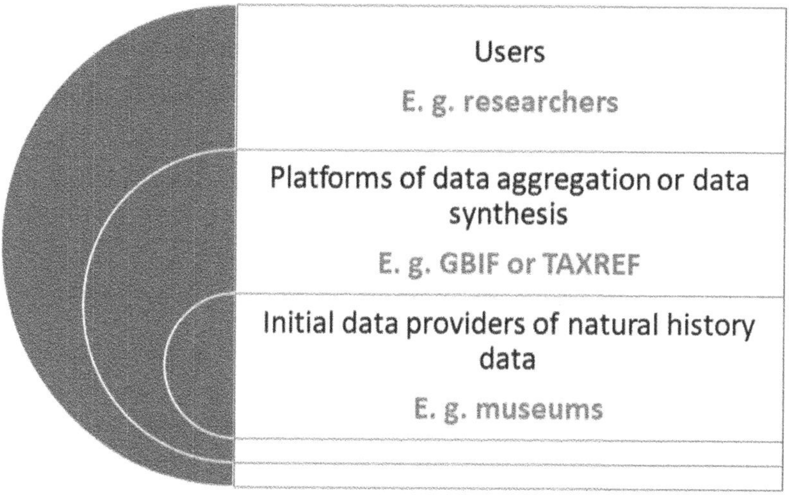

Figure 18.1. *The different actors, from data providers to data users through the data aggregation or synthesis platforms*

Some databases, such as *World Flora Online*[2], offer data resulting from a validation process of the whole database by a network of experts. Others like *Catalogue of Life*[3] offer a consensus classification from different sources: if some parts of the data are covered by different sources, the most relevant data are chosen and presented. It is therefore important to know whether the data used is the result of a process involving individual expert review, or an automated process of selection, correction and consensus. Sharing and open access to data not only makes it possible to deal with questions that could not be addressed before, but also to develop a new dynamic. New working methods based on the sharing of expertise and knowledge are being developed to deal with these new data and their advances, within the framework of open science (see note).

NOTE.– *Estimating the threat level of 345,000 plant species.*

A study carried out by our team aimed to estimate the threat level of all of the vascular plants (Haevermans et al. in preparation). To achieve this, the authors mobilized the knowledge provided by the *Muséum national d'Histoire naturelle's* herbarium collection, but not only this. In order to obtain all of the information necessary to answer their question, several data sources were gathered:

2 Available at: www.worldfloraonline.org.

3 Available at: www.catalogueoflife.org

– the data from the *Muséum national d'Histoire naturelle's* herbarium collection, providing information on the collection of more than six million specimens;

– IUCN Red List data[4], providing an assessment of the extinction risk of species based on multiple criteria;

– World Checklist of Selected Plant Families (WCSP[5]) data from the Royal Botanic Gardens, Kew, providing geographic information on species (national and sub-national occurrences), functional traits and life history;

– The Plant List 2.0[6], which is the most comprehensive list of plant species.

The next step was to evaluate the quality and relevance of the data for the question asked. Which taxonomic reference system should be used? Are the scientific names of species complete and correctly spelled? How much data are missing for the information needed? Do the codes given for geographical areas all match the standard used (e.g. ISO country code or TDWG code (Brummitt 2001))? On such large databases, these checks cannot be done visually.

Once these initial checks were completed, each of the databases used were matched against the checklist provided by Kew, which served as the taxonomic reference for this study.

Only then could the actual analysis work begin. Even though these steps are often well marked out by data managers, this data preparation work is often tedious and requires advanced technical skills. Therefore, the time involved should not be neglected when designing the project.

Finally, it is important for future readers or users of the study to be able to associate the study results with the versions of the databases used for the analyses. This good practice ensures the reproducibility of the analyses for potential new analyses or re-analyses. With the exception of *The Plant List*, which is published in successive versions, the databases used in this study are constantly updated. Because a DOI is not provided for these databases at the initial extraction stage, the best solution is to archive the databases used in the study in a directory linked to the publication.

4 Available at: www.iucnredlist.org.

5 Available at: www.wcsp.science.kew.org.

6 Available at: www.theplantlist.org.

18.3. Challenges for data providers

18.3.1. *Reading labels or directories*

Computerization and digitization mainly require making the museum specimen and the information associated with it available with labels or directories. The information traditionally concerns the conditions under which the specimens were collected, but sometimes also their subsequent uses. The data provider must ensure the quality and structure of the data and the appropriate documentation of the metadata. The information should be as close to reality as possible, understood by different users, and usable in a sustainable way by the computer algorithms of access portals.

The first challenge for data providers is to ensure that the information in the labels and field notebooks is correctly read and extracted. This may be easy with printed documents, but not always with older, handwritten documents with toponymies or rare names. This reading can lead to errors, and independent reading by at least two people allows for more reliable decision-making. The error rate or problems associated with one-person reading labels, the most common form of data entry in our institutions, are not known. This indicates the need to better address this problem and for institutions to invest in procedures that ensure a better way of entering the initial data. For example, in order to contribute to this task, the Paris Herbarium has set up "Les Herbonautes"[7]. This program aims to accelerate the reading of labels through citizen participation and proposes a repeatable method of reading. Each label is read independently by at least three people, and its information is cross-referenced to check the consensus rate. If there is no consensus, the tag is read again. If there are still problems, the case is taken over by the mission manager who looks for the cause of the problem and decides what to do next. The lack of consensus is itself an indicator of limitations/problems in the information and it has often been found that no one could extract more from these labels (M. Pignal, personal communication). The information in question is therefore considered to be lacking.

Typing errors, character inversion, as well as the diversity of special characters in foreign languages can give rise to problems that require a great deal of correction on the part of users. These problems can be minimized by creating a starting database and using a drop-down menu to enter taxon names, author names and place names. This approach is only valid if the starting database contains all of the possible choices. It is therefore necessary to associate a control system (such as an audit) of

7 Available at: www.lesherbonautes.mnhn.fr.

the quality of the digitization process, which can only take the form of a manual or semi-automated correction afterward.

A similar problem arises when transcribing, transliterating or Latinizing the names of localities, regions, authors' names and notes written in the original languages. One of the solutions adopted in the early days of computerization was to use only the symbols contained in the ASCII set corresponding to "basic English", thus excluding all accents, symbols and diacritics of other languages based on the Latin alphabet, as well as all other languages not using the Latin alphabet. Not all languages have a transliteration system that allows them to be transcribed using only unaccented Latin characters. This limitation linked to the character set used leads to a loss of information; a locality perfectly understandable in its original spelling can become uninterpretable when transcribed into a more restrictive character set. Only a return to the original label allows us to understand the locality or the collector's name. The advent of UTF-8 (Universal Character Set Transformation Format – 8 bits) as a character encoding standard compatible with almost all of the languages of the world has made it possible to faithfully transcribe the original spelling in the databases, and thus all of the data of the specimen labels. This standard makes it possible to exchange them without loss of information between database systems. Today, a large majority of websites use this standard. Thus, using a system that encodes characters natively in UTF-8 guarantees optimal encoding and data exchange between systems.

18.3.2. *Structure of the information related to the specimens*

In a natural history collection database, each specimen must have at least five pieces of information: an identifier, a name, information that allows it to be placed in space and time, in other words a reference to the place and collection data, and the collector's name. The latter is very important because it allows, among other things, the cross-referencing of information regarding the place and date, in order to give intellectual credit to the observations and associate them with specific sampling events. This allows, for example, an evaluation of sampling effort (Caesar *et al.* 2017). The level of precision and additional information on the tags may vary depending on the current practice of a period or collector, the purpose of the study that generated the collection and the information important for each taxonomic group (e.g. botanists often include soil type, vegetation type, while the collection method and time of sampling are very important for entomologists). In some cases, they may be complete enough to generate other information. For example, the geo-coordinates of a collection site can be reliably inferred from suitably filled-out locations. Conversely, properly informed geographic coordinates can be used to access any information about the environment where the specimen was sampled that was not included on the label or in the field notebooks and logs.

The format in which this information is presented, the level of detail and the association of several other pieces of information (e.g. soil, vegetation and host species) mean that the possible combinations of this information on the labels or field notebooks can be almost infinite. The second challenge in making the data available is therefore their structure, that is, the association of each piece of information on the label with a specific field in the database. This structure must be as developed as possible and use standard fields such as the Darwin Core standard.

18.3.3. *The taxonomic framework: moving information*

While place, date, environment and collection method are unchangeable parts of the information about a specimen, the name is assigned *a posteriori*; it may change with the evolution of knowledge, or even remain the same but change in meaning (see Chapter 2). For example, for mammals, the number of known species changed from 4,629 in 1993 (Wilson and Reeder 1993), to 6,495 in 2018 (Burgin *et al.* 2018). While new species have been discovered from new field sampling, a significant fraction of this is a result of renaming a portion of the specimens.

Taxonomic reference to specimens in collections is only possible if there is a set of taxon names placed in a classification, for which there is general agreement within the scientific community. A first major effort in this direction is the creation of taxonomic reference documents (e.g. *Catalogue of Life*[8], TAXREF for France (Gargominy *et al.* 2018)). These are databases resulting from the pooling of different reference documents that allow the identification of the valid name of a species, the name of the authors, the date of description and an exhaustive list of synonyms. This is an indispensable condition for the constitution of databases in order to know to which taxon a piece of data refers. Unfortunately, taxonomic reference systems do not always explicitly specify the criteria adopted to resolve taxonomic contradictions, and most often present a single classification. Some systems, such as Botalista[9], developed by the *Conservatoire et Jardin botaniques de la Ville de Genève*, allow for several competing classifications. Other approaches, such as *World Flora Online*[10], are based on a consensus classification developed and managed by groups of taxonomic experts who agree on the classification to be used. There is

8 Available at: www.catalogueoflife.org.

9 Available at: www.botalista.community.

10 Available at: www.worldfloraonline.org.

still a long way to go before we have an up-to-date list of valid species names, and especially an up-to-date identification of collection specimens.

18.3.4. *The importance of tracing the source of data*

Collection data is critical for scientific research because the data remain linked to a specimen to which we can always return to verify, add or compare information. The first piece of information required when building a database is therefore the assignment of a unique identifier for each specimen, to which the information can be associated and corrected over time. The practice of most institutions is to assign identification numbers from 1 to *N* for each collection set (mammals, insect group, plants, etc.). When these data are disseminated via the GBIF platform, they receive another unique identification number (the original number is kept in a separate field), similarly to the GenBank nucleotide sequence lists[11]. This universal practice in the database management community avoids the incorporation of duplicate specimens, if they were repeatedly made available by the data provider via different pipelines, for example. But this GBIF number must always be returned to the collection, otherwise no open access development of the collection is possible. For this reason, the creation of a unique identifier for each collection specimen is one of the major topics of discussion within the infrastructures for making collection data available (such as the pan-European infrastructure DiSSCo[12] – The Distributed System of Scientific Collections – or iDigBio[13] in the United States).

18.4. The role of access portals

The different data publishing institutions share their different local databases via data access portals, such as GBIF. The role of these portals is to allow the user to retrieve and download all the data needed for their work from a single access point. However, we have seen that data entry and preparation practices in each of these institutions can be very heterogeneous. A standardization and interpretation work is then carried out during the dissemination of the data from the institutions to the GBIF portal, an operation supervised by GBIF nodes (i.e. the GBIF teams in each country; e.g. GBIF France).

11 Available at: www.ncbi.nlm.nih.gov/genbank.

12 Available at: www.dissco.eu.

13 Available at: www.idigbio.org.

18.4.1. *The provision standards*

The GBIF secretariat has developed a free open source software tool used to publish and share biodiversity datasets from different institutions through the GBIF network called IPT (Integrated Publishing Toolkit). It is an open source software that allows databases to be transcribed according to the Darwin Core standard developed by the TDWG (Taxonomic Databases Working Group) (Wieczorek *et al.* 2012), a set of standards that make databases interoperable. One of the first steps in IPT, called mapping, is to match fields in an institution's database with those in the Darwin Core. For example, the field (or column in a table) containing the species name must be identified as such (*scientificName* in the Darwin Core standard), even if it is named something else in the initial table (e.g. *species*, *species-name* and *species_name*). This configuration is set up for each new database, then automated for their successive updates. Five fields are mandatory (*basisOfRecord*, *occurrenceID*, *scientificName*, *eventDate* and *countryCode*), 10 others are recommended, and up to 165 fields can be filled in.

Node teams are in charge of guiding institutions in this process. In general, large institutions host this software tool locally on their servers. The data and metadata (i.e. all information associated with the dataset as a whole) remain the property and responsibility of their provider (particularly in terms of data quality). The GBIF's role is to provide technical assistance to researchers and/or collection managers who wish to make their data available[14]. At the MNHN, for example, the data of the different local databases (e.g. Jacim and Sonnerat) are shared on the GBIF as soon as they are updated. When a new institution wants to publish data, the nodal point of the country concerned is informed in order to endorse this new institution and allow it to assure a follow-up to the providers.

During this step of data dissemination from institutional databases via the GBIF, an important verification is also carried out. The scientific names are compared to those of the GBIF Backbone Taxonomy, which serves as a taxonomic reference list. It is based on the *Catalogue of Life*[15], as well as on many other thematic and national taxonomic reference lists. They can either be validated or flagged as erroneous and a reference to an accepted name is suggested. Geographic information is checked and converted to the "decimal degrees" format used by the GBIF. Thus, many well-known errors or inaccuracies are flagged by data access portals such as the GBIF, which thus provide significant assistance to users. This is the case, for example, when latitude or longitude information is missing or inverted, or when the geographic coordinates are equal to zero. It is therefore in a data provider's interest to acquire data directly in

14 Available at: www.gbif.fr/page/contrib/connecter-des-donnees-au-gbif.

15 Available at: www.catalogueoflife.org.

the Darwin Core standard (with a universal character set such as UTF-8), to ensure that the information will be correctly interpreted later.

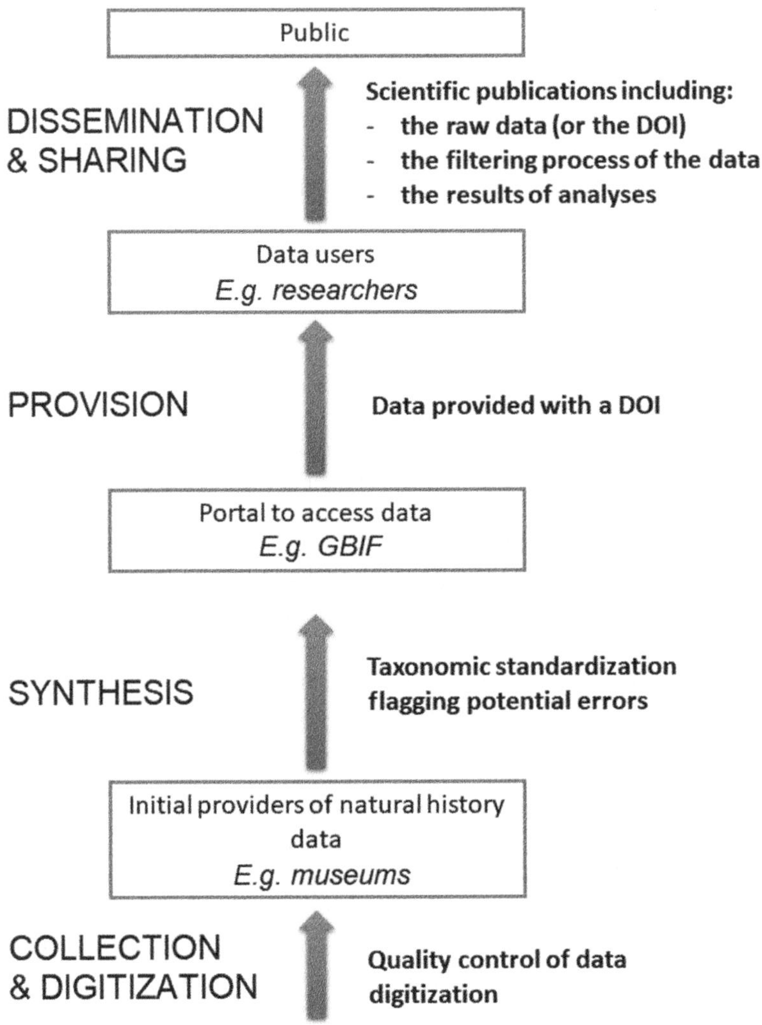

Figure 18.2. *The path from data collection to dissemination of scientific results: the verification and standardization processes undertaken by different actors. The data user has the final responsibility to ensure the reproducibility of the latest operations performed on the data. The DOI is a unique identification number that can be associated with data export*

18.5. The importance of scientific analysis design in appropriating the specificities of data from collections

Collection data often makes a considerable contribution compared to other datasets (Lavoie 2013). In particular, by providing access to historical data (Boakes *et al.* 2010), they document the distribution of organisms prior to landscape transformations by modern-day humans. Using historical data from natural history collections, Tóth *et al.* (2014), for example, showed that the current turnover of mammal species in protected areas in Kenya is lower than in the early 20th century. Furthermore, when compared to standardized vegetation surveys, herbarium data provide more complete species lists due to active searching for rare species by collectors (Garcillán and Ezcurra 2011). Museum data may also have broader spatial coverage than other data sources (e.g. atlas data or bird banding data focused on Western Europe; Boakes *et al.* (2010)). They represent a considerable opportunity for certain groups such as invertebrates, which are often neglected elsewhere (Ponder *et al.* 2001).

In most cases, natural history collections were constituted as a resource for taxonomy and systematics (see Chapter 2 and Chapter 20). Their use for questions of macroecology or macroevolution, which is more recent, goes along with the development of many methods to deal with the uncertainties inherent in the composite origins of data. But the first step is to make sure that the question asked can be answered through an appropriate analysis despite the sometimes heteroclite nature of collection data.

18.5.1. *Detecting the biases in collection data: advantages and opportunities for scientific analyses*

The opportunities offered by collection data should not make us forget their limitations and specificities. First, the availability of these data is often far from being completed. Some taxonomic groups are better studied and sampled than others, which produces differences between the number of species represented in the collections and their real diversity. This is obvious, for instance, when comparing the completeness of representation in the databases of vertebrates versus non-vertebrates (Troudet *et al.* 2017), or butterflies and large beetles versus small flies (Ponder *et al.* 2001). In addition, digitization capabilities also cause huge disparities in data availability. For example, in large collections, plants are almost completely digitized (Le Bras *et al.* 2017) through chain digitization of herbarium pieces. At the other extreme, digitization of most insect groups is proceeding at a very slow pace due to the fragility of the specimens, their three dimensional nature, and the need for label manipulation.

Geographic biases have often been noted in data from collections. Collections near roads, academic centers and in the most accessible locations (Pautasso and McKinney 2007; Daru *et al.* 2018) tend to be more frequent and complete. Depending on the geographic scale considered, other factors, such as economic and academic differences between countries, also play an important role (Araújo 2003). At the global scale, differences in sampling effort coupled with the ability of wealthy countries to make their data available quickly lead to significant distortions in the picture of the distribution of biodiversity on the planet. For example, the picture of the occurrences accessible via the GBIF shows that the amount of data available for the most developed countries with a strong academic tradition is several times greater than that for the Amazon or the Congo Basin, the places with the greatest biological diversity on the planet.

Despite the high value of historical data for diverse studies and for tracking biodiversity change through time (see Chapter 16), few studies have focused on assessing their temporal biases. Nevertheless, the study by Daru *et al.* (2018) indicates that these are also temporally biased, following peaks in the activity of large collectors and significant historical moments in the three countries examined. Boakes *et al.* (2010) showed a reduction in the number of species harvested between 1930 and 1940, the period of the Great War in Europe.

18.5.2. *Toward a good balance between the question and the available data*

As in any research project, the major intellectual challenge is to properly formulate the question being studied and to assess the availability and relevance of data that can help answer that question. It is then necessary to appropriate these data by resampling them and selecting the adequate spatio-temporal window. It is also necessary to identify other available resources that can facilitate the control of the data and correct possible biases.

The data users can then face different situations. They can either meet questions for which data will not be available or relevant, or questions for which validation steps and appropriate statistical processes will make it possible to resample the data, obtaining some response elements, even if they are limited.

A good example of a positive situation is provided by Delisle *et al.* (2003), who were able to understand the invasion dynamics of invasive plants in southern Quebec from historical herbarium data. How did they manage to overcome the biases inherent in the collection process? Initially, the plants were not collected with the aim of measuring the expansion of their distribution. Sampling effort is not uniform over time or space, and thus the increased collection of exotic species in collections themselves can portend an expansion phenomenon, as well as an increase in sampling effort. The authors corrected this bias by jointly studying the sampling of native species collected simultaneously: they compared the cumulative space occupied by each exotic species over time to that occupied by native species. When the proportion (of exotic to native species) increases over a period of time, it strongly suggests that the area occupied by exotic species is actually increasing, because it is increasing more rapidly than if it were strictly the result of improved spatial coverage of herbarium sampling (Delisle *et al.* 2003).

18.5.3. *Playing the advantage of multiple spatial scales*

The geographic localities associated with collection specimens represent an enormous amount of information, but with great disparities in accuracy, especially with historical data that often have low accuracy. These data are often discounted statistical analyses that may require high precision occurrence data acquired with GPS. Yet, these data, despite their low accuracy, provide information that can be critical. Researchers have seized this issue and built models for estimating spatial distributions using data at fine and low resolution (Keil *et al.* 2014). Synergy between the different types of data is often crucial because of their complementarity: historical data can help us to check the completeness of estimated distributions made with limited but accurate recent data, for example.

18.6. Moving from raw data to sorted data that can be used for scientific analyses

The pooling and increasing size of datasets has led to an increase in their complexity and analyses. New skills are therefore needed to build and assemble these databases, and also to analyze them.

The first step in a scientific analysis, once the question has been well defined, is therefore to detect residual errors in the datasets and to discard or correct them (Figure 18.3). However, because of the large amount of data that has become available in recent decades, not every piece of data can generally be checked and corrected individually. Errors are inevitable in a large dataset. They are of a multiple nature and from multiple sources, and their presence increases with the number of data sources in the same database or with the size of the database.

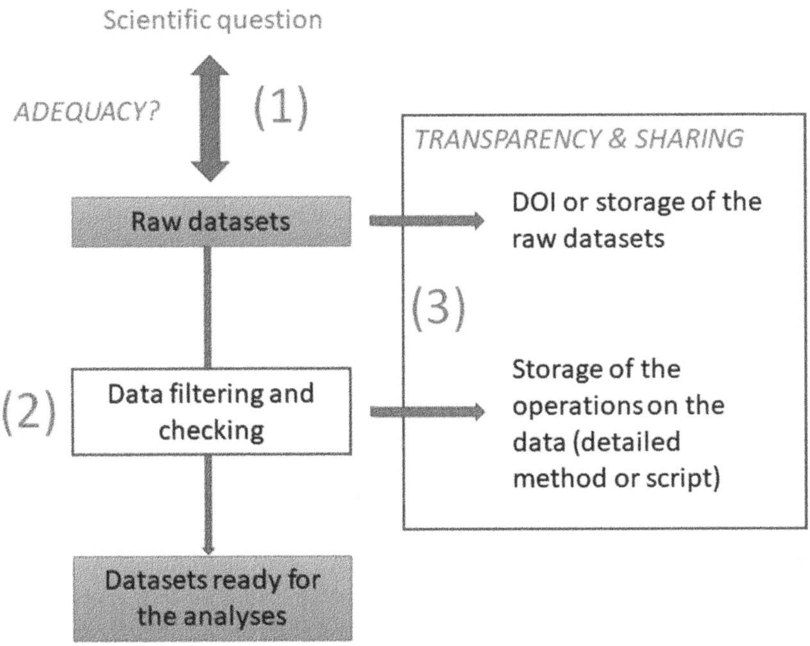

Figure 18.3. *From downloaded data to data prepared for analysis: 1) verification of the adequacy of the data to the question asked, 2) operations regarding the data and 3) responsibility of the user to ensure traceability of the data used and the operations performed*

The first type of error is in the taxonomic information: the names assigned to the taxa may have been misspelled or synonyms may have been used instead of valid names. These errors are usually filtered out at the computerization stage thanks to the standardization methods developed by the data providers. However, some fall through the cracks. Zermoglio *et al.* (2016) analyzed the errors found in VertNet[16], a data aggregator bringing together several hundred vertebrate collections. They showed that only 47% of the taxa had valid names, but that 97% of the taxa could be associated with a valid name when the names contained in the database were compared to a taxonomic backbone. The solution was to compare the extracted list of taxon names to a taxonomic reference system. It is therefore crucial to be able to refer to an up-to-date authoritative reference. Technical tools exist to automate the

16 Available at: www.vertnet.org.

matching of a taxon name with the accepted name (see, for example, Cayuela *et al.* (2012), or the Biodatascreen webservice produced by our research team[17]).

The second type of errors frequently encountered are the errors inherent in geo-spatialized data. These are recurring but generally well-known errors. In 2019, Zizka *et al.* (2019) detected 3.4 million potentially problematic occurrences (3.7%) among 90 million occurrences of flowering plant. Again, automated assessment for error detection exists and is often reported by data providers. The coordinates of a terrestrial plant that would fall into the ocean or dubious coordinates (e.g. with latitude and/or longitude equal to zero, or coordinates not converted to decimal degrees) are easy to detect and discard. However, it is much more complicated to detect coordinates that appear normal, but fall outside the known and validated range of a species. It is then essential to be able to distinguish a misidentification from a valid station, either by expert validation or by comparing the data with an independent dataset. Some databases provide presence data on a sub-national or national scale that are very useful to validate occurrence data. For example, Veron *et al.* (2019) used the *e-monocot*[18] database to help exclude non-native occurrences of species in their analyses. Similarly, Monnet *et al.* (2021) discarded dubious occurrences of Mediterranean trees by confronting the data with the most recent species inventory conducted at the scale of countries and their islands (Médail *et al.* 2019).

18.6.1. *From open data to open science, a responsibility for the traceability of data and operations*

The data users must be able to detect residual errors in order to obtain a working dataset whose quantity and quality they are able to assess (Figure 18.2, note "Some good practices for the data user"). However, since the user does not have access and authority to modify the raw data, most of the time, data with errors or incomplete information are only discarded from the analysis.

It is then critical to store the resampled data and the changes made for future analyses or re-analyses. This usually requires recording all of the processes performed on the data by coding using a programming language. These filters must be transparent to improve the reproducibility of the analyses. Moreover, in order to replicate these analyses, a potential new user must have access to the same export of the database on which the first users worked. Thus, a condition for the reproducibility of the analyses implies communicating the DOI associated with the dataset provided

17 Available at: www.bloom.snv.jussieu.fr/biodatascreen/.

18 Available at: www.about.e-monocot.org.

by some data providers (as the GBIF does), or failing that, storing the raw database in a durable archive. In several decades, a researcher should be able to access the same database used for the analyses performed today.

The DOI provided by the GBIF for a dataset also allows the data that have been used to link to a publication: when consulting data on the GBIF, a user can see how they were previously used. The challenge is also to make the work more collaborative: the effort put into understanding the data and cleaning it up is not to be done anymore and the work can be shared and improved. Open data is thus rapidly leading us to open science, with the sharing of more analyses in addition to data. More and more tools are available for data processing, they are becoming more user-friendly and more efficient. The SQL (Structured Query Language) programming language dedicated to the manipulation of relational databases and the R programming language (R Core Team 2019) are widely used by scientists. Recently, the OpenRefine[19] software, a tool designed "for working with messy data", enables a step toward automation for people not familiar with programming for the detection and correction of spelling differences in databases. This software allows manual modifications in a dataset without formulas, while keeping track of these modifications, thus offering the possibility of repeating the operations carried out the first time.

Corrections made by the user are therefore not made directly on the raw data, but only *recorded*. Indeed, in many cases the data user is not the data supplier. This shows the advantage of establishing a supplier/user link in order to set up a positive feedback loop from the user to the data supplier, which can accelerate the correction of errors at the source.

18.6.2. *Toward a necessary reorganization of collaborative work*

An additional challenge lies in the heterogeneity in the nature of data needed for the scientific analysis. Each dataset brings different types of errors to be detected, and discarded or corrected. Thus, the same scientific question may require the collection of systematic data on occurrence, genetics, phylogenetics, environment, etc. Research questions therefore require interoperability between the different types of data and their validation tools. Scientific analysis must be based on a well-structured workflow.

---

19 Available at: www.openrefine.org.

This synthesis may also require new collaborative tools to enable interoperability between different expertise cells, especially when they are managed by different people. For example, the Functional Webs project of the *Centre de synthèse et d'analyse sur la biodiversité* (CESAB, France) has set up a package[20] on R software offering each collaborator access to the constantly evolving database. Thus, while continuing to collect and update data, the various members of their consortium could already work simultaneously on the latest updated version of the database.

NOTE.– *Some good practices for the data user.*

Ideally, to make it reproductible, all data processing should be done in a script (R or other language) and include all of the modifications made to the raw data. The following is a summary of good practices for each major step in data processing and verification.

– *As soon as the data are uploaded*: the information needed for the publication of the results and data can be collected in advance:

- store a raw version of the export for archiving;
- report how the dataset is cited (e.g. citation and date of data access);
- report the DOI if available.

– *Design of the scientific project – check the relevance of the data mobilized for the question:* the following questions can be asked to query the database and check the adequacy between the data and the scientific question asked:

- How much data are provided? For how many taxa?
- How many unique coordinates are there?
- Are all of the fields required for the analysis available? How much data are missing from each required field?
- Have any data been flagged as dubious by the data providers? If so, how many?

These preliminary questions make it possible to assess whether the data meet the expectations of the scientific question posed. If not, additional data can be mobilized and/or the question must be redefined (in its taxonomic, spatial and/or temporal dimension).

20 Available at: www.github.com/SrivastavaLab/fwdata.

– *Data filtering operations: subsampling data to reappropriate it*:

- Are the taxon names accepted names or do they need to be mapped to the accepted name (synonymy) according to a current taxonomic standard (taxonomic standardization according to a taxonomic backbone)?

- Have the most frequent errors in geolocated data (Zizka *et al.* 2019) been previously identified by the data provider or do they need to be detected by the user (removal of dubious data identified by the user and/or provider)?

- Can additional errors be detected by viewing the distribution of data for each taxon (or higher group) or by comparing the resulting data with an independent dataset or expert opinion (eliminating potential misidentifications)?

– *Publication*:

- Are the various databases properly cited and *directly* accessible to the public, either on an archive directory or via a DOI provided by the source database?

- Is the filtering process performed on the data reproducible thanks to a detailed description of the methods or the archiving of the script?

18.7. Conclusion

Studies using specimens and data from natural history collections have multiplied since the early 1990s (Lavoie 2013; Suarez and Tsutsui 2013). They address a very large number of different research topics and mobilize an increasing number of computerized or digitized specimens. These data, emblematic of open science, offer both an access to the universe of biodiversity and a change at the taxonomic, geographic or temporal scale. Thanks to them, crucial questions for science or society can be asked, embracing a large part of the Living World of the globe or relying on reference points ("zero points") to define trends. All of these questions would otherwise be impossible to address without huge investments or long waits to acquire poor equivalents of datasets.

However, using these data requires particular expertise; they must be mixed, assembled, validated and resampled, and these steps represent a significant and sophisticated task. However, the scientific community gains expertise and develops new methods and user-friendly tools to help the data users to sort their data.

In doing so, we benefit from open data and it is up to us to use these data and to perform our analyses with the same ethics, ensuring the reproducibility of the operations performed. Open data is thus rapidly moving us toward open science, encouraging increasing sharing of analysis in addition to data. Reinserting natural history collections into scientific and societal issues implies not only the synthesis of different datasets, but also the synthesis of different expertise and knowledge, calling for a rethinking of the organization of scientific analysis work toward more collective work. The specificities of collection data (e.g. the existence of historic data) represent incredible opportunities to open up the field of knowledge, but the associated biases should not be neglected in the analyses. On the contrary, a better knowledge of these biases can guide the choices of spatial and temporal resolution of the analyses that can be conducted, as well as the preparation of future expeditions for new data collection.

In this perspective, several methodological aspects are currently being developed at the global scale. Taxonomic reference systems are under construction and are permanently changing and must reflect the state of knowledge on biodiversity. They are based on thousands of works concerning more than two million known species of organisms. This progress requires that unique identifiers be defined and used for each preserved specimen, or for each mobilized dataset. These are issues that may seem simple, but require consensus and organization at the international level.

But the progress we need the most is to raise awareness that biodiversity data – including those we currently collect – must remain open and usable for many years to come. The *sine qua non* for such availability is the reference to a material sample or at least to rich digital data (Grandcolas 2017, 2019; Troudet *et al.* 2018). Only this reference can help us to overcome inaccuracies, errors in the work that generated the biodiversity data, but also the evolution of taxonomies and names. The availability of data does not go without the availability of analyses and intermediate data processing steps that also must be available for future reference.

While facing the biodiversity crisis, the scientific community keeps discovering and using the treasure and heritage represented by natural history collections and the biodiversity data that reside there brilliantly, but should not forget that we must not only make scientific "profits" from this heritage, but also conserve it and enable it to bear fruit for future generations.

18.8. References

Araújo, M.B. (2003). The coincidence of people and biodiversity in Europe. *Global Ecology and Biogeography*, 12(1), 5–12.

Boakes, E.H., McGowan, P.J.K., Fuller, R.A., Chang-Qing, D., Clark, N.E., O'Connor, K., Mace, G.M. (2010). Distorted views of biodiversity: Spatial and temporal bias in species occurrence data. *PLoS Biol.*, 8, e1000385.

Brummitt, R.K. (2001). World Geographical Scheme for Recording Plant Distributions, Edition 2. Document, Biodiversity Information Standards (TDWG), Geneva.

Burgin, C.J., Colella, J.P., Kahn, P.L., Upham, N.S. (2018). How many species of mammals are there? *J. Mammal.*, 99, 1–14.

Caesar, M., Grandcolas, P., Pellens, R. (2017). Outstanding micro-endemism in New Caledonia: More than one out of ten animal species have a very restricted distribution range. *PLoS ONE*, 12(7), e0181437.

Cayuela, L., Granzow-de la Cerda, Í., Albuquerque, F.S., Golicher, D.J. (2012). Taxonstand: An R package for species names standardisation in vegetation databases. *Methods Ecol. Evol.*, 3, 1078–1083.

Daru, B.H., Park, D.S., Primack, R.B., Willis, C.G., Barrington, D.S., Whitfeld, T.J.S., Seidler, T.G., Sweeney, P.W., Foster, D.R., Ellison, A.M., Davis, C.C. (2018). Widespread sampling biases in herbaria revealed from large-scale digitization. *New Phytol.*, 217, 939–955.

Delisle, F., Lavoie, C., Jean, M., Lachance, D. (2003). Reconstructing the spread of invasive plants: Taking into account biases associated with herbarium specimens. *J. Biogeogr.*, 30, 1033–1042.

Garcillán, P.P. and Ezcurra, E. (2011). Sampling procedures and species estimation: Testing the effectiveness of herbarium data against vegetation sampling in an oceanic island. *J. Veg. Sci.*, 22, 273–280.

Gargominy, O., Tercerie, S., Régnier, C., Ramage, T., Dupont, P., Daszkiewicz, P., Poncet, L. (2018). TAXREF v12, référentiel taxonomique pour la France : méthodologie, mise en oeuvre et diffusion. Report, Muséum national d'Histoire naturelle, Paris.

Grandcolas, P. (2017). Loosing the connection between the observation and the specimen: A by-product of the digital era or a trend inherited from general biology? *Bionomina*, 12, 57–62.

Grandcolas, P. (2019). The rise of "Digital Biology": We need not only open, FAIR but also sustainable data! *Biodiversity Information Science and Standards*, 3, e37508.

Haevermans, T. *et al.* (in preparation). A majority of plant life globally at risk as predicted by machine learning threat analysis.

Keil, P., Belmaker, J., Wilson, A.M., Unitt, P., Jetz, W. (2013). Downscaling of species distribution models: A hierarchical approach. *Methods Ecol. Evol.*, 4, 82–94.

Lavoie, C. (2013). Biological collections in an ever changing world: Herbaria as tools for biogeographical and environmental studies. *Perspect. Plant Ecol. Evol. Syst.*, 15, 68–76.

Le Bras, G., Pignal, M., Jeanson, M.L., Muller, S., Aupic, C., Carré, B., Flament, G., Gaudeul, M., Gonçalves, C., Invernón, V.R., Jabbour, F., Lerat, E., Lowry, P.P., Offroy, B., Pimparé, E.P., Poncy, O., Rouhan, G., Haevermans, T. (2017). The French Muséum National d'histoire Naturelle vascular plant herbarium collection dataset. *Scientific Data*, 4, 1–16.

Médail, F., Monnet A.-C., Pavon D., Nikolic T., Dimopoulos P., Bacchetta G., Arroyo J., Barina Z., Cheikh Albassatneh M., Domina G., Fady B., Matevski V., Mifsud S., Leriche A. (2019). What is a tree in the Mediterranean Basin hotspot? A critical analysis. *Forest Ecosystems*, 6(17).

Monnet, A.-C., Cheikh albassatneh, M., Arroyo, J., Bacchetta, G., Bagnoli, F., Barina, Z., Cartereau, M., Casajus, N., Dimopoulos, P., Domina, G., Doxa, A., Escudero, M., Fady, B., Hampe, A., Matevski, V., Misfud, S., Nikolic, T., Pavon, D., Roig, A., Santos Barea, E., Spanu, I., Strid, A., Vendramin, G.G., Leriche, A. (2021). WOODIV, a database of occurrences, functional traits, and phylogenetic data for all Euro-Mediterranean trees. *Sci. Data*, 8(89).

Pautasso, M. and McKinney, M.L. (2007). The botanist effect revisited: Plant species richness, county area, and human population size in the United States. *Conservation Biology*, 21(5), 1333–40.

Ponder, W.F., Carter, G.A., Flemons, P., Chapman, R.R. (2001). Evaluation of museum collection data for use in biodiversity assessment. *Conserv. Biol.*, 15, 648–657.

R Core Team (2019). R: A language and environment for statistical computing. R Foundation for Statistical Computing, Vienna [Online]. Available at: https://www.R-project.org/.

Rouhan, G., Chagnoux, S., Dennetière, B., Schäfer, V., Pignal, M. (2014). The herbonauts website: Recruiting the general public to acquire the data from herbarium labels. In *Botanists of the Twenty First Century: Roles, Challenges and Opportunities*. UNESCO, Paris.

Suarez, A.V. and Tsutsui, N.D. (2004). The value of museum collections for research and society. *Bioscience*, 54, 66–74.

Tóth, A.B., Lyons, S.K., Behrensmeyer, A.K. (2014). A century of change in Kenya's mammal communities: Increased richness and decreased uniqueness in six protected areas. *PLoS One*, 9(4), e93092.

Troudet, J., Grandcolas, P., Blin, A., Vignes-Lebbe, R., Legendre, F. (2017). Taxonomic bias in biodiversity data and societal preferences. *Scientific Reports*, 7(1), 1–14.

Troudet, J., Vignes-Lebbe, R., Grandcolas, P., Legendre, F. (2018). The increasing disconnection of primary biodiversity data from specimens: How does it happen and how to handle it? *Systematic Biology*, 67(6), 1110–1119.

Veron, S., Mouchet, M., Govaerts, R., Haevermans, T., Pellens, R. (2019). Vulnerability to climate change of islands worldwide and its impact on the tree of life. *Scientific Reports*, 9(1).

Wieczorek, J., Bloom, D., Guralnick, R., Blum, S., Döring, M., Giovanni, R., Robertson, T., Vieglais, D. (2012). Darwin core: An evolving community-developed biodiversity data standard. *PLoS One*, 7(1), e29715.

Wilson, D.E. and Reeder, D.M. (1993). *Mammal Species of the World: A Taxonomic and Geographic Reference*, 2nd edition. Smithsonian Institution Press, Washington, DC.

Zermoglio, P.F., Guralnick, R.P., Wieczorek, J.R. (2016). A standardized reference data set for vertebrate taxon name resolution. *PLoS One*, 11, 1–20.

Zizka, A., Silvestro, D., Andermann, T., Azevedo, J., Duarte Ritter, C., Edler, D., Farooq, H., Herdean, A., Ariza, M., Scharn, R., Svantesson, S., Wengström, N., Zizka, V., Antonelli, A. (2019). CoordinateCleaner: Standardized cleaning of occurrence records from biological collection databases. *Methods Ecol. Evol.*, 10, 744–751.

19

Is There a Need for Biocultural Collections? State of the Art and Perspectives[1]

Serge BAHUCHET
EA – CNRS, Muséum national d'Histoire naturelle, Université Paris Diderot, Musée de l'Homme, Paris, France

19.1. Introduction

In the vast field of human sciences, the MNHN holds several collections of various compositions which result from the history of laboratories and present different modes of evolution. Some of them date back to the origins of the *Jardin du Roi* in the 17th century (physical anthropology, then "anatomy"), others appeared in the 19th century when their science was born (prehistory), and others are much more recent, created in the 20th century (ethnology and ethnobotany). In this chapter, we propose to briefly describe these last two elements, which directly concern the lifestyles of modern-day humans, by tracing their origins.

For a color version of all figures in this chapter, see www.iste.co.uk/pellens/natural.zip.

1 This chapter has benefited from the help of several people whom I would like to thank, especially Pauline Rameau, for her collaboration in the collection and management of the collections both in the field and in the laboratory, and for the numerous discussions about this, as well as Flora Pennec for her help in the drawing up of the maps, Simon Juraver for the information and the photographs on the ethnobotanical collection, and Nathalie Lasgleyzes for the editing of the photographic plates.

Natural History Collections in the Science of the 21st Century,
coordinated by Roseli PELLENS. © ISTE Ltd 2021.

The primary question is indeed to decide what to do with the collections we inherit, which are the result of different trajectories and have been built up independently of each other. The links between them have never been made. It is doubtful whether it would be useful to continue this segmentation. How can we revive collections that have been virtually abandoned for three decades (ethnobotany)? Is it really necessary to develop collections of objects again, post-2002, and if so, in which directions? Starting from the principles that were behind their existence, we will expose the principles of a scientific and functional reorganization and then some lines of enrichment and research.

Rather than calling these collections "cultural anthropology", which could restrict them to the field of ethnographic objects and would maintain a segmentation, we prefer the *biocultural* notion, which has the advantage of including the ethnobiological approach, which takes into account the elements of the natural environment involved in human activities:

> Biocultural collections are ethnobiological specimens, artifacts and documents – plant, animal and cultural – that represent dynamic relationships among people, biota, and environments. (Salick *et al.* 2014, p. 1)

It is in the use of this term that the ambition of the profound restructuring of the collections for which we are responsible lies.

19.2. Origin of these collections

19.2.1. *Ethnobotany*

Beginning in the 1920s, in the former laboratories of colonial and then tropical agronomy (Prof. Auguste Chevalier, from 1911), then of ethnobotany (Prof. Roland Portères, from 1962; (Portères 1965; Bahuchet and Lizet 2003)), the ethnobotany collection is the oldest and most imposing, with about 100,000 items. It consists of useful plants, known, named and used by human societies in various geographical areas. These are plant specimens in various forms: herbaria, seeds, fruits, fibers and flours, and wood samples. Added to this are various objects made of plant material (farming tools, calabashes, beaten bark cloth, African masks).

At the end of the 1960s, the laboratory extended to ethnozoology, and built up a modest collection of animal samples (insects, fish, reptiles in alcohol, and mammal skins or skulls), collected by ethnologists or linguists for identification purposes, as well as objects related to hunting or made with animal elements and traditional beehives.

19.2.2. *Ethnology*

This collection is a continuation of the collections of the former ethnology laboratory of the *Musée de l'Homme*. After the transfer to the *Musée du quai Branly* of the non-European ethnology collections of the MNHN registered before 2002, a few hundred objects remained the property of the *Muséum*, either because they were not incorporated into the registered collections or because they were acquired after the 2002 decree. From this point on, it was decided (a) to maintain the existence of an ethnological collection at the MNHN, and (b) to limit its scope to themes appropriate to the *Muséum*'s new program, focused on biodiversity and natural environments.

The ethnological collection now includes over 6,000 ethnographic objects.

19.3. Collection principles and the function of collections

19.3.1. *The role of objects in "Maussian" ethnology*

In 1928, when Paul Rivet took over as director of the *Musée d'ethnographie du Trocadero*[2] and reorganized it, following the creation of the *Institut d'ethnologie* with Lucien Lévy-Bruhl and Marcel Mauss in 1925, he trained young ethnologists in new principles of collecting objects for ethnographic research (Bahuchet 2015, pp. 125–139).

The best document explaining the role of objects, both as a manifesto and as a practical guide, is the booklet "*Instructions sommaires pour les collecteurs d'objets ethnographique*" (Leiris 1931)[3]. Here we reproduce the most characteristic passages,

---

2 The MET was created in 1879 by Ernest Théodore Hamy within the framework of the Ministry of Public Education, but outside of the *Muséum*, to house all of the State's ethnographic collections, which had been dispersed in various public institutions following the short-lived provisional Museum of Ethnography of the Scientific Missions of the 1878 World's Fair. It was not attached to the chair of anthropology at the MNHN until 1928 when Paul Rivet was elected professor.

3 The *Instructions* were written by Michel Leiris at the start of the great Dakar-Djibouti mission organized by Marcel Griaule (1931–1933), based on the teachings of Marcel Mauss at the *Institut d'ethnologie*. See Jolly (2016) for the development of these instructions and their application by the missions of this period in Africa. Marcel Mauss's lectures were transcribed by his students and later published (Mauss 1967).

which specify the place of material culture in ethnographic study and lay down the principles of collecting (see Leiris 1931).

> What is ethnography? Ethnography [...] has as its essential object the study of material civilization: food and housing, clothing and adornment, weapons and instruments, hunting, fishing, culture and industry, means of transport and exchange, ceremonial attributes, games, artistic productions, etc. (p. 5)
>
> *It is urgent to constitute collections of objects.* Almost all the phenomena of collective life are likely to be translated by given objects, because of this need which has always pushed men to print on matter the trace of their activity. A collection of objects systematically collected is thus a rich collection of "pieces to convict", whose meeting forms archives [...]. (p. 6)
>
> *Choice of the object.* A collection of ethnographic objects is neither a collection of curiosities, nor a collection of works of art. The object is no more than a witness, which must be considered according to the information it brings about a given civilization, and not according to its esthetic value. It is therefore necessary to get used to collecting all kinds of objects and to get rid first of two prejudices, that of the purity of style and that of rarity. [...] *2°. Prejudice of rarity.* The most common objects are those that tell us the most about a civilization. A tin can, for example, characterizes our societies better than the most sumptuous jewel or the rarest stamp. (p. 8; italic text was in bold in the original French text)

In his ethnography course, Marcel Mauss insisted on the fact that "any object must be studied: 1° in itself; 2° in relation to the people who use it; 3° in relation to the totality of the observed system" (Mauss 1967, p. 34).

These recommendations by Mauss found their development in the department of comparative technology that Rivet had created in his *Musée de l'Homme*, with André Leroi-Gourhan (1971, 1973). To study the object in itself, comparative series were constituted. The second axis ("... in relation to the people who use them") led to insisting on the importance of the gesture and the handling of the objects, as Leroi-Gourhan affirmed: "[...] the tool exists only in the operative cycle; it is a good witness of it [...]. [...] the tool is really only in the gesture which makes it *technically effective*" (Leroi-Gourhan 1965, p. 35, our use of italics)[4]. Consequently, it is of extreme importance to support the collected object with precise documentation[5] (Figure 19.1).

Figure 19.1. *Handling of an agricultural knife (DRC): blade pushed tangentially to weed from the tip, blade thrown to cut grass on the curved side, blade pulled towards oneself, to dig before planting. Source: S. Bahuchet (1990)*

19.3.2. *Ethnobotanical collections*

Roland Portères defined ethnobotany as follows when he founded his laboratory in 1963 by transforming the former chair of tropical agronomy:

> *Ethnobotany* is an interpretative and associative discipline that researches, uses, links and interprets the facts of interrelations between Human *Societies* and Plants in order to understand and explain the birth and progress of civilizations [...]. It studies, in themselves, neither Plants, nor Human Societies, nor Humans. [...] Ethnobotany will only rely on [agronomy, agriculture, economic botanics], having

4 As early as 1940, Haudricourt had already affirmed: "The *material civilization* of a determined human grouping is *all of the traditional muscular movements with technical effectiveness*" (author's use of italics), and, further, "the tool is adapted to the gesture and not conversely" (Haudricourt 1940, p. 760).

5 More recently, François Sigaut demonstrated how only observations on the way the knife is held and wielded allows us to understand its effectiveness (Sigaut 2012).

> in view only *relations with human Societies*. [...] *Ethnobotany can be classified in the Human Sciences*. (Portères 1961, pp. 102–104; author's use of italics)

And Portères adds: "Ethnobotany belongs more to the group of human societies than to that of plant biology, where it has no business" (Portères 1961, p. 104).

In 1969–1970, Roland Portères wrote a handout to accompany his course in ethnobotany at the University of Paris V (Portères 1969). He devoted a significant section to ethnobotanical objects and collections, and described the "ethnographic object" as:

> A concrete, material witness of Civilization. Its presence gives an account of the mentality, the ingenuity, the technical skill and the technological degree of an operator in the shaping or transformation of a raw product into something elaborate. [An object that has been created] in order to play a role, to assume one or more functions. (Portères 1969, p. 117)

He goes on to define the *botano-ethnographic object* as "any ethnographic object made from materials of plant origin" (p. 118). He clarifies, however, that these are not just materials, but properties:

> For the Ethno-botanist, as for the Ethno-zoologist, the Object, part of a living Being, holds the Life of the Being from which it comes. This fact gives the Object a Life and this is transferred to the Ethnographer's Object.

And to continue:

> The Object tends to be only Life, which is integrated with the Function; the matter itself becomes secondary, insofar as one considers only the Form, conceived to allow the achievement of the Function. (p. 120)

He then describes all of the "botano-ethnographic objects" which are whole objects, but also any vegetable fragment worked, triturated, "as well as what it evokes, or suggests". He insists on the need to take into account the plant itself as an object, that is to say, what allows the identification of the taxonomic status of the plant that provided the material. Portères then gives detailed indications on the methods, starting with the obtaining and preparation of ethnobotanical samples and then on "field notes", that is, investigations, which must deal with the plant, the raw and then transformed product, the motivations, the rituals of collection, the names, the technique and the tools of transformation, etc. (p. 127), all accompanied by photographs and sketches. Indeed, he insists that:

> Ethnobotanical description goes beyond botanical description and much more information is needed. The notation is part of the Object collected in the field. (Portères 1969, p. 127)

It is on this basis that the laboratory's collections were constituted: "The Ethnobotanist must systematically collect samples of the Plants to which he will refer elsewhere" (Portères 1961, p. 106). In addition to the collections made by the researchers of the laboratory and by himself for their own research, Portères had set up a real ethnobotanical service to help ethnologists and linguists in their research and descriptions of human communities, which consisted of giving them indications on the collections to be made, and even providing them with material, and then determining their herbaria on returning from the field.

Later, the ethnozoologist Raymond Pujol created a similar service for animals.

19.3.3. *Biocultural collections*

19.3.3.1. *Ethnobiological specimens*

Ethnobiology is the branch of human sciences that studies the interrelations of human societies with their environment, through knowledge and know-how; it includes, in particular, ethnozoology (relations with the animal world) and ethnobotany (relations with the plant world). Like ethnology, ethnobiology takes place in the field, within human communities. The ethnobiologist investigates technical practices through direct observation, called participatory observation, and through variously organized interviews (free, semi-directive, by questionnaire, etc.). For all the activities, they collect the technical and naturalistic vocabulary, transcribing it into phonetics. They collate the information through audio recordings, photos and videos detailing the techniques.

Their main concern is to understand the meaning of local terms, especially with regard to species. They must also try to identify the principles of local taxonomy and the groupings of species that are made by the speakers of the language being studied (categorization and classification processes). It is at this point that they need to find the relationship between the local terms designating natural elements and the scientific nomenclature concerning species. In addition to Portères' recommendations cited above, we will rely on the more recent article by Robert Bye (1986) concerning the status of specimens, which are essential for the ethnobiologist's work, because they are necessary for identifying the organism: "With an accurate identification, the scientific name becomes the crucial link between people with folk knowledge and people trained in western sciences" (Bye 1986, p. 2). It is the bridge between two cultures, facilitating the exchange of mutually beneficial information.

As often as possible, the ethnobiologist collects the specimens described by vernacular terms in order to obtain later the determination. For this, they must ask the help of systematicians in the appropriate scientific institutions. They must also follow the regulations in force concerning biological collections in the countries where they work (which may forbid them from repatriating their herbarium to their home country), as well as international regulations (CITES convention, Nagoya protocol). It is important that the specimens are deposited in reference collections (museums, universities) in order to be conserved permanently and made available to researchers, thus becoming reference material.

The ethnobiological specimens essentially correspond to a research stage, and they are not an end in themselves. They carry cultural information and bear witness to local knowledge. They must therefore be linked to documents containing this information (handwritten field notes and publications, but also drawings, photos and films showing the techniques of use). From the ethnobiologist's point of view, what matters is the information; once this has been collected, the specimens have only an episodic use. Ethnobiological collections have, first of all, the function of obtaining the scientific determination of a specimen that bears a local name. The specimens will then essentially serve as a reference for later reference when a monograph has been published by the researcher. However, duly informed, they remain an important source for taxonomists for the places and periods of collection; conversely, the ethnobiologist's specimens, re-examined by naturalists, benefit from the most recent verifications and modifications of nomenclature.

The specimen depends on the field conditions, it is obtained from the members of the community studied, and it is of extremely variable quality: horns, pieces of animal skin, shells, seeds or dried fruits, in the best of cases, parts of herbs, but are most often sterile! The goal is to collect a relevant piece that is sufficient to distinguish the species. Unfortunately, specimens with all the characters of a "clean" taxonomic collection are the exception. However, both Portères and Bye insist that the specimen, even if "unacceptable" by the standards of natural history collections, should nevertheless be preserved, either in general taxonomic collections or possibly in particular areas outside natural history collections.

19.3.3.2. *Material culture in ethnobiology*

Ethnobiology involves the study of techniques (know-how) in relation to the elements of the natural environment (plants, animals). These practices concern predation (the taking of spontaneous species by hunting, fishing, collecting), agriculture and breeding (including species and varieties resulting from human selection). They take into account the actions of transformation (manufacturing techniques), and food, which is the primary intermediary between humans and living species.

As such, it addresses the question of the tools that are associated with these practices that are linked to the natural environment. It thus joins the field of material ethnology, as described above (section 19.3.1). What may distinguish it is the emphasis that is placed on the local knowledge that is associated with these techniques and the living beings involved in them, as well as the stages of material transformation. From this point of view, the object is a medium, a witness to this ethnobiological knowledge, in the same way as the previously-defined specimen. The object is accompanied by detailed surveys on the activity in which it is used and on the conditions under which it was obtained by its user. Examples of survey questionnaires can be found in Bahuchet (2014, pp. 169–182). This fact underlines the indispensability of linking objects with cultural information.

In other words, the objects are intended to document the uses of biodiversity, *hic et nunc*, here and now.

Therefore we have two types of collections in the biocultural field, with different purposes[6]:

– reference specimens (plant or animal samples) to scientifically determine what has a local name, and to preserve the testimony;

– objects to document the uses of biodiversity.

These two types of collections each rely on associated documentation, both for ethnobiological specimens and for material culture (artifacts).

19.4. Principles for the articulation of sets

The justification of biocultural collections in a natural history museum necessarily implies a coherence of these two types of elements and to create links

6 Examples of ethnobotanical collections made during ethnographic surveys can be found in Brousse (2017).

between them. Indeed, the principle of biocultural collections involves linking the animal or plant species to the cultural object: which species are used to make a utensil, and for the obtaining of which species is this object intended (Bahuchet 2014, pp. 169–182)?

The most useful concept in this setting of linking elements is that, created by the cultural technology, of an operational chain (Lemonnier 1976, 1992; Balfet 1991). Any object is the result of a process of manufacture, of transformation of matter, by the implementation of energy, based on this simple notion explained by François Sigaut (published in 1978 and republished in 2010) (Figure 19.2).

Material state 1 → *activity* → **Material state 2**

Figure 19.2. *Transformation of matter following Sigaut (2010)*

The "activity" is a technical act, according to Bertrand Gille's conception (1978): the technique is an action which has an effect, resulting from an "obligatory couple: matter–energy whose two elements are linked together by the technical act which, most often, needs a medium" (Gille 1978), and thus, this energy can possibly be transmitted by an object. Pierre Lemonnier specifies that each technique is made up of five elements that are always linked: (1) a material, (2) an energy, (3) tools (means of action), (4) gestures, (5) knowledge (ideas, i.e. a mental process) (Lemonnier 1992, pp. 5–6). I would tend to put gestures before tools, because there are always gestures even when one does not use utensils, for example, when only the hand is used.

The gestures follow one another in a logical series of operations that depend on one another to achieve a result. This is what we call an operational chain: "an operational chain is the series of operations that transform a raw material into a product, whether it is an object of consumption or a tool", according to Robert Cresswell's minimalistic definition (1976, p. 26)[7]. If we consider a basket, for example: the operational chain will consist of starting with a plant, say a bamboo, which is cut in the bush, brought back to the village, and cut into long strands. These are thinned, tapered, smoothed, then intertwined to make the basket. The bamboo is a specimen, the tapered strands, a material and to end up the basket, the object (Figures 19.3 and 19.4).

7 Lemonnier (1976) and Balfet (1991) have commented on the work program defined by the operational chain concept, showing both its interest in providing a detailed description of techniques and its complexity of application.

Each of these transformations can be carried out with the use of utensils, which may be one for each stage or for several different stages (Figure 19.5).

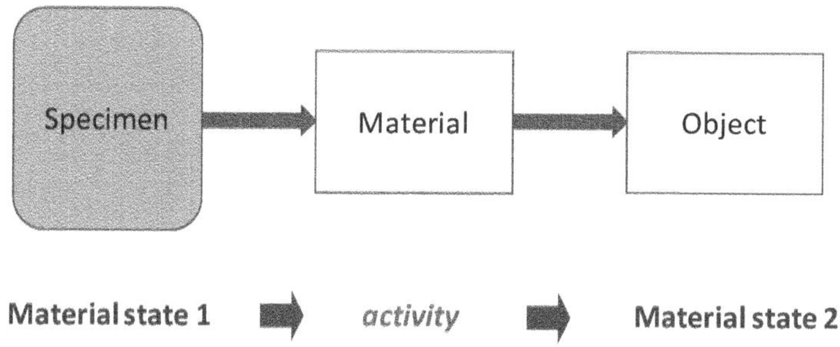

Figure 19.3. *The operational chain*

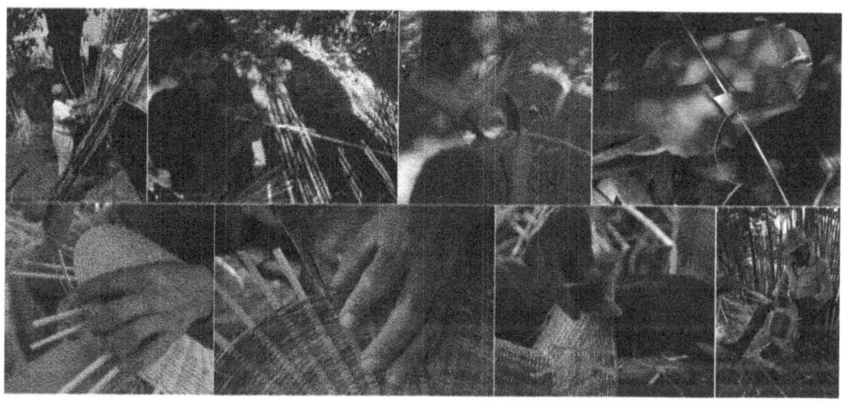

Figure 19.4. *Example of an operational chain for making an* Otatea *bamboo hat. The same knife is used to split the stem, thin and taper the strands, and spread them apart to thread them into the basket shape. Don Lauro and his family, Chiquilistlán (Jalisco, Mexico). Source: Pauline Rameau (2017)*

Finally, each object is, itself, the result of a transformation chain similar to that which has just been defined. Moreover, it can be considered as being situated at the intersection of four systems. First, the object is manufactured; it is the product of the operational chain which has just been described. This is the *production system*. Second, it is a tool that is used to do something; this is the *system of use*. Third, this object can be given, borrowed, bought or exchanged, it is a commodity; this is the *system of consumption*. Finally, the object is compared to other objects, it is an element inserted in a *system of objects* (Figure 19.6).

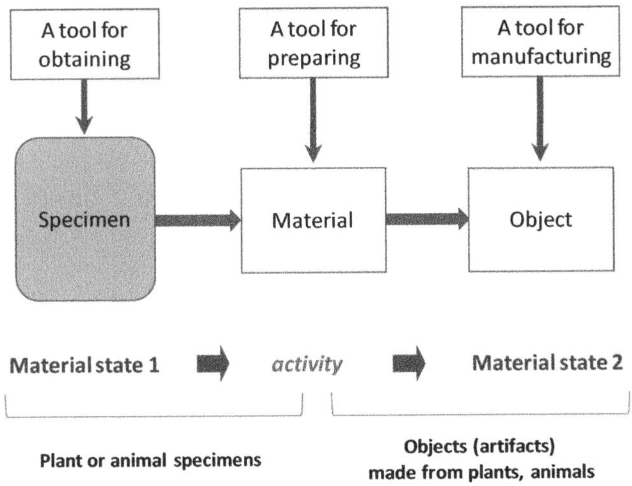

Figure 19.5. *Conceptual relationships between specimens and objects*

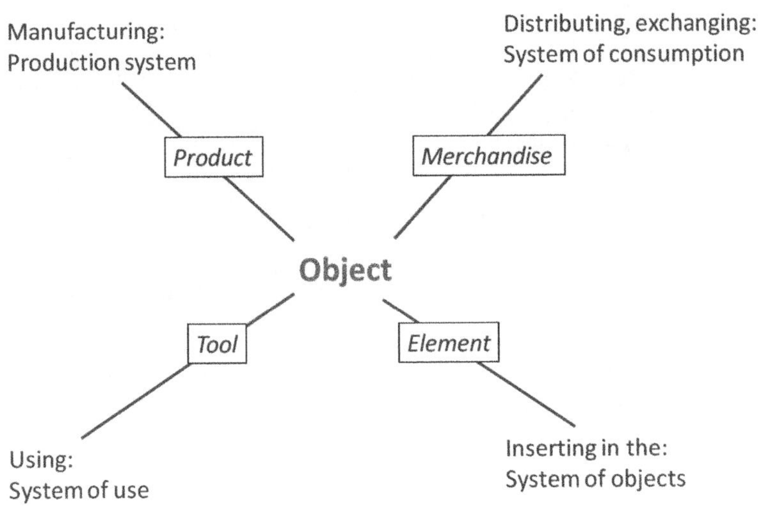

Figure 19.6. *The system of objects*

Using this principle, a different functional distribution of the elements of the two previous collections will be made, no longer taking into account their institutional origin. Thus, on the one hand, ethnobiological (ethnobotanical and ethnozoological) specimens, including cultivated plants, will be grouped together, and, on the other hand, all artifacts, including transformed materials, whether technical or food-related (Figure 19.7).

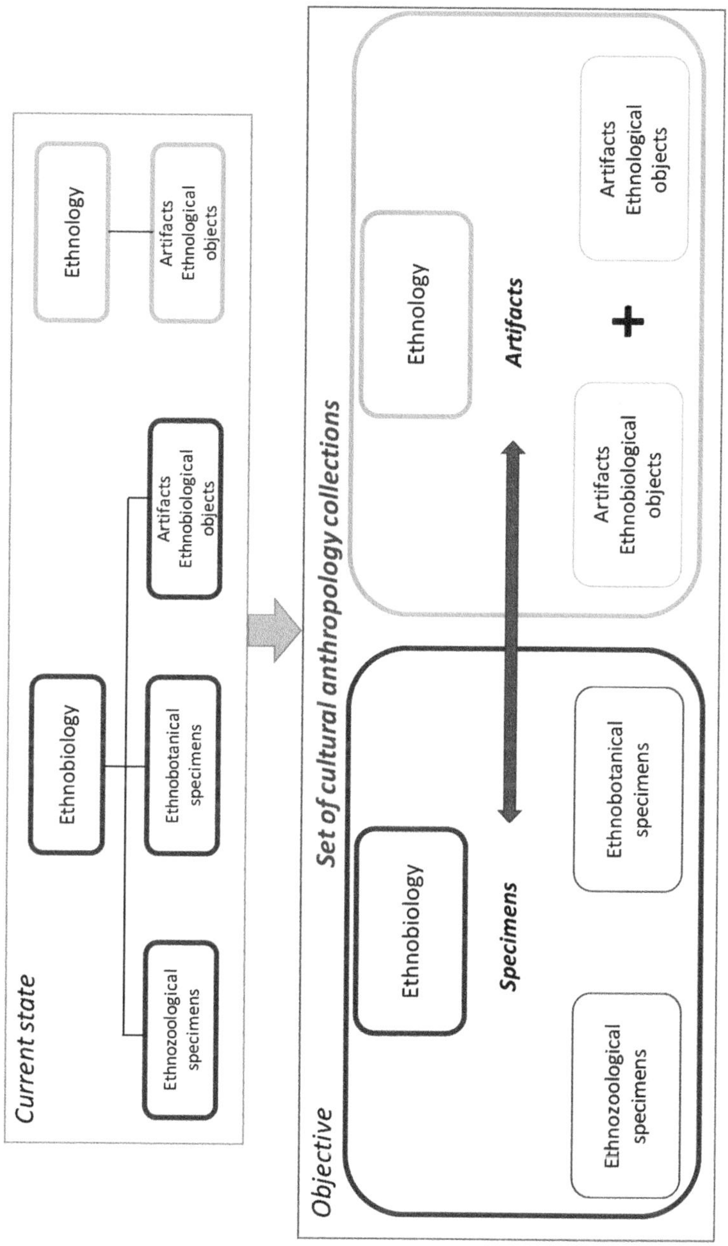

Figure 19.7. *Diagram of the evolution of the organization of groups*

a) b)

Figure 19.8. *Two objects and associated species: (a) ridging hoe for yam cultivation (*Dioscorea *spp., Dioscoreaceae), Benin, collection R. Pujol; (b) duiker skin hunting quiver (*Cephalophus *sp., Bovideae), Mbuti Pygmies, DRC, collection S. Bahuchet. Source: J.-C. Domenech – MNHN*

An ear of corn or bean kernels, for example, would be considered ethnobiological specimens, but a commercial bag of millet semolina, or dried chuño potatoes, would be an artifact, since it is the result of human transformation.

However, what gives the term "biocultural collections" its full meaning, and what is most important, is to link these two sets to each other so that they form a single scientific whole, beyond the technical constraints of conservation. It is imperative to create the link between the object and the specimen, and to pool the documents that concern them: the plant that provided the wood to make this object; the prey for which this weapon was intended; the hoe used for this cultivated plant (Figures 19.8 and 19.9).

The documents include notes and descriptions from the field, publications (theses, monographs, articles, etc.) and, all image media, sketches, photos or videos concerning the gesture and the object, its use, its manufacture and its location *in situ*. In other words, it is a question of constituting an ambitious database (which we had named "ethnobiological information system" in an earlier stage; a sketch of the base diagram can be found in (Pennec *et al.* 2014 , p. 182)).

19.5. Description of the collections

It is time to describe briefly what the *Muséum*'s biocultural collections are made up of. Here we take into account the new division in progress, presented above, which distinguishes specimens from objects, whatever their institutional origin.

Figure 19.9. *A biocultural set, the betel quid: the tongs for breaking areca nuts (Areca catechu) which will be wrapped in Piper betel leaves with lime. Here we have gathered various transformations of areca nuts, whole, sliced, and broken into fragments sold in bags. Source: Nathalie Idalie*

19.5.1. *Ethnobiological specimens*

From the beginning of the 20th century, these collections have concerned the relations of human societies with the plant and animal world. A rich documentation is associated with them (archives, field notes, iconography and specialized publications).

As far as *ethnobotany* is concerned, several important sets make up the bulk of the collection, nearly 100,000 in number:

– sets of plants, wild and cultivated, named and used in human societies, collected in the field and accompanying the ethnobotanical study of the same society: Andes

(Friedberg, Girault, etc. collections), Southeast Asia (Haudricourt, Barrau, Friedberg, Macdonald, Revel, etc. collections), and Central Africa (Thomas, de Garine, Bahuchet, Breyne, Peeters, Pujol, etc. collections). This represents about 25,000 herbarium parts (Figure 19.10);

– voluminous historical herbaria constituted by the founders of the laboratory, Auguste Chevalier, Roland Portères, Jean-François Leroy and Hubert Gillet. Some concern wild species of domesticated plants (coffee trees in particular), others herbaceous plants consumed by tropical livestock (agrostology, work of H. Gillet);

– significant collections of cultivated plants, in the form of herbaria or seeds in bags or vials. The most important collection concerns African cereals: sorghum *Sorghum*, fountaingrasses *Pennisetum*, fonio *Digitaria* and African rice *Oryza glaberrima* and *barthii* (work of Portères). In addition, there are varieties of Asian rice *O. sativa* and maize, and samples of coffee cultivars (work of Chevalier and Leroy). This set represents about 28,000 herbarium parts (Figure 19.11);

– samples of plant material: seeds and fruits, fibers and flours (in dry or wet jars, boxes and bags); wood chips and billets from the five continents. The core of this collection comes from various colonial exhibitions of the 19th and early 20th centuries. The set represents about 10,000 bags and jars and 20,000 wood samples (which should join the *Muséum*'s xylotheque) (Figure 19.12).

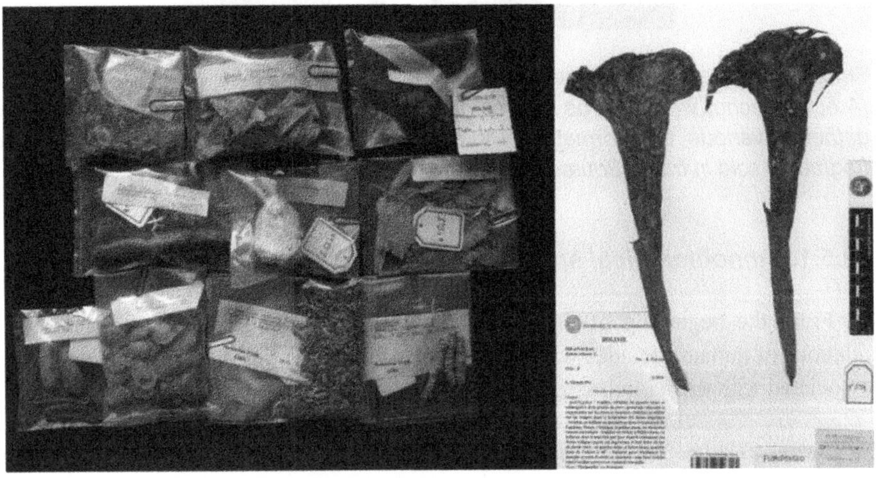

Figure 19.10. *Louis Girault Collection of medicinal plants from Bolivia. Source: S. Juraver*

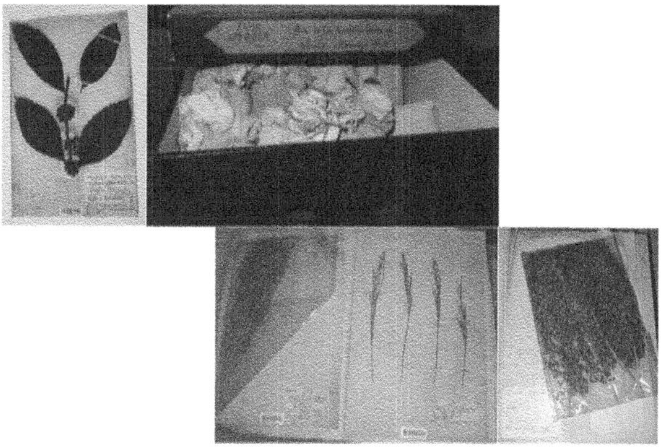

Figure 19.11. *Cultivated plants: herbaria of wild coffee (1905) and sorghum (1933) collected by Chevalier,* Oryza glaberrima *collected by Portères (1942), bags of O. sativa sent by Morange to Chevalier (1914). Source: J. Larco-Laurent and S. Bahuchet*

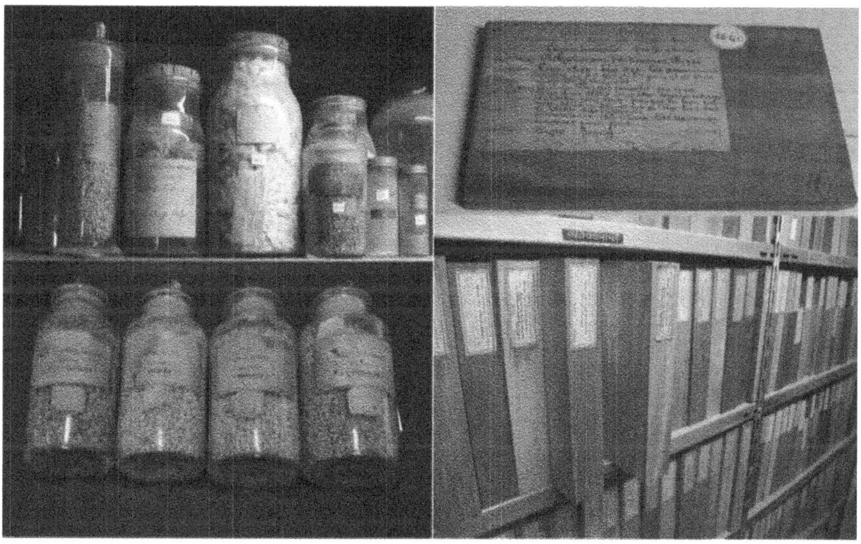

Figure 19.12. *Grain, fiber, wood samples. Source: J. Larco-Laurent and S. Juraver*

These specimens come from about 40 countries in Europe, Africa, East Asia, and Latin America, with Peruvian, Indonesian, and Central and West African specimens dominating (Figure 19.13).

Figure 19.13. *Geographic origin of ethnobotanical collections. Map drawn by F. Pennec*

The small ethnozoology collection was initiated in the late 1960s by Raymond Pujol, when he opened this field in the laboratory at the request of R. Portères. It includes about 1,000 specimens entrusted by researchers in the humanities to obtain the determination of animals named and used by humans.

These are whole animals, dry or in liquid preservative (insects, mollusks, reptiles, fish), and fragments (skulls, horns, dry skins, shells). Coming mainly from Central Africa, Amazonia and France, there are about 240 birds, 200 mammals and 300 mollusks.

19.5.2. *Objects and artifacts*

After 2002 and the transfer of the ethnology laboratory's collections to the new *Musée du quai Branly*, we defined five themes for a new collection program:

- objects testifying to the relationship with nature;
- objects that mark cultural identity;
- objects that witness technical changes;
- objects representing nature;
- objects marking the treatment of the human body.

Centered, without exclusivity, on the relationships of societies with their environment, these themes also take into account the human species itself, as inserted, at the same time, in society and in nature, and the technical transformations at work, as visible through material culture (whether by circulation of objects or by technical adaptation).

Thus defined, these themes obviously met those that had presided over the constitution of the "ethnobiological" collections of the ethnobotany laboratory (these "botano-ethnographic objects" that Portères had described), the systematic collection of which had been initiated by Raymond Pujol, professor of ethnozoology, among his correspondents and students. These objects all bear witness to techniques and uses concerning plants and animals.

There are now nearly 10,000 items that have been collected, some as early as the 1970s, and the majority during the past 15 years. They come from 115 countries, on all continents (Figure 19.14).

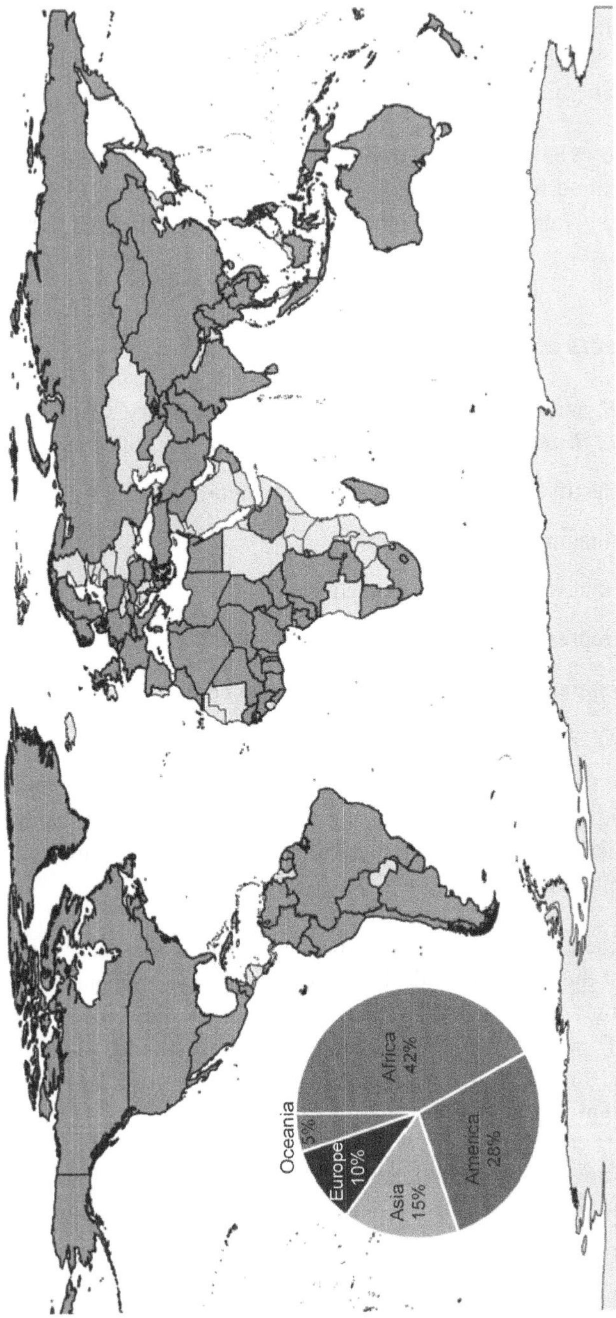

Figure 19.14. *Geographical origin of the objects. Map drawn by F. Pennec*

COMMENT ON FIGURE 19.14. – *Major regions and countries within the continents from which the majority of objects originate: Central Africa (Cameroon, Gabon, CAR, DRC), West Africa (Burkina Faso, Liberia, Senegal), North Africa (Algeria), Central Asia (Kyrgyzstan, Tajikistan, Uzbekistan), East Asia (China, Japan, Laos), Latin America (Brazil, Colombia, Guyana, Mexico, Peru, Venezuela), Oceania (Indonesia, Papua New Guinea, Philippines, Polynesia), Eastern Europe (Russia), and Western Europe (France).*

The objects are mostly made of vegetable and animal materials, and the main themes[8] are:

– tools, weapons and utensils for hunting and fishing;

– farming tools, utensils related to agriculture and breeding (nearly 200);

– utensils related to the preparation and consumption of food;

– basketry (more than 500);

– traditional hives (60, Figure 19.15);

– musical instruments (more than 120);

– accessories related to body care and protection (cosmetics, ornaments and ornaments, clothing, amulets, etc.).

Figure 19.15. *Peasant hive from Diois, made of oak wood, for rearing* Apis mellifera *bees (Apideae), France, collection D. Chevalier ETB-FR-DC-2011-002. Source: J.-C. Domenech – MNHN*

8 We have begun publishing thematic catalogs on fisheries (Bahuchet and Rameau 2016; Bahuchet *et al.* 2018) and on hens and roosters (Rameau *et al.* 2017).

This dynamic collection is regularly enriched by the missions of researchers. In fact, for the past 12 years, all field missions by researchers and students (doctoral students and master's degree internships) have resulted in the collection of objects documenting the relationship with the natural environment. We also benefit from donations and bequests by private individuals, collectors and amateurs, often through the intermediary of the *Société des amis du musée de l'Homme*, as well as from researchers outside the *Muséum*. Finally, from 2013 to 2015, specific collections were organized from our fields in anticipation of the showcases in the *Galerie de l'Homme* (Bahuchet *et al.* 2016).

19.6. What changes?

In this context, we have already started some developments concerning the collections. For example, based on elements already present, we are continuing in the same direction: for ethnobotany, specimens of cultivars, peppers (*Capsicum* spp.), corn, beans (*Phaseolus* spp., *Vigna* spp.), in all regions where we circulate. To realize the showcase on food plants in the Botanical Gallery at the *Jardin des Plantes* in 2013, we collected dry plant-based foods available in the stores of migrant populations in Paris, which proved to be particularly promising since products from more than 40 species, from 14 different families, were inventoried. Joining the collections made during our missions of the last few years, where we find a significant range of domestic utensils for the inhabitants of these places, this fact convinced us of the interest of markets and small local stores as a target for surveys of daily life (which has already been proposed by the Mexican ethnobotanists Bye and Linares (1983)). In the context of large multicultural metropolises such as Paris, repeated surveys in markets are likely to provide interesting information on the evolution of the community and the mixing of populations.

Another theme that naturally imposed itself on us in the continuation of ethnozoology is that of food insects and the utensils related to the world of insects and first hives, continuing the enrichment of the collection formerly gathered by Raymond Pujol. In the same way, we have continued his collections and studies on featherwork (Lecuyer and Pujol 1975, pp. 205–216), with objects made of feathers[9], objects composed of animal materials, but, more broadly, with new materials for new uses, for example, pouches for urban use, made up of tanned *Tilapia* fish skins (from Brazil), or even varnished teak tree leaves (from Thailand)!

We were also interested in object transformations or innovations (Figure 19.16).

9 A few examples can be found in our chicken catalog (Rameau *et al.* 2017).

Figure 19.16. *Basket for shellfish harvesting on foot. It is made from recycled oyster bags and was invented by Mohamed Teffahi (St. Nazaire, Loire-Atlantique). Collection S. Le Bomin, ETB-FR-SLB-2016-001. Source: S. Bahuchet*

However, today, local traditions are confronted with globalization. The context of the constitution of "archives of humanity", advocated in 1930 when the tropical world was tipping over, is no longer the case. Nowadays, if there are still artifacts made directly by their users, or obtained from local craftsmen who master the entire chain of manufacture, the majority of objects are acquired on the market and are industrial products, from small mechanized industries or, more often, from very large series, of standardized mass production. In the words of Lewis Mumford, "we live in the machine age and are proud of it" (Mumford 2015)! Our crucial question is then: how can we document the contemporary era, to bear witness to it for future generations, in line with the MNHN's approach – that is, humans' relations with the environment and biodiversity? In other words, what collections are there for the 21st century? We are not a museum of technology, focusing on engines, machines and industry. We will also recall the analysis of Muraoka and Okamura, on Japanese traditional arts: "The basic attribute of folk crafts is utility; the objects have practical value. One can sense the beauty of folk crafts by seeing utensils in actual use" (Muraoka and Okamura 1981, pp. 103 and 121).

Without eliminating the collection of what remains of locally-made artisanal artifacts, we can see that the description of increased manipulation of industrialized objects for environmental relations will have to take on greater importance in the building of biocultural collections.

The overall objective is to document the current uses of biodiversity, knowledge and know-how. The second objective is to analyze the series of objects thus realized by a comparative and typological method to highlight similarities, convergences, transmissions and similarities. An innovative approach consists of applying calculation methods specific to phylogenetic research, which is still not very widely applied to material culture[10]; this approach seems to us to be necessary to study cultural diversity and the processes of cultural evolution. Finally, we believe that making the collections available to communities other than those involved in research in the humanities, such as those involved in arts and crafts and design, is a very promising way to enhance the value of this heritage.

19.7. References

Bahuchet, S. (2014). Curating ethnographic information for biocultural collections. In *Curating Biocultural Collection. A Handbook*, Salick, J., Konchar, K., Nesbitt, M. (eds). Royal Botanical Gardens, Kew.

Bahuchet, S. (2015). Hommes, natures et sociétés : la construction de l'interdisciplinarité. In *Le Musée de l'Homme ; histoire d'un musée laboratoire*, Blanckaert, C. (ed.). Artlys-MNHN, Paris.

Bahuchet, S. and Lizet, B. (2003). L'ethnobotanique au Muséum national d'Histoire naturelle. Les hommes, les idées, les structures. In *Plantes, sociétés, savoirs, symboles. Matériaux pour une ethnobotanique européenne*, Lieutaghi, P., Musset, D. (eds). Les cahiers de Salagon, Les Alpes de lumière, Forcalquier.

Bahuchet, S. and Rameau, P. (2016). Quelques engins de pêche en eau douce d'Afrique centrale. *Revue d'ethnoécologie*, 10.

Bahuchet, S., Kozlowski, V., Merlin, M., Parent, A. (2016). Exposer l'ethnoécologie au Musée de l'Homme. *Revue d'ethnoécologie*, 9.

Bahuchet, S., Rameau, P., Reyes, N. (2018). Engins de pêche dans les collections du Musée de l'Homme. *Revue d'ethnoécologie*, 13.

Balfet, H. (ed.) (1991). *Observer l'action technique. Des chaînes opératoires, pour quoi faire ?* Éditions du CNRS, Paris.

10 The problem of the phylogenetic approach and cultural evolution has been addressed by Mace and Holden (2005), Collard *et al.* (2006), Tëmkin and Eldredge (2007), and applied notably to basketry (Jordan and Shennan 2003) weaving (Buckley 2012) and music (Le Bomin *et al.* 2016).

Brousse, C. (2017). De l'objet à caractère végétal à l'objet ethnobotanique. Penser les collectes de plantes et objets en plante. *In Situ*, 32.

Buckley, C.D. (2012). Investigating cultural evolution using phylogenetic analysis: The origins and descent of the southeast Asian tradition of warp ikat weaving. *PLoS ONE*, 7(12), e52064.

Bye, R.A. (1986). Voucher specimens in ethnobiological studies and publications. *Journal of Ethnobiology*, 6(1), 1–8.

Bye, R.A. and Linares, E. (1983). The role of plants found in the Mexican markets and their importance in ethnobotanical studies. *Journal of Ethnobiology*, 3(1), 1–13.

Collard, M., Shennan, S.J., Tehrani, J.J. (2006). Branching, blending, and the evolution of cultural similarities and differences among human populations. *Evolution and Human Behavior*, 27(3), 169–184.

Cresswell, R. (2010). Techniques et culture, les bases d'un programme de travail. *Techniques & Culture*, 54–55, 23–45.

Gille, B. (1978). *Histoire des techniques*. Gallimard, Paris.

Hamy, E.-T. (1988). *Les origines du Musée d'ethnographie*. J.-M. Place, Paris.

Haudricourt, A.-G. (1940). Les moteurs animés en agriculture. Esquisse d'une histoire de leur emploi à travers les âges. *Revue de botanique appliquée et d'agriculture coloniale*, 230–231, 759–772.

Jolly, É. (2016). Les collectes d'objets ethnographiques. À la naissance de l'ethnologie française. Les missions ethnographiques en Afrique subsaharienne (1928–1939) [Online]. Available at: http://naissanceethnologie.fr/.

Jordan, P. and Shennan, S. (2003). Cultural transmission, language, and basketry traditions amongst the California Indians. *Journal of Anthropological Archaeology*, 22(1), 42–74.

Le Bomin, S., Lecointre, G., Heyer, E. (2016). The evolution of musical diversity: The key role of vertical transmission. *PLoS ONE*, 11(3), e0151570.

Lecuyer, J. and Pujol, R. (1975). L'oie plumassière du Poitou, utilisation des peaux et des plumes. In *L'homme et l'animal, premier colloque d'ethnozoologie*. Institut international d'ethnosciences, Paris.

Leiris, M. (1931). *Instructions sommaires pour les collecteurs d'objets ethnographiques*. Musée d'ethnographie, Muséum National d'Histoire Naturelle et mission scientifique Dakar-Djibouti, Paris.

Lemonnier, P. (1976). La description des chaînes opératoires : contribution à l'analyse des systèmes techniques. *Techniques et culture (Bulletin)*, 1, 100–151.

Lemonnier, P. (1992). *Elements for an Anthropology of Technology*. Museum of Anthropology University of Michigan, Ann Arbor, MI.

Leroi-Gourhan, A. (1965). *Le geste et la parole II : la mémoire et les rythmes*. Albin Michel, Paris.

Leroi-Gourhan, A. (1971). *L'homme et la matière (Évolution et techniques 1)*. Albin Michel, Paris.

Leroi-Gourhan, A. (1973). *Milieu et techniques (Évolution et techniques 2)*. Albin Michel, Paris.

Mace, R. and Holden, C.J. (2005). A phylogenetic approach to cultural evolution. *Trends in Ecology & Evolution*, 20(3), 116–121.

Mauss, M. (1967). *Manuel d'ethnographie*. Payot, Paris.

Mumford, L. (2015). *Art et technique. Six conférences inédites*. La lenteur et la roue, Paris.

Muraoka, K. and Okamura, K. (1981). *Folk Arts and Crafts of Japan*. Weatherhill, New York, and Heibonsha, Tokyo.

Pennec, F., Juraver, S., Quiñe, M., Bahuchet, S. (2014). Draft structure for the ethnobiological database in the Museum national d'Histoire naturelle, Paris. In *Curating Biocultural Collections: A Handbook*, Salick, J., Konchar, K., Nesbitt, M. (eds). Royal Botanic Gardens, Kew.

Portères, R. (1961). L'ethnobotanique : place – objet – méthode – philosophie. *Journal d'agriculture tropicale et de botanique appliquée*, VIII(4–5), 102–109.

Portères, R. (1965). Le laboratoire d'ethnobotanique du Muséum National d'Histoire Naturelle. *Journal d'agriculture tropicale et de botanique appliquée*, 12(1–3), 1–4.

Portères, R. (1969). Cours d'ethno-botanique générale. Course, laboratoire d'ethnobotanique et d'ethnozoologie, MNHN, Paris

Rameau, P., Brabant, D., Bahuchet, S. (2017). La poule et le coq dans les collections du Musée de l'Homme. *Revue d'ethnoécologie*, 12.

Salick, J., Konchar, K., Nesbitt, M. (eds) (2014). *Curating Biocultural Collections: A Handbook*. Royal Botanical Gardens, Kew.

Sigaut, F. (2010). Des idées pour observer. *Techniques & Culture*, 54–55(1), 87–97.

Sigaut, F. (2012). *Comment* Homo *devint faber*. CNRS, Paris.

Tëmkin, I. and Eldredge, N. (2007). Phylogenetics and material cultural evolution. *Current Anthropology*, 48(1), 146–154.

20

Why Preserve?

Véronique ROUCHON
Centre de Recherche sur la Conservation, Muséum national d'Histoire naturelle, CNRS, Ministère de la Culture, Paris, France

The natural sciences are based on the observation of nature in all its forms and are, primarily, descriptive. The naturalists of the 19th century, as well as current researchers, collected samples during their field trips. These samples eventually became collections. Their conservation depends not only on the environmental conditions of the storage places, but also on the methods used, which have largely evolved with technological progress and are often poorly documented. They were designed to preserve, at a certain time, a certain form of information but may leave lasting traces on the specimens.

However, science progresses. Collections have been passed on from teacher to student or have been expanded by acquisition, until they constitute collections of several hundreds, thousands or even millions of specimens. These collections are considered today as a heritage to be preserved. But in what way? To answer this question, we need to consider the values of this heritage, which are multiple and largely dissociated from commercial values. We can qualify them in different ways: esthetic value, historical value, affective value, etc. The most important remains, without any doubt, the scientific value: the specimens have above all the value of being witnesses to a history that we are trying to record, the history of the Earth, of life, of humankind. This scientific value is also declined according to several facets, depending on whether this history has been written or remains to be written.

For a color version of all figures in this chapter, see www.iste.co.uk/pellens/natural.zip.

Natural History Collections in the Science of the 21st Century, coordinated by Roseli PELLENS.

20.1. The museum's collections: between study and heritage

The museum, in the current sense of the term, has its origins in the Enlightenment philosophy of the 18th century, which advocated universal knowledge and a shared understanding. In France, the Revolution of 1789 and the numerous confiscations that followed led to the creation of collections, which were very quickly assimilated into national property and had to be administered in the common interest. As early as 1793, two great museums were created, each very different from the other: the Louvre, whose purpose was to give the nation access to the royal collections of fine arts, and the *Muséum national d'Histoire naturelle* (referred to hereafter as the *Muséum*), which, under the impetus of Daubenton and Lakanal, transformed the *Jardin du Roi* into a place of public education and scientific research.

Twelve professorships were created at the *Muséum*, covering various disciplines such as natural history drawing, anatomy (human and animal), chemistry, botany, agronomy, zoology, mineralogy and geology. Today, the research conducted at this institution brings together the work of about 500 researchers and covers a much broader scope, with biochemistry, paleontology, human paleontology, archeology and ethnology. In the 19th century, the development of experimental research in chemistry and physics contributed to technological progress in all spheres of society. It benefited naturalists as it provided new means of investigation, such as microscopy. However, experimental and natural sciences follow very different approaches: experimental sciences aim to describe universal behaviors, which are supposed to be reproducible by everyone, as long as the experiment is described in detail. The natural sciences, on the other hand, aim to describe nature in all its forms, in order to give a better understanding of its mechanisms and evolution, over all time spans, from the creation of the solar system to today. Observation of samples is essential: this may be performed either in the field or at the laboratory. These samples constitute the basis of this research. For the most part, they remain in the *Muséum's* premises, contributing to the constitution of collections throughout the 19th and 20th centuries. These collections are regularly enriched by donations or purchases from enlightened amateurs or confirmed researchers. For example, the famous naturalist Jean Henri Fabre (1823–1915) worked extensively in the garden of his property in Sérignan-du-Comtat (Vaucluse), and in the surrounding countryside, collecting specimens, observing them, drawing them and inventorying them, thus constituting subsequent material for research purposes (Figure 20.1). After his death, this collection was bought by the State, along with his house, called the Harmas de Fabre, which is now open to the public.

These considerations show that natural history collections are the foundations of natural sciences and therefore have a major societal role, as they help us to understand where we come from and where we are going. They must therefore be considered as heritage assets of public interest. This is one of the reasons why the status of "*Musée de France*"[1] applies to the *Muséum* with, in particular, the principle of inalienability of its collections.

a) b)

Figure 20.1. *a) Jean Henri Fabre observing specimens, photo: Ville d'Avignon, Palais du Roure, Fondation Flandreysy-Espérandieu; b) Jean Henri Fabre's office visible today. MNHN/A. Iatzoura*

20.2. Disrupting the equilibrium

The natural sciences aim to explore all time spans. As soon as we are interested in materials dating from several thousands, millions or even billions of years ago, it is tempting, given their age, to imagine that their characteristics are everlasting. Nothing could be further from the truth! These testimonies to the past remain likely to evolve with time because their removal from the ground is often accompanied by a radical change of their environment. The ground in which they were buried is often depleted of oxygen. Depending on the level of the water table, it can be wet or dry. It can also be a site of bacterial activity. It is generally considered that the samples have reached us because they have reached a state of quasi-equilibrium with the soil that surrounds them. But excavation breaks this equilibrium in several ways, by exposing the collected pieces to untimely handling, but also to oxygen in the air, to new humidity and temperature conditions as well as to new biological contaminants.

1 Law n° 2002-5 of January 4, 2002 relating to the museums of France.

A certain number of more or less visible transformations follow, depending on the materials considered.

These transformations can be spectacular. They can be related to the deformation of the material during its drying. Organic fossil materials, bones and most organic remains (leather, skin, etc.) are subject to this type of phenomenon, which causes cracking and loss of cohesion. The alterations are not only physical, but can also involve chemical reactions or biological activity. This is particularly true of sulfide-rich specimens, which are chemically very reactive in the presence of water and oxygen (Rosso and Vaughan 2006). Their conservation is often a challenge. This is the case for so-called "pyritized" fossils, which take their name from pyrite, an iron sulfide (formula FeS_2), often found at fossil sites. When they are buried, in a reducing environment, these fossils can be preserved for millions of years. But conservation difficulties begin at the time of their excavation when these waterlogged materials are exposed to the oxygen in the air. Sulfides undergo oxidation reactions leading to sulfate efflorescence that damage the material esthetically and mechanically (Figure 20.2) (Rouchon *et al.* 2012).

These alterations are even less predictable because pyrite is often present at a micro-nanoscopic scale, in the form of framboids, a name inspired from the raspberry aspect of the structure (the word for raspberry in French is *framboise*), made up of more or less organized micro-nanograins (Figure 20.3). It is then impossible to detect their presence without the use of analytical tools.

Beyond this spectacular damage, many materials entering the *Muséum's* collections evolve under ambient conditions, due to temperature variations, presence of humidity and oxygen, or presence of atmospheric pollutants. These changes can remain, up to a certain point, hardly perceptible to the eye, but they nevertheless mark the intrinsic composition of the material and must be taken into account because they can bias the results of physicochemical analyses.

One will then wonder about the necessity of breaking the equilibrium between the object and its environment of origin. For the physical chemist, it is quite obvious that the best way to preserve the various archeological and paleontological remains would be to leave them undisturbed. To isolate them from their original environment with which they were in equilibrium in order to expose them to a completely different environment is an error. It is equally obvious that archeologists and paleontologists are of the opposite point of view, because they need access to evidence. In the present state of knowledge, buried specimens are, so to speak, non-existent because they are ignored. Their conservation is therefore not only a question of physico-chemical stability, but it is also intimately linked to the meaning we give them, and which is the result of a meticulous work of interpretation, motivated by the will to write or re-write the history of the Earth, of life, of humankind. It is clear, through this trivial example, that the conservation of collections generates antagonisms and requires compromises.

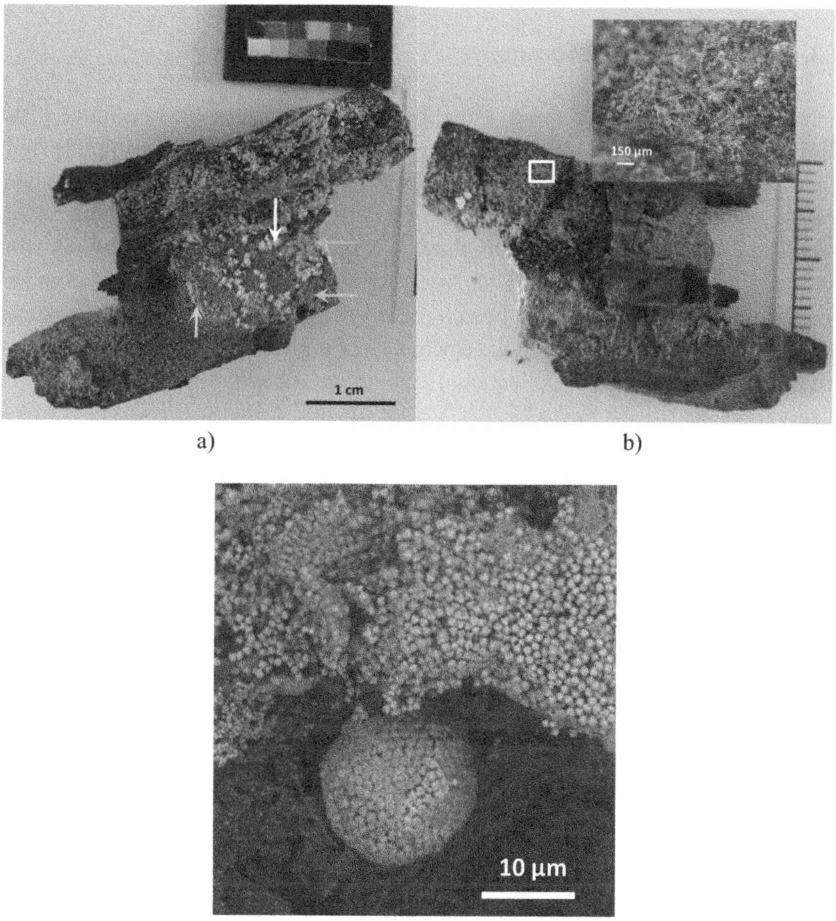

c)

Figure 20.2. *Fossil wood fragment, Eocene, La Plaine Chevrière. © MNHN/V, Rouchon*

COMMENT ON FIGURE 20.2. – *a) and b) This fragment is rich in pyrite, and its alteration has led to the emergence of several sulfate phases: coquimbite [$Fe^{III}_2(SO_4)_3{\cdot}9H_2O$, pink arrow]; metavoltine [$K_2Na_6Fe^{II}Fe^{III}_6\,O_2\,(SO_4)_{12}{\cdot}18H_2O$, orange arrow]; ferrinatrite [$Na_3Fe^{III}(SO_4)_3{\cdot}3H_2O$, white arrow]; magnesiocopiapite*

[$MgFe^{III}{}_4(SO_4)_6(OH)_2{\cdot}20H_2O$, green arrow]; gypsum [$CaSO_4{\cdot}2H_2O$, white needles in the zoom on the back]. c) Example of framboidal pyrite grains (scanning electron microscopy, backscattered electrons mode).

20.3. Preparation and storage

Collecting living matter raises a number of concerns specific to natural history collections, particularly related to the conservation of tissues. How can putrefaction be avoided? The term conservation takes on a very specific connotation here, which is not so much linked to the preservation of a heritage for future generations as to the preservation of biological tissues for study. Without intervention, these tissues would degrade in a few hours or days. Thus, preserving the appearance of dead organisms in order to better study them has been a recurrent preoccupation of curiosity seekers, and, later, of naturalists. It allowed the emergence of a great variety of skills associated with the preparation of collections, including taxidermy, embalming, fluid conservation, etc.

Figure 20.3. *Illustration of the variety of sealants present in fluid collections. © MNHN/V, Rouchon*

These skills are the results of more or less successful experiments, sometimes shared in an academic context. Thus, in 1662, William Croone (1633–1684) presented two dog embryos preserved in alcohol for several days to the Royal Society in London. These tests were widely taken up by his contemporaries, such as Frederik Ruysch, to supply curiosity shops. The following generations of naturalists widely used different types of alcohol, most of them from everyday life (rum, brandy, etc.) (Simmons 2014), as well as salt solutions with a wide variety of ingredients. Beginning in the 1890s, the commercialization of formaldehyde, also known as "formalin", had a large impact on practices because this product had the ability to "fix" tissues, thus ensuring better preservation of the morphology of specimens. The quality of the sealing of the jars was also an essential parameter of preparation. Here again, many products were used: resins, waxes, bitumens, etc. These preparation practices, which were multiple and not well documented except through a few professional manuals (Boitard 1859), have led to the existence, in the *Muséum's* collections, of a great variety of possible preparations, the visual aspects of which are very diverse (Figure 20.3). These preparations have influenced the composition and morphology of tissues and organs and, in a sense, are attached to them. It is important to learn more about them to optimize the conservation of specimens (from the heritage viewpoint).

Moreover, the *Muséum's* collections have always been designed to be exhibited, whether to a large audience (since its creation, the *Muséum* has had a teaching and dissemination vocation) or to a more restricted sphere of researchers and enlightened amateurs. The methods of preparation have thus inherited from the cabinets of curiosity a certain theatrical tendency. Thus, naturalized specimens are often shown in postures that are judged to be the most suitable to please an audience and one will not hesitate to enhance the color of certain parts that have become a little dull. The presentations lend themselves to the estheticism of the moment. For example, a mounting of paleontological specimens, frequent in Gaudry's time, involves mounting the fossil plates on plaster structures, themselves surrounded by a wooden frame. Here also, one will not hesitate to paint the specimens so that they have better appearance. In the case of pyritic-bearing fossils, such as those from the fossiliferous shale of Igornay, near Autun (Permian), these preparations age very badly: the alteration of the pyrite generates crystalline growths of white sulfates under the layer of paint, causing a weakening of the structure and a substantial deformation of the specimen visible when one compares it to its plaster cast (Figure 20.4). This example shows that preparation can cause a great loss of authenticity of the specimen: it is very difficult today to distinguish, between repaints and plaster, where the original material is. Moreover, the morphology of the bones is no longer reliable: in this respect, the plaster cast will be of more scientific interest to a paleontologist today than the original. On the other hand, the original may be of new scientific interest to specialists who are interested in the history of preparation methods and alterations in museum collections.

Another intervention, common in preparation but which compromises the authenticity of the original material, is varnishing or impregnation with resins. It is not always clear whether these varnishes and resins, whose composition is often unknown, are intended to embellish the object or to protect it. Probably both, because their use is very common on many specimens: snakes, crocodiles, turtles, bones, fossil remains, etc. (Figure 20.5).

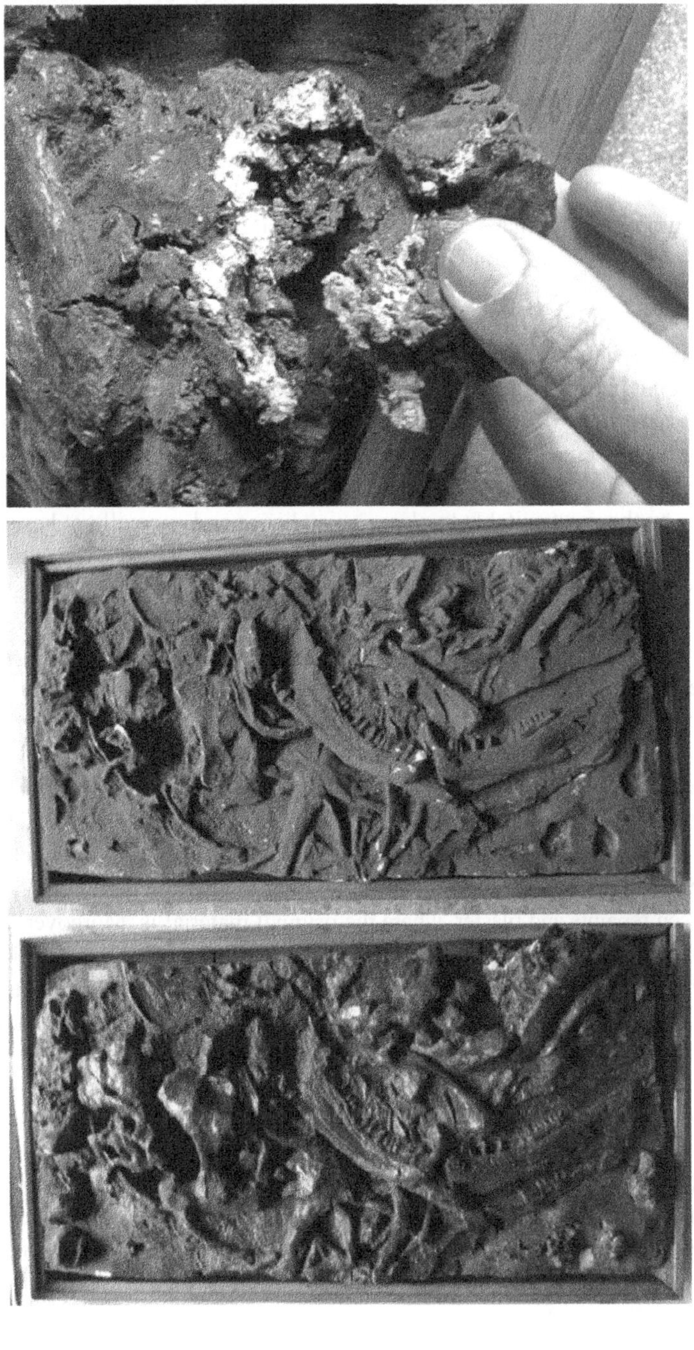

a) b) c)

Figure 20.4. *Example of a paleontological preparation that has evolved unfavorably: remains of a* Stereoachis dominans *skeleton (A. Gaudry, 1880) © MNHN/V, Rouchon*

COMMENT ON FIGURE 20.4. – *The actual specimen (a), photographed with the same light as the plaster cast (b), shows completely blistered areas due to visible crystal growth on the most fragile external parts (c).*

In general, since the 1970s, a wide variety of synthetic polymers have been tested with varying degrees of success (Wayne Smith 2003). One example is plastination techniques that employ silicone resins for the preservation of biological tissue. Their use is far from systematic in the museum environment. It focuses mainly on a few examples of specimens that are difficult to maintain by other techniques, such as large underwater specimens like the giant squid.

Figure 20.5. *Examples of varnished specimens:* Caiman crocodilus *(Linnaeus, 1758, Cr11-8, 7542) and partially varnished bone on the right side that has yellowed (*Cylindrapsis triserrata*, Mare aux songes, Mauritius, Günther, 1873, 1905.0476-0481, To.0415-1). © MNHN/V, Rouchon*

In the field and in the workshops, many of the resins currently used belong to the polyacrylate family, which have a strong adhesive power and are relatively stable over time. The most widely used in France, Paraloid B72, is an ethyl methacrylate copolymer manufactured by Rhom and Haas. It is usually diluted in organic solvents and applied to objects after they have been dried. But there is no question here of detailing the use of synthetic polymers on specimens, especially since the names of the products, and even their composition, often change according to the brands and patents of the major manufacturers, which does not facilitate their traceability. It is however crucial to underline their importance, both by the number of objects concerned and by their impact on future research. Indeed, the application of these products corresponds to the exogenous intrusion of a chemical compound likely in the long term to interact with the object, to degrade or even to modify the information contained in the material, thus compromising future analyses.

20.4. The main principles of conservation

An awareness of the need to better define conservation strategies in museums began to emerge at the end of the 19th century. This was illustrated by the first professional colloquia of curators, such as the colloquium in St. Gallen (1898, Switzerland), organized for the heads of libraries, during which several recurring concerns about the conservation of archival records were addressed. However, it was not until the second half of the 20th century that a systematic reflection was carried out, in particular through the publications of Cesare Brandi (1906–1988) (Brandi 1963) and the establishment, around the 1980s, of heritage conservation courses. This discipline, modest and pragmatic, was inspired by an ambitious and absolute view that obeys several major principles listed below (Berducou 1990):

– The first principle corresponds to the need to establish a condition report and a diagnosis allowing interventions carried out on the objects or the specimens to be justified.

– The second principle is correlated to the first one and concerns the scrupulous recording of information relating to the interventions, to ensure a traceability which will make it possible to better understand the future of the objects or the specimens, akin to personal health records.

– The third principle refers to the reversibility of interventions. This notion aims to guarantee the possibility of intervening in the future by undoing what has been done previously, either to replace certain products with others deemed more efficient, or to rectify a perceived error in the initial treatment.

– The fourth principle refers to the compatibility of the materials used (especially for repair or mounting) with those to be preserved. This principle aims to avoid further alteration of the objects by exogenous material.

– The fifth principle is that of the legibility of the intervention, which must remain clearly identifiable so as not to “deceive” future generations. In particular, a repaired fracture must remain visible on an object, because this fracture is an element of the object’s life that must be respected. To conceal it by filling and retouching would be to erase this element of the object’s history and thus to falsify it.

– Finally, the sixth principle is that of minimal intervention, which aims to modify the object as little as possible in its composition and structure.

These broad principles were generally defined in response to past mistakes. For example, the compatibility of materials used reflects concerns with some materials that deteriorate (wood, paper, polymer films, etc.) by emitting volatile organic compounds. When these materials are used to house sensitive collections, they cause damage. For example, oak furniture, in spite of its beauty and strength, is not recommended for collections rich in calcareous materials, such as malacological collections. Indeed, oak is a wood that has a strong tendency to emit organic acidic vapors. In the presence of humidity, these vapors are adsorbed in the porosity of the specimens, causing a dissociation of carbonates and a precipitation of organic calcium salts (Figure 20.6). This phenomenon, also known as Byne decay, is named after Loftus Byne (1872–1947) who was one of the first to report this type of damage in 1899 (Tennent and Baird 1985).

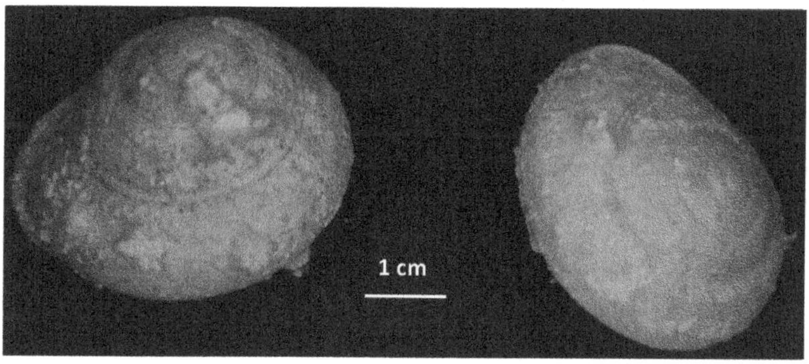

Figure 20.6. *Examples of malacological specimens affected by Byne decay © MNHN/V, Rouchon*

20.5. The main principles of conservation being undermined

Some of the main principles, widely promoted today in the field of heritage conservation, do not withstand the practical realities (Richmond and Bracker 2009). Thus, the recording of information on interventions is often fragmented due to a lack of time or resources. The first element of information, which corresponds to the position of the object or specimen at the time of collection, may only be partially recorded. In addition, it is sometimes necessary to "prepare" certain specimens at the time of collection. For example, animals and fish that are to be preserved in fluid must be fixed as soon as possible after death in solutions designed to preserve tissue. These operations, which are generally done in the field, with the available resources, remain difficult to document in detail.

Diagnosis is another principle that is not always easy to apply under time constraints. Its relevance depends largely on the knowledge one has of the object, and this knowledge is not always acquired instantaneously. The most stable objects or those with classic types of alterations can be dealt with according to well-defined operating chains, but for complex or very unstable objects whose behavior is not predictable, there is often an urgency and little time to think, which seriously compromises the quality of the diagnosis. Moreover, when interventions take place in the field, their recording is often imprecise. It is generally not possible to produce a complete photographic file with precise information on the interventions. We are therefore often contented with a summary of the type of product used.

The reversibility of interventions is also a principle that quickly reaches its limits. Over the last 30 years, the term reversibility has been widely used to designate interventions that are "respectful" of the object. This is often wrong, because the interventions carried out, however respectful they may be, are generally irreversible. However, the notion of reversibility becomes meaningful when we look at polymers used to repair fractured objects or to consolidate materials that have lost their mechanical properties. In suppliers' catalogs, polymers (resins, glues, adhesives, etc.) described as "reversible" are those that can dissolve in a solvent, even after alteration. This property is interesting when one considers reprocessing objects that have been restored in the past. However, it is not sufficient to guarantee the reversibility of the intervention: resins often impregnate porous materials to the core, and it is very difficult to extract them afterward. Moreover, it is sometimes the extraction solvent that can cause concern, because of its capacity to dissolve some of the constituents of the treated material. For these reasons, the term "reversible" has evolved to the concept of "retreatable" which seems more adapted to qualify the interventions that have been carried out.

These great principles of conservation, although undermined by economic and practical realities, have the recognised merit of existing and defining a framework for reflection. For it is indeed a question here of thinking before acting, so as to make the most pertinent choices. A complementary practical approach, sometimes presented as a principle, corresponds to preventive conservation. Widely developed over the last 30 years, it does not aim so much to intervene on the object so that it resists its new environment as to conceive an environment that will engender the least possible alteration of the object. In this approach, it is no longer the object that is adapted to its environment, but the opposite, which means focusing more on the causes than the effects of aging. This approach is often more efficient and less expensive on a collection scale. It focuses mainly on the control of environmental conditions in the widest sense: temperature, hygrometry, volatile organic compounds, oxygen content, packaging materials, etc.

20.6. Multiple values

Thinking about the value of the collections and how this evolves according to the choice of intervention is an effective approach for defining conservation priorities. But what value should be placed on the specimens? The market value is the one that might first come to mind. However, this notion is reductive because it is based on the law of supply and demand, which is often completely independent of the common interest. Moreover, it assumes that there is a market in which the specimen can be traded, which is not always the case.

In general, the notion of value is difficult to grasp because it is multiple and is both qualitative and quantitative. In the field of conservation, it is common to define several types of value. First of all, let us mention the esthetic value, which is probably the most likely to seduce the eye during an exhibition. It is probable that a collection of gems, showing precious stones in multiple colors, will hold the attention of an audience more easily than a collection of gray clay sedimentary rocks. This does not mean that the latter have no value, quite the contrary, but this value is of a different nature. In the field of archeology, at the time of the first explorations carried out in the 18th and 19th centuries, the excavations aimed primarily at exhuming objects of art, such as antique statues, to the detriment of the testimonies to everyday life. The value of objects was then often reduced to their estheticism. This reductive approach has long since been abandoned and many other types of value have been defined.

There is, for example, the historical value, which corresponds to the anchoring of an object in history (in the broadest sense). The example of the lead cast of Louis XIV's blue diamond is, in this respect, emblematic of an object, unnoticed for several centuries, and whose value became clear when it was placed in its historical[2] context. An object that belonged to a personality or that witnessed a significant event will also have a great value. For example, most of the fossil specimens assembled and described by Georges Cuvier (1769–1832) have a great historical value, because of Cuvier's fame for his works of comparative anatomy. Some have an even greater historical value because their discovery was a milestone in the discipline.

This is the case of the Montmartre opossum, *Peratherium cuvieri* (specimen GY679a-b), which includes the fossil remains of a small marsupial mammal, dated to about 35 million years ago (Eocene). This specimen, extracted from the gypsum quarries of Montmartre, was used as demonstration material when Cuvier, in 1804, presented his theory on the correlation of parts to a select audience (Figure 20.7). Cuvier had carefully examined the shape of the teeth, which he interpreted as the teeth of insectivores because he found them to be very similar to those of extant

2 For details, see Chapter 3 by Farges dealing with the blue gems of Louis XIV.

marsupials living in South America. In front of the assembly, he then announced that the rest of the skeleton, not visible because it was buried in the block, must present marsupial characteristics, such as the pelvic bones, which he verified by freeing these bones in front of his audience. This brilliant demonstration, published in the annals of the *Muséum d'Histoire naturelle* (Cuvier 1804), ensured the fame of Cuvier and his theories. It constitutes a significant event in the history of paleontology, and gives a particular value to the specimen currently preserved.

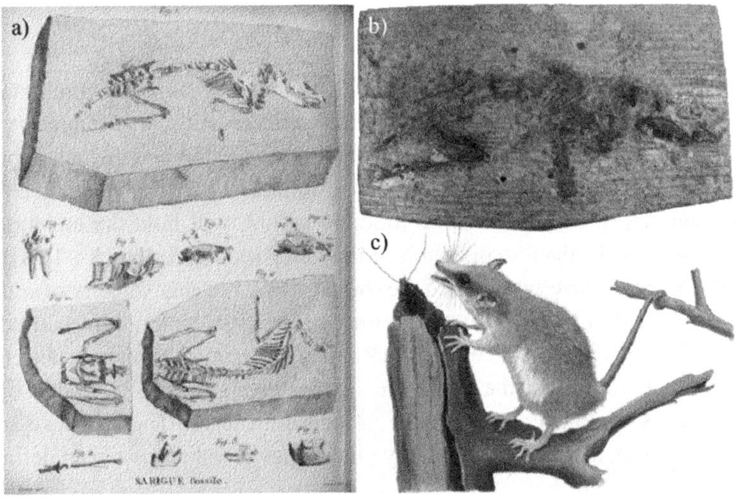

Figure 20.7. *Cuvier's opossum. a) Plate published by Cuvier in the annals; b) present appearance of the specimen © MNHN/L. Cazes; c) reconstruction by Bernard Duhem, MNHN*

Value of an affective nature can also be defined when the object or specimen attracts a certain empathy, as in the case of the naturalized specimen of Kiki the tortoise. This male individual of the giant Seychelles tortoise, *Dipsochelys dussumieri*, lived in the menagerie of the *Jardin des Plantes* until it was 146 years old, inspiring children's books and eliciting public sympathy. In a similar way, the remains of the giraffe Zarafa, which in its time was a huge success, are capable of provoking a certain emotion.

Use is also an aspect to be considered, especially in the case of ethnographic collections[3]: the marks of blows on a sword, wear on a tool, the remains of food,

3 Significant examples of these objects, marked by their use, are mentioned in "Is There a Need for Biocultural Collections? State of the Art and Perspectives" (Chapter 19 of this book).

perfume, cosmetics in pottery are inseparable from the life of the objects. In the same vein, the traces of sampling, disassembly and reassembly of a natural history specimen are precious testimonies to a use that clearly participates in its value.

Finally, authenticity is also an important criterion. It is linked both to the documentation of the context of the discovery and to the fact that the materials that make up the object have not been modified by the addition of exogenous compounds, such as paints, resins or varnishes.

In general, the typology of the different values that a heritage collection can have depends largely on the authors and the collections (de la Torre 2002). For example, the collections of the *Muséum* which are used as support for researchers will clearly have scientific value, which will not be, *a priori*, the case for a collection of contemporary paintings.

The current tendency to favor minimal intervention (Bradley 2004) aims not only to save resources, but also to limit the errors caused by an underestimation of the values of an object. These errors are all the easier to make because the exhaustive awareness of the different values of an object is generally not instantaneous, but results from a long process of considering all the facets of the object. In this respect, the consultation of the different actors of the collection (curators, restorers, public, researchers, etc.) is essential. The difficult part, which generally falls to the person in charge of the collection, consists of evaluating, in the most judicious way possible, these different values in order to estimate whether an intervention envisaged corresponds globally to a gain or a loss. In the English-language conservation community, this approach, initially thought of in the case of specific interventions, has been widely developed, until it has become part of the general policies of collection management. Institutional guides are offered to help museums define, as exhaustively as possible, the right value criteria for their collections and to carry out a global evaluation (Cultural Heritage Agency 2014). These tools are seen as decision aids, offering the possibility to calculate the evolution of the value of a collection according to different scenarios. They are also effective communication tools that allow for greater credibility and persuasiveness when it comes to defending a valuation project.

20.7. The scientific value of the collections

The scientific value of natural history collections is a multiform notion that can be broadly defined in the plural sense. The first of the scientific values is that of evidence, a decisive notion in natural sciences. For example, comparative anatomy is entirely built around the description of specimens[4]. In a more general way, the classification of species (taxonomy) is a discipline where the attributions are regularly re-interrogated and it is essential to have access to the specimens having been the support of the reference publications to re-examine, even refine the established classifications. These specimens, called "types", are linked to their first published description. Zoological and botanical taxonomies have defined a wide variety of terms describing different kinds of types, depending on the nature of the link between these specimens and the described species: holotypes, isotypes, neotypes, paratypes, etc. In natural history museums, the type specimens and materials are often gathered and regarded as having the greatest scientific value. This attachment to evidence corresponds to the fact that a previously defended interpretation can be reconsidered at any time in light of a new discovery. Re-examining the specimen is then necessary to go back to the original source of the information. Thus, the preservation of types, and also of all specimens collected, is essential for the accuracy and rigor of scientific work concerning present and past biodiversity, mineralogy and geology.

These notions of evidence also apply to archeological collections, but with a less systematic approach. Access to collections remains essential to researchers because it is not uncommon to re-examine a set of objects. However, the notion of types, in the taxonomic sense of the term, does not exist in archeology. It is rather the dating that serves as a reference: dated objects constitute essential reference points for a good number of studies and therefore have a great scientific value.

In general, the scientific value of a specimen or object will be largely related to the quality or quantity of information it carries. In the field of archeology, the complexity of an object is often perceived as a criterion of value, because a complex object generally carries more information than a simple object. In the same way, the finesse of the realization participates in the scientific value (in addition to the esthetic value) because it testifies to an elaborate know-how.

Another decisive aspect in defining scientific value is the notion of group (or assemblage). For example, dissociated from its excavation context, the archeological object loses a large part of its scientific value (a bit like an amnesiac person loses his social identity). When considered as part of an ensemble, the object can be representative of this ensemble if it presents the same characteristics as its neighbors, or, on the contrary, atypical if it presents different characteristics. All intermediate situations can occur, and it is therefore understandable that the way in

4 International code of zoological nomenclature, 5th edition, available at: http://iczn.org. International code of botanical nomenclature, available at: https://www.iapt-taxon.org/icbn/main.htm.

which the object is positioned in relation to the group is a source of information and contributes to its value. The same is true for most natural history specimens. The collections often include several specimens of the same species and the whole, far from being redundant, makes it possible to estimate the intraspecific variability (i.e. between individuals). When the collection of specimens is rare, to constitute sets is difficult and the collections of several individuals are all the more precious. The collection of coelacanths in the *Muséum* is, in this respect, quite remarkable. This species, identified in the 19th century through fossil specimens, was said to be extinct since the end of the Cretaceous. In fact, the coelacanth, far from being extinct, has evolved up to our times, living in the deep waters of the Indian Ocean, beyond a depth of 100 meters. The fishing of a specimen, in December 1938, near the coasts of South Africa, caused a great stir in the scientific community, especially since this species, of the class of sarcopterygians, is distinguished from fishes (class of actinopterygians) by the monobasal structure of its fins which presents several homologies with the members of the tetrapods that have conquered the terrestrial environment. The third specimen, caught near the Comoros Islands in 1952, joined the collections of the *Muséum* in Paris. Thanks to the influence of Jacques Millot, then director of the laboratory of comparative anatomy, a great number of specimens caught thereafter were transferred to France for study. As coelacanth specimens are both rare and scattered in many countries, their inventory was necessary (Nulens *et al.* 2011). It shows that the *Muséum national d'Histoire naturelle* now hosts 15 whole specimens and more than 300 samples from 35 different specimens, which, compared to the 237 specimens located around the world, represents the most extensive collection. This aspect is particularly interesting for researchers who can thus more easily study the morphological diversity and the various stages of development of the species (Figure 20.8).

a) b)

Figure 20.8. *Specimens of coelacanth,* Latimeria spp. *Embryo with its yolk sac (specimen CCC29.5, MNHN AC 2012-22, from a pregnant female caught in 1962 in the Comoros) (a); adult specimen (CCC8, MNHN AC 2012-6) captured in 1974 (b). Collection of the Muséum national d'Histoire naturelle. © MNHN/V, Rouchon*

Similarly, all information relating to the provenance of the object or specimen and its history contribute to its scientific value. Thus, on a paleontological or archeological excavation site, it is essential to record as precisely as possible (photographs, handwritten records, theodolite measurements, 3D measurements, etc.) the location in the ground of the pieces collected, because these data are likely to provide a great deal of information, both temporal (stratigraphy) and structural (organization of activity, sudden events, etc.). If they are not collected at the time of the excavation, they are lost forever.

A crucial question in the evaluation of collections that are the subject of scientific research is whether one should evaluate the value of the items in relation to the knowledge acquired or the knowledge to be acquired. When the motivation is focused on the knowledge to be acquired (which is often the case for researchers), there is a temptation to give major importance to the publication and less importance to the objects that inspired it, as long as one considers that they no longer have any information to deliver. These objects then have mainly evidential value. In this approach, the major interest is given to the new objects discovered, not yet interpreted (thus still unpublished), but whose potential for study is sensed. This is not the opinion of the sociologists, who reason in terms of added value for society, and therefore in terms of acquired knowledge. For them, knowledge that is not yet formulated does not exist and therefore has no scientific value for society. The visitor to a museum wants to be told a story and therefore to be shown evidence. They may be occasionally interested in questions and uncertainties, but the heart of the process remains the story. We can see here again that the notion of value is a question of point of view, which leads some authors (Jeram 1996) to consider two types of scientific value: a real scientific value correlated to the existing story (i.e. to the publications) and a potential scientific value, linked to future stories. An object or a specimen that has just been discovered and that we wish to study will not have a real scientific value, but a potential scientific value. In the same way, an object or a specimen that has only been studied on one aspect, and that would deserve a study on another aspect, would have a scientific value that is both real and potential.

The potential scientific value of an object or a specimen is not easy to estimate because it inevitably evolves with time. Oxygen in the air, humidity, pollution, but also human interventions cause changes in composition that can seriously compromise its future study. For example, limestone matrix extraction techniques that use carboxylic acids can substantially alter the isotopic signature of extracted pieces (Hellawel and Chris 2002) and thus bias their analysis. When objects are impregnated with an exogenous substance (e.g. consolidation resin) or subjected to chemical "stabilization" treatments, the shape and external appearance of the object are preserved, allowing the study of its morphology. However, the materials are largely transformed and their study potential is substantially reduced.

The usefulness of preventive conservation policies and the principle of minimal intervention is therefore understandable. However, one should not deduce from this that the potential scientific value of the *Muséum's* collections is inevitably bound to diminish with time. On the contrary! With the development of analytical tools, what was not possible yesterday becomes possible today and what is not possible today may be possible tomorrow. For example, advances in tomography now make it possible to study the fine structure of archeological and paleontological material without having to resort to invasive extraction techniques (Sutton 2008). In particular, X-ray tomography measurements have recently been performed on Cuvier's opossum (Figure 20.7) to refine the classification of this species (Selva and Ladevèze 2017).

On another note, we can see today that the natural history collections gathered before the 20th century constitute a formidable source of information for a better understanding of climate change and its effects on the environment. In light of this research, these collections now have a high potential scientific value, even though they were initially constituted and studied for a completely different purpose (Winker 2004). We can see here that the scientific use of collections evolves according to societal concerns. Moreover, many of these collections are irreplaceable because the habitats and localities from which the specimens were sampled have disappeared.

Sometimes, it is the species themselves that have disappeared from their natural environment, either because these environments have evolved or because they have been overhunted. Some of them owe their survival only to the existence of zoological parks which manage the surviving individuals by avoiding consanguinity. When the species are completely extinct, the specimens are the only remains that are accessible to us today (Figure 20.9). They are all the more precious, and their value is often not only scientific but also historical: thus, the *Muséum* houses the naturalized specimen of a dwarf emu, *Dromaius baudinianus*, which has an immense scientific value, because it is a type and the only naturalized specimen in the world of this species, extinct since the 19th century. This specimen has in addition a great historical value because it was brought back, under Napoleon I, by the expedition of Captain Baudin to Australia.

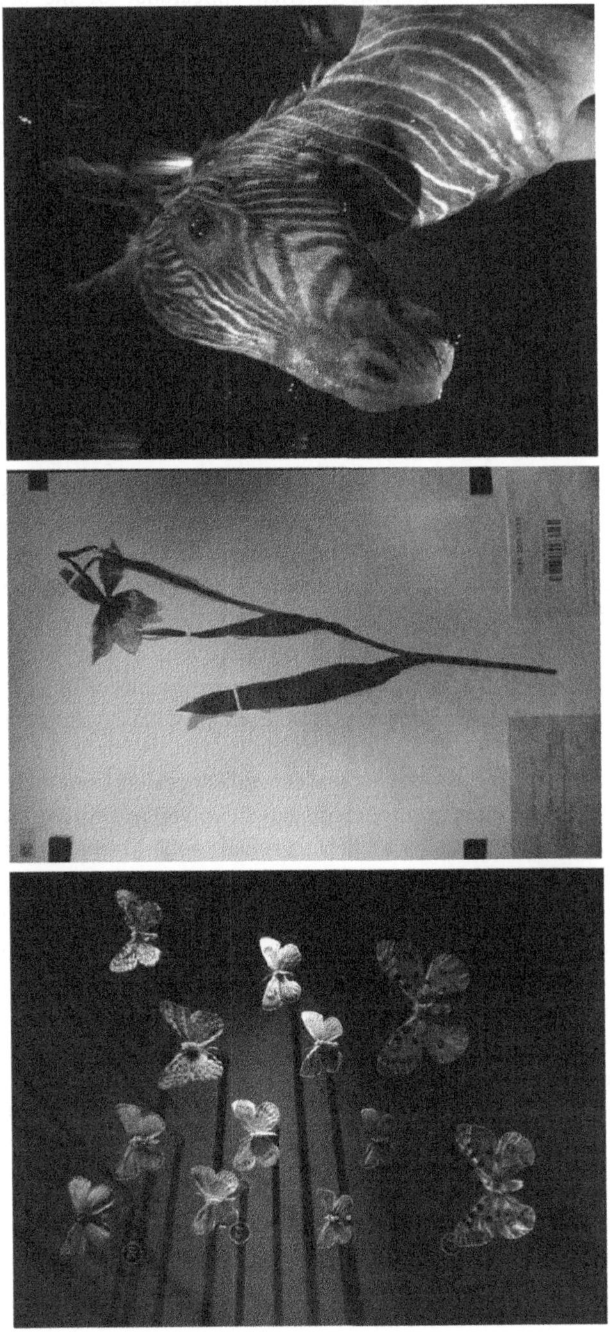

a) b) c)

Figure 20.9. *Examples of specimens corresponding to extinct species. © MNHN/V. Rouchon*

COMMENT ON FIGURE 20.9. – *a) Lepidoptera of different genera (*Lycaena, Phalanta, Zerinthia, Coenonympha, Maculinea, Glaucopsyche *and* Parnassio*); b) Garden tulip (*Tulipa didieri*, extinct in the wild); c) quagga (*Equus quagga quagga*; this specimen also has a historical value: brought back during an expedition to Africa, it joined the royal menagerie in Versailles in 1784, then the menagerie in 1794 where it died 4 years later).*

These different examples show that the scientific value of natural history collections has multiple facets, depending largely on past, ongoing and foreseen studies. This value evolves with time, scientific tools and the concerns of society. In order to evaluate it, we need to project into the future and imagine what tools will be available and what the challenges of tomorrow's research will be. Some of these projections will come true, others will not. Our ability to anticipate and imagine the future is nonetheless crucial to carry out a conservation policy that will be useful for future generations.

20.8. Conclusion

The different aspects we have just developed show that the conservation of natural history collections is far from being obvious, even when the specimens or objects are very old. Their future is linked, not only to the interest they arouse, but also to the respect we show them, which depends largely on the awareness we have of their values. The definition of collections that group together coherent sets of objects or specimens contributes to give them value and meaning, but it also generates a particular practical organization because of the large number of items concerned. As alterations are generally irreversible, conservation will be greatly facilitated by the anticipation of difficulties, and therefore by a preventive policy. It is true that the conservation of certain materials still constitutes a technological challenge. However, the tree should not hide the forest: most objects and specimens have a satisfactory stability. Their good conservation depends to a large extent on their entry into the collection and its wise management.

20.9. References

Berducou, M. (1990). *La Conservation en Archéologie*. Masson, Paris.

Boitard, M. (1859). *Nouveau manuel du naturaliste préparateur*. Roret, Paris.

Bradley, A.R. (2004). *The Archaeologist's Manual for Conservation: A Guide to Nontoxic, Minimal Intervention Artifact Stabilization*. Kluwer Academic, Plenum Publishers, New York.

Brandi, C. (1963). *Teoria del Restauro*. Edizioni di Storia e Letteratura, Rome.

Cultural Heritage Agency (2014). Assessing museum collection: Collection valuation in six steps. Document, Cultural Heritage Agency.

Cuvier, G. (1804). Mémoire sur le squelette presque entier d'un petit quadrupède du genre des sarigues, trouvé dans la pierre à plâtre des environs de Paris. *Annales du Muséum national d'Histoire naturelle*, 5(an XIII), 277–292.

Hellawel, J.O. and Chris, J.N. (2002). Acid treatment effects on the stable isotopic signatures of fossils. *Palaeontology*, 55(1), 1–10.

Jeram, A.J. (1996). Criteria for establishing the scientific value. In *Value and Valuation of Natural Science Collections*, Nudds, J.R. and Pettitt, C. (eds). Geological Society, Bath.

Nulens, R., Scott, L., Herbin, M. (2011). An update inventory of all specimens of the coelacanth, Latimeria spp. *Smithonian Special Publication*, 3.

Richmond, A. and Bracker, A. (eds) (2009). *Conservation: Principles, Dilemmas, and Uncomfortable Truths*. Victoria and Albert Museum, Butterworth-Heinemann, London.

Rosso, K.M. and Vaughan, D.J. (2006). Reactivity of sulfide mineral surfaces. In *Reviews in Mineralogy and Geochemistry: Sulfide Mineralogy and Geochemistry*, Vaughan, D.J. (ed.). Mineralogical Society of America, Chantilly.

Rouchon, V., Badet, H., Belhadj, O., Bonnerot, O., Lavédrine, B., Michard, J.G., Miska, S. (2012). Raman and FTIR spectroscopy applied to the conservation report of paleontological collections: Identification of Raman and FTIR signatures of several iron sulfate species such as ferrinatrite and sideronatrite. *Journal of Raman Spectroscopy*, 43, 1265–1274.

Selva, C. and Ladevèze, S. (2017). Computed microtomography investigation of the skull of Cuvier's famous "opossum" (Marsupialiformes Herpetotheridae) from the Eocene of Montmartre. *Zoological Journal of the Linnean Society*, 180(3), 672–693.

Simmons, J.E. (2014). *Fluid Preservation: A Comprehensive Reference*. Rowman & Littlefield, Plymouth.

Sutton, M.D. (2008). Tomographic techniques for the study of exceptionally preserved fossils (review). *Proceedings of the Royal Society B*, 275, 1587–1593.

Tennent, N.H. and Baird, T. (1985). The deterioration of mollusca collections: Identification of shell efflorescence. *Studies in Conservation*, 30, 73–85.

de la Torre, M. (2002). *Assessing the Values of Cultural Heritage: Research Report*. Getty Conservation Institute, Los Angeles, CA.

Wayne Smith, C. (2003). *Archaeological Conservation Using Polymers: Practical Applications for Organic Artifact Stabilization*. Texas A&M University Press, College Station, TX.

Winker, K. (2004). Natural history museum in a postbiodiversity era. *BioScience*, 54(5), 454–459.

21

Collections for Scientific Research in the 21st Century and Beyond

Roseli Pellens

ISYEB – CNRS, Muséum national d'Histoire naturelle, Sorbonne Université, École Pratique des Hautes Études, Université des Antilles, Paris, France

21.1. Collections in the quest for knowledge

The studies presented in this volume show how much collections have supported humanity in its quest for knowledge about its history and its environment. Collections used to be focused on the discovery of the exotic, the precious or the commercial, with a perspective on cataloging and listing. Today, they are the source of the new views that we have on Life and on our place in the Universe. Collections also represent a material basis – a collected vision of biodiversity – that confronts us with the world around us. We can hardly imagine the emergence of many concepts in modern biology without the many observations, deposits and re-examinations of specimens. For example, in *The Voyage of the Beagle*, Charles Darwin makes the examinations and comparisons that allowed him to find numerous examples to develop the theory of natural selection visible (Darwin 1845, 1859). By comparing and reviewing the same specimens, defining the criteria that bring them together, we realize the existence of the range of possibilities about the emergence of this diversity.

For a color version of all figures in this chapter, see www.iste.co.uk/pellens/natural.zip.

Natural History Collections in the Science of the 21st Century,
coordinated by Roseli Pellens.

This book shows us that as we work with collections, we can see other information and theories emerge that have remained uninformed or unnoticed until now. Open science traditions inherent to the history of systematics continue to chart this course. New communities of scientists are discovering this resource and are immersing themselves in this know-how and modality of re-examining the data as many times as needed.

21.2. Three main kinds of new uses for collections

These studies suggest the existence of three major kinds of new uses for collections.

21.2.1. *Enriching the life sciences, human sciences and the sciences of the universe with new technologies*

The first kind of use gathers the research that uses new technologies to pursue the discovery of Life, Human diversity and culture, rocks, meteorites and minerals. These forms of research represent, in different degrees, the contemporary development of systematics, evolutionary biology, comparative anatomy, mineralogy and geology. The innovation lies in the new ability to study specimens that could not have been analyzed without the new technologies. Access to DNA, promoted by new sequencing technologies (Chapter 13), and the use of green-yellow fluorescence to study fossil crustaceans (Chapter 7) are good examples of these developments. The new information obtained is not only molecular, but also anatomical (Giribet 2010). Thanks to these new technologies and after more than a century of controversy, the existence of an additional vein on the wings of insects has been confirmed (Schubnel *et al.* 2020). The same is true for the study of giraffe diversity, which has been resolved, among other things, by sequencing specimens from collections belonging to extinct populations (Petzold *et al.* 2020; Petzold and Hassanin 2020). Chapters 3, 4 and 5 also provide us with examples of how these new approaches are enriching several scientific fields.

21.2.2. *A pool of information on the environment*

The second type of use assembles the research intended to clarify and/or refine knowledge of the characteristics of past environments. Their purpose is obviously to understand the impacts of environmental changes on organisms, mainly as a result of human actions. The value of chemical contents or physical parameters can be analyzed as proxies of the environment. The occurrence of parasites, commensals and symbionts can be studied because it informs about the origin and transmission routes of pathogens ravaging crops, or diseases affecting humanity. The information may be directly in the tissues of the specimens, in their stomach contents, in their feces, in their associated organisms or in their interactions (which can be explored through simple co-occurrences, or through changes in frequencies over time). For example, a study on herbivory using items from herbaria has highlighted relationships between herbivory rates and global changes (Meineke and Davies 2018; Meineke *et al.* 2019). Environmental changes in the oceans over two centuries can be explored through the earplugs of blue whales (see note "Whale earplugs: indicators of ocean quality" in this chapter). These research works are still based on the analysis of collection specimens, but the prospect of finding environmental indicators/markers opens up a whole universe of very innovative uses (Chapters 6 and 11–16 offer several examples). The information on sets of specimens (batches of samples grouped by date, collector and locality, event sampling) and the conservation treatments that have been applied to them acquire an important place in this type of research. The access to their successive past states and the establishment of reference points represent its main interests.

The development of genetics, physiology and population biology also contributes to the use of collection specimens for purposes not envisioned at the time of their collection. In some studies, collections are used to answer questions on ecological restoration, one of the great technological challenges of our time. For example, the reintroduction of individuals developed from herbarium-preserved seeds and spores can enable the reintroduction of lost alleles and contribute to the enrichment of genetic diversity in small populations (Nakahama *et al.* 2015). Studies with the goal of bringing extinct species to life are another very specific example. As demonstrated in Chapter 10, this challenge requires experiments on germination and developmental physiology in closely related species to reduce the impact on the target material, which is present in small quantities in the collections. The final result is nevertheless very limited by the storage conditions of this material, which was not intended for this use. However, the knowledge gained from recent studies allows us to intervene at the source of the problem when designing future collections. New criteria are being added for the selection of species to be included in a seed bank (e.g. rare and endangered species) and new technologies are being developed to preserve their germination potential (Liu *et al.* 2018; Royal Botanic Gardens – Kew 2020).

21.2.3. *The era of digital data*

The third kind of use comprises the research based on virtual data (transcription of information from labels, field notebooks, images, molecular sequences, sounds, etc.) made available by many institutions on specific websites or through aggregator portals. What they have in common is a focus on the data and, all too frequently, a certain disinterest in the original specimens or collections (Grandcolas 2017; Chapter 2 of this book), including when their origin is indeed a specimen (Troudet *et al.* 2018). The impacts of this approach are multiple: (1) our understanding of the colossal size of the biodiversity data available via collections (Page *et al.* 2015); (2) the use of this information/data by scientific communities much larger than those familiar with collections; (3) the possibility of studying life at unprecedented spatiotemporal scales (Veron *et al.* 2019); and (4) the possibility of including collection data in various protocols. Nonetheless, this Big Data consists of billions of data assembled from a multitude of projects, each with their own goals, histories, limitations and methods. Understanding what these datasets contain and the impact of our ignorance on their completeness and biases in our reading of the living world are today's great challenges (Maldonado *et al.* 2015; Troudet *et al.* 2017; Daru *et al.* 2018; Dias Tarli *et al.* 2018). This new use is promoting great transformations in statistical analyses (Graham *et al.* 2004), in the interactions between users, and in the entire data delivery chain (see Chapter 18). The goal is to be able to include as many data as possible in the studies, despite their uneven levels of precision, uncertainties about their completeness and various biases. More recently, these data have also been analyzed with machine learning or e-learning, which look for patterns within their heterogeneity. These methods thus enable the exploration of problems that require the use of huge datasets (see note "Estimating threat status of 345,000 plant species", Chapter 18).

These last two kinds of uses are particularly innovative. They reveal the power of approaches based on huge datasets, even composites, or for the search for information present in minute quantities in specimens. The specimen is no longer considered only for itself, but as a witness of a bygone era and of the environmental changes that have occurred since. This use is particularly important in our time, in view of the global changes caused by human activities.

21.3. Lessons from these new uses

It is essential to identify and understand the multiple ways in which we look at collections today. They teach us how to enrich and manage collections so that they can offer the maximum content, which is possibly far beyond what we can presently predict or conceive of.

21.3.1. *The importance of richness and diversity*

The first lesson of this innovative body of research is on the importance of richness and diversity in collections. It is impossible to predict which species, specimen, object or piece of information will be of crucial importance to knowledge in the short and long term.

But new uses show us the importance of the huge pool where we look for information to test our hypotheses (see note "Whale earplugs: indicators of ocean quality" in this chapter). The diversity of collections represents multiple possible options for researchers now and in the future. This option value (Faith 2016; Chapter 20 of this book) is central to the contribution of collections to scientific approaches. Therefore, all collection development and curation strategies should contribute to improving both the diversity and quality of sampling. Focusing, for example, on a single species or taxonomic group of interest or, conversely, incorporating only a few specimens of a maximum number of species are strategies that are equally unlikely to increase the number of possible options. As with the characterization of the diversity of life, the phylogenetic diversity (Faith 1992; Pellens and Grandcolas 2016) of a collection can be an indirect measure of its option value (Liu *et al.* 2018).

21.3.2. *Information at the heart of new research*

The second lesson concerns the information linked to the specimen, one of the aspects that has become extremely important for innovative uses. For systematics, a specimen with very imprecise information about locality and date of collection (e.g. "North America, 18th century") – a common situation for specimens collected during the 18th and 19th centuries – is still usable as long as its morpho-anatomical structures and, in the best case, its DNA are in good condition.

For many innovative uses, whether related to specimen characterization or large computerized datasets, accurate information about the specimen, its location or its date of collection is crucial. Specimens with incomplete or poorly confirmed collection or identification information are of reduced value for most uses.

Moreover, the quality of the information is not necessarily guaranteed by the use of new and powerful technological tools (e.g. geolocation and digitization); the reliability and durability of this information must be absolutely taken into account.

NOTE.– *Whale earplugs: indicators of ocean quality.*

Earplugs are composed of a mixture of lipid and keratin, which is deposited in thin layers of alternating light and dark colors. These layers accumulate semi-annually in a systematic way throughout the individual's life and, for this reason, are used as a proxy to determine the age of individuals.

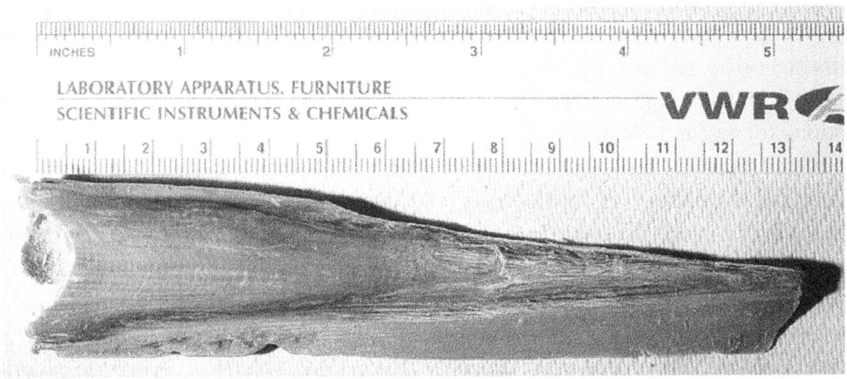

Figure 21.1. *Transverse section of an earplug from a fin whale (Balaenoptera physalus). Source: Stephen J. Trumble, Baylor University, Waco, Texas, USA*

COMMENT ON FIGURE 21.1.– *The alternating light and dark laminae of cerumen provide information on the chemical composition of lipids every half year and represent an indicator of stress and ocean chemistry throughout the individual's life.*

Recent studies show that each of these laminae captured lipophilic substances (hormones and chemicals) represent levels in the individual's body for each half year of life (Trumble *et al.* 2013). The analysis of cortisol (the mammalian stress hormone), mercury and organic pesticide concentrations in these laminae thus allowed for the assessment of periods of stress and the variation of the concentration of these chemicals in the environment where the individuals lived (Trumble *et al.* 2018). Because whales have very large ranges, live for a very long time (average life expectancy for many species is on the order of 200 years) and are very sensitive to changes in their environment, they are considered marine sentinels. The information collected in the earplug is thus very useful to date and marks certain aspects of the oceans on a planetary scale and with an unprecedented temporal depth (from the 1870s to the present day).

21.3.3. *Good conservation and good practices*

The third lesson concerns the state and means of conservation. The techniques for obtaining information on specimens are improving and becoming more refined every day. The ability to measure chemical contents in minute quantities is increasing; new sequencing technologies allow access to the DNA of almost all specimens held in collections; a CT-scan (computed tomography scanning) allows the examination of internal parts of specimens without recourse to dissection. At the same time, research on conservation techniques is intensifying and many solutions are being made available (see the many examples given in Chapter 20). Nevertheless, the more specimens can be used for various purposes, the more necessary it is that there is proper preservation and archiving of conservation methods and possible treatments. Gaudeul and Rouhan (2013) emphasized this with respect to DNA conservation for genetic and molecular studies in plants. Also, for example, Peñuelas and Matamala (1990) were able to observe the change in nitrogen and sulfur contents in the leaves of vascular plants between 1750 and 1990, using specimens from a collection whose treatment was known and well documented; it had been treated with DDT, a pesticide that does not contain any of these elements, thus ruling out any possibility of contamination.

A second point related to conservation is that uses, both old and new, can be invasive and/or destructive. The effects of some uses are well known, such as tissue sampling, coating structures with gold for scanning electron microscopy, slide preparation and mounting, and insertion into resins and staining (Espinosa and Pinedo Castro 2018; Chapter 20 of this book). But other uses have more subtle impacts, invisible to the naked eye, and even variable depending on the organism or type of structure. Therefore, special attention must be given to them to ensure proper preservation of specimens and their subsequent uses. For example, X-ray tomography (CT-scan) can damage the DNA of hard mineralized tissues (bones and teeth), while its direct impact on soft tissues remains poorly understood. In addition, this technology applied to soft tissues requires the use of contrasting products with diverse characteristics and specific effects (Hall *et al.* 2015).

To ensure the conservation and integrity of specimens, it is therefore important to detect the possible effects of different methods and make good practice guides available to mitigate their impacts (such as the guide produced by Keklikoglou *et al.* (2019)). *Ad hoc* working groups and expert panels identifying new uses and their impacts on specimens at different time scales are also highly desirable. This is especially important as the number of potential uses increases, especially with the computerization of collections and the availability of information acquired on specimens (barcodes, sequences, isotopes and various two- and three dimensional

(2D, 3D, 4D) images – see Chapter 17). In this context, the best-studied specimens will tend to be used more often for other research, which can build on some of the knowledge already acquired.

Today, another facet of this topic concerns the preservation of data. As noted in Chapter 18 (Monnet *et al.*), computerization is important as it links the entire chain, from acquisition to final use of the data. Like everything else related to information technology, this subject is evolving very rapidly. Many institutions ensure that data remain available in the future, as defined by the FAIR (Findable, Accessible, Interoperable and Reusable)[1] principles. In this respect, it is essential that data be stored in formats and on media that are resistant to computer obsolescence and to crashes of large operational systems. Files in .txt, .rtf, .csv format, which can be opened by a wide range of software, and open source repositories of data and metadata can contribute to the durability of information.

21.3.4. *The importance of sets*

The search for information on the past environment and on the impacts of environmental changes on organisms (interactions, phenology, etc.) from collection specimens, highlights the importance of event sampling. This is a new dimension that adds to the recurrent debate on the ideal arrangement of specimens in collections. Batches of specimens from the same collection event may represent, as a whole, samples or replicates of an interaction or environmental feature at known locations and dates (GBIF 2017).

The dispersion of sets in several storage units greatly reduces this potential because it generates information losses. Even in a situation where all specimens would be fully computerized and easily reassociated, which is the case for a minority of taxonomic groups in the collections, some information that are not computerized *a priori* may escape notice simply because the individuals are not grouped. For example, samples collected later in the year do not show evidence of consumption by herbivores. The sets also offer opportunities to set up time series to monitor and understand changes in the environment over the long term.

It is obvious that the management of collections is constrained by various logistical problems and that taxonomic arrangement is inseparable from the existence of collections. Nevertheless, if this option were possible, it should not be assumed, *a priori*, that all of the information residing in the collections could be recovered with computerization. For this, the specifications for computerization should

1 Available at: https://www.go-fair.org/fair-principles.

include a much wider range of information to be taken into account than is currently the case.

21.4. Collections in 21st century science and beyond

This diversity of new uses shows us how central information richness is to the option value of collections. This places new responsibilities on today's collectors. Traditional collections with information only regarding dates and locations have thus become quite limited considering the scientific know-how that has developed in this field (Schindel and Cook 2018). We have the means to document everything that can be seen on the specimen through 2D, 3D or even 4D digital images (Chapter 17). Information on behaviors, interactions, micro-habitat details, as well as collecting associated species (predators, prey, host plants, endosymbionts, etc.) is also valuable. Consideration of diversity at all observation levels should be encouraged; several specimens of a species with respect to intraspecific diversity, several samples from the same individual to understand intra-organism genetic and phenotypic diversity (e.g. elements of herbaria corresponding to samples located at different heights in a tree) and so on. This concept, which some call the extended specimen, has now become inseparable from natural history.

Our responsibility in new collections is all the more important because we are going through a major biodiversity crisis. To face this crisis, we need information on species with low population numbers (rare or endemic) that are frequently vulnerable to extinction. In this context where multiple irreversible environmental transformations take place every year, collecting events represent unique opportunities to keep track of heritage for the generations that will follow us. Of course, this crisis situation also calls for greater rigor in the access and benefit sharing (ABS) and other ethical issues related to the uses of collections. The availability of collections and resident information is not only a matter of good scientific ethics, but also of respect for political agreements established between different parties at the international level. This is the case, for example, with the Nagoya Protocol for biodiversity, or with the restitution agreements for human remains or artifacts.

21.5. Conclusion

We can conclude this book by asking ourselves the question of the role of collections in the sciences of the future. Can we envisage life sciences without collections? Can we envisage a harmonious relationship between Humans and Nature without knowledge based on collections?

Understanding Life and Humankind essentially has the challenge of integrating their diversity, this incredible product of the evolution of life. Nevertheless, our mind tends to dismiss everything that is not already known (Kuhn 2012). Collections represent a fantastic tool to counter this confirmation bias and combat our natural anthropocentrism, by becoming aware of diversity. Consequently, the first risk of a life science without a collection is that of focusing on the human species and a few model organisms, while overlooking the diversity that is essential for understanding essential human or environmental issues.

Not having or losing collections can lead to a second extinction, that is, erasing the existence of the enormous diversity that our generation still perceives from our memory (Raven and Miller 2020). Recent history shows us that some species that were previously widespread with huge populations, such as the migratory pigeon in North America, became extinct within a few decades. In the same short period of time, vast forests have been replaced by megacities, and only collections can still account for the diversity that existed there. The fire that destroyed the Museu Nacional do Rio de Janeiro in September 2018 is a tragic example of this kind of loss. It did not only destroy a large number of types of organisms (especially insects) or artifacts described in Brazil during the 20th century. With the fire, we lost the witnesses of the diversity of an extremely rich tropical forest that no longer exists (Pellens and Grandcolas 2018).

Collections are also an extraordinary source of inspiration for seeking solutions to multiple questions or problems. The field of bioinspiration or nature-based solutions, central to our industrial societies, relies heavily on them for its full redeployment (Benyus 1997; Chapelle and Raskin 2020). Losing them or not having them would deprive us of an essential tool for this inspiration and of this particular view that we have of Life.

More generally, collections are a precursor and yet living example of open science, in which not only the results, but also the research data should be available to everyone (Schilthuizen *et al.* 2015). They should continue to inspire the whole of science in the implementation of an open science policy, going beyond the simple electronic availability that is too often equated with open data.

Finally, this openness must also be toward the public and civil society, well beyond the borders, for that matter imaginary, of the scientific community (Bakker *et al.* 2020). The collections must also be a tool to help pedagogy, museography and all other disciplines related to knowledge within our societies.

21.6. References

Bakker, F.T., Antonelli, A., Clarke, J.A., Cook, J.A., Edwards, S.V, Ericson, P.G.P., Faurby, S., Ferrand, N. (2020). The Global Museum: Natural history collections and the future of evolutionary science and public education. *PeerJ*, 8225, 1–40.

Benyus, J.M. (1997). *Biomimicry: Innovation Inspired by Nature*. William Morrow, New York.

Chapelle, G. and Raskin, K. (eds) (2020). *Humanité bio-inspirée ; une autre approche*. Le Cherche Midi/Cité des Sciences et de l'Industrie, Paris.

Daru, B.H., Park, D.S., Primack, R.B., Willis, C.G., Barrington, D.S., Whitfeld, T.J.S., Seidler, T.G., Sweeney, P.W., Foster, D.R., Ellison, A.M., Davis, C.C. (2018). Widespread sampling biases in herbaria revealed from large-scale digitization. *New Phytologist*, 217(2), 939–955.

Darwin, C. (1845). *Journal of Researches into the Natural History and Geology of the Countries Visited During the Voyage of H.M.S. Beagle Round the World, under the Command of Capt. Fitz Roy, R.N.*, 2nd edition. John Murray, London.

Darwin, C. (1859). *On the Origin of Species by Means of Natural Selection, or the Preservation of Favoured Races in the Struggle for Life*. John Murray, London.

Dias Tarli, V., Grandcolas, P., Pellens, R. (2018). The informative value of museum collections for ecology and conservation: A comparison with target sampling in the Brazilian Atlantic forest. *PLoS ONE*, 13, e0205710.

Espinosa, F. and Pinedo Castro, M. (2018). On the use of herbarium specimens for morphological and anatomical research. *Botany Letters*, 165(3–4), 361–367 [Online]. Available at: http://doi.org/10.1080/23818107.2018.1451775.

Faith, D.P. (1992). Conservation evaluation and phylogentic diversity. *Biological Conservation*, 61, 1–10.

Faith, D.P. (2016). A general model for biodiversity and its value. In *The Routledge Handbook of Philosophy of Biodiversity*, Garson, J., Plutynski, A., Sarkar, S. (eds). Routledge, Abingdon-on-Thames.

Gaudeul, M. and Rouhan, G. (2013). A plea for modern botanical collections to include DNA-friendly material. *Trends in Plant Science*, 18(4), 184–185.

GBIF (2017). Best practices in publishing sampling-event data [Online]. Available at: https://github.com/gbif/ipt/wiki/BestPracticesSamplingEventData.

Giribet, G. (2010). A new dimension in combining data? The use of morphology and phylogenomic data in metazoan systematics. *Acta Zoologica*, 91(1), 11–19.

Graham, C.H., Ferrier, S., Huettman, F., Moritz, C., Peterson, A.T. (2004). New developments in museum-based informatics and applications in biodiversity analysis. *Trends in Ecology and Evolution*, 19(9), 497–503.

Grandcolas, P. (2017). Losing the connection between the observation and the specimen: A by-product of the digital era or a trend inherited from general biology? *Bionomina*, 12, 57–62.

Hall, A.C., Sherlock, E., Sykes, D. (2015). Does micro-CT scanning damage DNA in museum specimens? *Journal of Natural Science Collections*, 2, 22–28.

Keklikoglou, K., Faulwetter, S., Chatzinikolaou, E., Wils, P., Brecko, J., Kvaček, J., Metscher, B., Arvanitidis, C. (2019). Micro-computed tomography for natural history specimens: A handbook of best practice protocols. *European Journal of Taxonomy*, 1–55.

Kuhn, T.S. (2012). *The Structure of Scientific Revolutions*. University of Chicago Press, Chicago, IL.

Liu, U., Breman, E., Cossu, T.A., Kenney, S. (2018). The conservation value of germplasm stored at the Millennium Seed Bank, Royal Botanic Gardens. *Biodiversity and Conservation*, 27(6), 1347–1386 [Online]. Available at: https://doi.org/10.1007/s10531-018-1497-y.

Maldonado, C., Molina, C.I., Zizka, A., Persson, C., Taylor, C.M., Albán, J., Chilquillo, E., Rønsted, N., Antonelli, A. (2015). Estimating species diversity and distribution in the era of Big Data: To what extent can we trust public databases? *Global Ecology and Biogeography*, 24(8), 973–984.

Meineke, E.K. and Davies, T.J. (2018). Museum specimens provide novel insights into changing plant–herbivore interactions. *Philosophical Transactions of the Royal Society of London B*, 374(2017.0393).

Meineke, E.K., Classen, A.T., Sanders, N.J., Davies, T.J. (2019). Herbarium specimens reveal increasing herbivory over the past century. *Journal of Ecology*, 107, 105–117.

Nakahama, N., Hirasawa, Y., Minato, T., Hasegawa, M., Isagi, Y., Shiga, T. (2015). Recovery of genetic diversity in threatened plants through use of germinated seeds from herbarium specimens. *Plant Ecology*, 216, 1635–1647.

Page, L.M., Macfadden, B.J., Fortes, J.A., Soltis, P.S., Riccardi, G. (2015). Digitization of biodiversity collections reveals biggest data on biodiversity. *BioScience*, 65(9), 841–842.

Pellens, R. and Grandcolas, P. (eds) (2016). *Biodiversity Conservation and Phylogenetic Systematics: Preserving our Evolutionary Heritage in an Extinction Crisis*. Springer, Cham [Online]. Available at: http://link.springer.com/10.1007/978-3-319-22461-9.

Pellens, R. and Grandcolas, P. (2018). Après l'incendie du Museu Nacional de Rio : que faire maintenant ? *The Conversation* [Online]. Available at: http://theconversation.com/apres-lincendie-du-museu-nacional-de-rio-que-faire-maintenant-104115.

Peñuelas, J. and Matamala, R. (1990). Changes in N and S leaf content, stomatal density and specific leaf area of 14 plant species during the last three centuries of CO_2 increase. *Journal of Experimental Botany*, 41(9), 1119–1124.

Petzold, A. and Hassanin, A. (2020). A comparative approach for species delimitation based on multiple methods of multi-locus DNA sequence analysis: A case study of the genus Giraffa (Mammalia, Cetartiodactyla). *PLoS ONE*, 0217956, 1–28.

Petzold, A., Magnant, A., Edderai, D., Chardonnet, B., Rigoulet, J., Saint-Jalme, M., Hassanin, A. (2020). First insights into past biodiversity of giraffes based on mitochondrial sequences from museum specimens. *European Journal of Taxonomy*, 703, 1–33.

Raven, P.H. and Miller, S. (2020). Here today, gone tomorrow. *Science*, 370(6513), 149.

Royal Botanic Gardens – Kew (2020). A global resource for plant and fungal knowledge: Science strategy 2015–2020. Report, ORBIS, United Kingdom.

Schilthuizen, M., Vairappan, C.S., Slade, E.M., Mann, D.J., Miller, J.A. (2015). Specimens as primary data: Museums and "open science". *Trends in Ecology & Evolution*, 30, 237–238.

Schindel, D.E. and Cook, J.A. (2018). The next generation of natural history collections. *PLoS Biology*, 1–8 [Online]. Available at: doi.org/10.1371/journal.pbio.2006125.

Schubnel, T., Desutter-Grandcolas, L., Legendre, F., Prokop, J., Mazurier, A., Garrouste, R., Grandcolas, P., Nel, A. (2020). To be or not to be: Postcubital vein in insects revealed by microtomography. *Systematic Entomology*, 45(2), 327–336.

Troudet, J., Grandcolas, P., Blin, A., Vignes-Lebbe, R., Legendre, F. (2017). Taxonomic bias in biodiversity data and societal preferences. *Scientific Reports*, 7(1), 1–14.

Troudet J., Vignes-Lebbe, R., Grandcolas, P., Legendre, F. (2018). The increasing disconnection of primary biodiversity data from specimens: How does it happen and how to handle it? *Systematic Biology*, 67, 1110–1119.

Trumble, S.J., Robinson, E.M., Berman-Kowalewski, M., Potter, C.W., Usenko, S. (2013). Blue whale earplug reveals lifetime contaminant exposure and hormone profiles. *Proceedings of the National Academy of Sciences of the United States of America*, 110(42), 16922–16926.

Trumble, S.J., Norman, S.A., Crain, D.D., Mansouri, F., Winfield, Z.C., Sabin, R., Potter, C.W., Gabriele, C.M., Usenko, S. (2018). Baleen whale cortisol levels reveal a physiological response to 20th century whaling. *Nature Communications*, 9(4587), 1–8.

Veron, S., Haevermans, T., Govaerts, R., Mouchet, M., Pellens, R. (2019). Distribution and relative age of endemism across islands worldwide. *Scientific Reports*, 9(1), 11693 [Online]. Available at: https://doi.org/10.1038/s41598-019-47951-6.

List of Authors

Arnaud ANSART
Consultant

Anne-Sophie ARCHAMBEAU
PatriNat
IRD – OFB – CNRS
Muséum national d'Histoire naturelle
Paris
France

Éric BADEL
PIAF
INRAE
Université Clermont Auvergne
Clermont-Ferrand
France

Serge BAHUCHET
EA – CNRS
Muséum national d'Histoire naturelle
Université Paris Diderot
Musée de l'Homme
Paris
France

Claudia BAIDER
The Mauritius Herbarium
Ministry of Agro-Industry and Food Security
Port-Louis
Mauritius

Annachiara BARTOLINI
CR2P – CNRS
Muséum national d'Histoire naturelle
Sorbonne Université
Paris
France

Lucile BECK
LMC14 – LSCE/IPSL
CEA – CNRS – UVSQ
Université Paris-Saclay
Gif-sur-Yvette
France

Nathalie BECKER
ISYEB – CNRS
Muséum national d'Histoire naturelle
Sorbonne Université
École Pratique des Hautes Études
Université des Antilles
Paris
France

Cécile BERNARD
MCAM – CNRS
Muséum national d'Histoire naturelle
Paris
France

Christophe BOU
PACEA
UFR Sciences Odontologiques
CNRS – Université de Bordeaux
Pessac
France

Stéphane BUORD
Conservatoire botanique national de Brest
France

Paola CAMPOS
PVBMT
CIRAD
INRAE
Université de la Réunion
Saint-Pierre
and
ISYEB – CNRS
Muséum national d'Histoire naturelle
Sorbonne Université
École Pratique des Hautes Études
Université des Antilles
Paris
France

Sylvain CHARBONNIER
CR2P – CNRS
Muséum national d'Histoire naturelle
Sorbonne Université
Paris
France

Katia COMTE
MCAM – CNRS
Muséum national d'Histoire naturelle
Paris
France

Bruno DAVID
Muséum national d'Histoire naturelle
Paris
France

Emmanuelle DELQUE-KOLIC
LMC14 – LSCE/IPSL
CEA – CNRS – UVSQ
Université Paris-Saclay
Gif-sur-Yvette
France

Sébastien DUPERRON
MCAM – CNRS
Muséum national d'Histoire naturelle
Paris
France

Bernard DUTRILLAUX
ISYEB – CNRS
Muséum national d'Histoire naturelle
Sorbonne Université
École Pratique des Hautes Études
Université des Antilles
Paris
France

Charlotte DUVAL
MCAM – CNRS
Muséum national d'Histoire naturelle
Paris
France

Guillaume ECHEVARRIA
LSE
INRAE
Université de Lorraine
Nancy
France

François FARGES
IMPMC – CNRS
Muséum national d'Histoire naturelle
Institut universitaire de France
Paris
France

Éric FEUNTEUN
BOREA – IRD
CNRS
Muséum national d'Histoire naturelle
Sorbonne Université
Université des Antilles
Université de Caen-Normandie
Dinard
France

Marie-Béatrice FOREL
CR2P – CNRS
Muséum national d'Histoire naturelle
Sorbonne Université
Paris
France

Martin FRIESS
UMR 7206 "Éco-Anthropologie"
Muséum national d'Histoire naturelle
Université Paris Diderot
France

Lionel GAGNEVIN
PHIM
CIRAD – IRD – INRAE
Institut Agro
Université Montpellier
France

Manon GALLAND
UMR 7206 "Éco-Anthropologie"
Muséum national d'Histoire naturelle
Université Paris Diderot
France

Romain GARROUSTE
ISYEB – CNRS
Muséum national d'Histoire naturelle
Sorbonne Université
École Pratique des Hautes Études
Université des Antilles
Paris
France

Myriam GAUDEUL
ISYEB – CNRS
Muséum national d'Histoire naturelle
Sorbonne Université
École Pratique des Hautes Études
Université des Antilles
Paris
France

Michèle GERBAULT-SEUREAU
ISYEB – CNRS
Muséum national d'Histoire naturelle
Sorbonne Université
École Pratique des Hautes Études
Université des Antilles
Paris
France

Philippe GRANDCOLAS
ISYEB – CNRS
Muséum national d'Histoire naturelle
Sorbonne Université
École Pratique des Hautes Études
Université des Antilles
Paris
France

Thomas HAEVERMANS
ISYEB – CNRS
Muséum national d'Histoire naturelle
Sorbonne Université
École Pratique des Hautes Études
Université des Antilles
Paris
France

Sahima HAMLAOUI
MCAM – CNRS
Muséum national d'Histoire naturelle
Paris
France

Jean-Bernard HUCHET
PACEA
CNRS – Université de Bordeaux
Pessac
and
AASPE
ISYEB – CNRS
Muséum national d'Histoire naturelle
Sorbonne Université
École Pratique des Hautes Études
Université des Antilles
Paris
France

Vanessa R. INVERNÓN
ISYEB – CNRS
Muséum national d'Histoire naturelle
Sorbonne Université
École Pratique des Hautes Études
Université des Antilles
Paris
France

Pierre JOUANNAIS
I2BC – CEA – CNRS
AgroParisTech
Université Paris-Saclay
Gif-sur-Yvette
France

Romain JULLIARD
CESCO – CNRS
Muséum national d'Histoire naturelle
Sorbonne Université
Paris
France

Véronique LABORDE
DGDcoll
Muséum national d'Histoire naturelle
Paris
France

Line LE GALL
ISYEB – CNRS
Muséum national d'Histoire naturelle
Sorbonne Université
École Pratique des Hautes Études
Université des Antilles
Paris
France

Jean-Michel LETT
PVBMT
CIRAD
INRAE
Université de la Réunion
Saint-Pierre
France

Samuel MERIGEAUD
Clinique du Parc
Castelnau-le-Lez
France

Sylvain MERLOT
I2BC – CEA – CNRS
AgroParisTech
Université Paris-Saclay
Gif-sur-Yvette
France

Anne-Christine MONNET
ISYEB – CNRS
Muséum national d'Histoire naturelle
Sorbonne Université
École Pratique des Hautes Études
Université des Antilles
Paris
France

Eva MORENO
LOCEAN
CNRS
IRD
Muséum national d'Histoire naturelle
Paris
France

Serge MULLER
ISYEB – CNRS
Muséum national d'Histoire naturelle
Sorbonne Université
École Pratique des Hautes Études
Université des Antilles
Paris
France

Dulce M. NAVARRETE GUTIÉRREZ
LSE
INRAE
Université de Lorraine
Nancy
France

Roseli PELLENS
ISYEB – CNRS
Muséum national d'Histoire naturelle
Sorbonne Université
École Pratique des Hautes Études
Université des Antilles
Paris
France

Adrien PERRARD
iEES Paris
CNRS – INRAE – IRD
Université de Paris
Sorbonne Université
Paris
France

Yohan PILLON
LSTM
IRD – INRAE – CIRAD
Institut Agro
Université Montpellier
France

Valérie PRIOLET
Muséum national d'Histoire naturelle
Paris
France

Éloïse QUETEL
Pôle Patrimoine
Sorbonne Université
Paris
France

Anita QUILES
IFAO
Cairo
Egypt

Adrien RIEUX
PVBMT
CIRAD
INRAE
Université de la Réunion
Saint-Pierre
France

Marine ROBUCHON
ISYEB – CNRS
Muséum national d'Histoire naturelle
Sorbonne Université
École Pratique des Hautes Études
Université des Antilles
Paris
France

Quentin ROME
PatriNat
IRD – OFB – CNRS
Muséum national d'Histoire naturelle
Paris
France

Véronique ROUCHON
Centre de Recherche sur la Conservation
Muséum national d'Histoire naturelle
CNRS
Ministère de la Culture
Paris
France

Germinal ROUHAN
ISYEB – CNRS
Muséum national d'Histoire naturelle
Sorbonne Université
École Pratique des Hautes Études
Université des Antilles
Paris
France

Philippe ROUMAGNAC
PHIM
CIRAD – IRD – INRAE
Institut Agro
Université Montpellier
France

Florence ROUSSEAU
ISYEB – CNRS
Muséum national d'Histoire naturelle
Sorbonne Université
École Pratique des Hautes Études
Université des Antilles
Paris
France

Boris SZUREK
PHIM
CIRAD – IRD – INRAE
Institut Agro
Université Montpellier
France

Aline THOMAS
EA – CNRS
Muséum national d'Histoire naturelle
Université Paris Diderot
France

Romane TISSERAND
LSE
INRAE
Université de Lorraine
Nancy
France

Claire VILLEMANT
ISYEB – CNRS
Muséum national d'Histoire naturelle
Sorbonne Université
École Pratique des Hautes Études
Université des Antilles
Paris
France

Claude YÉPRÉMIAN
MCAM – CNRS
Muséum national d'Histoire naturelle
Paris
France

Index

D

E

F

G

H

I

K, L

M

N

O

P

Q, R

S

T

U, V

W, X, Y, Z

Printed and bound by CPI Group (UK) Ltd, Croydon, CR0 4YY

13/09/2026

14960994-0008